The Use of Video Technologies in Refereeing Football and Other Sports

For a long time, various different lobbying sectors have claimed that the use of video technology is an effective aid in decision-making. Now the IFAB has taken a historic step in the approval of experiments on the use of video to provide support to football refereeing.

The Use of Video Technologies in Refereeing Football and Other Sports analyses the capacity of audio-visual technology from different perspectives to help understand the best implementation of the Video Assistant Referee (VAR) system in football and, more generally, in other sports. This book addresses in-depth interdisciplinary viewpoints on the need and the opportunity of the implementation procedures regarding how to use it, considering that it could lead to very important changes.

The book goes on to examine various approaches to the most interesting topics for players, amateurs, coaches, referees and referees' coaches. Offering viewpoints from both academics and professionals, this new volume addresses the VAR issue in a multidisciplinary way, analysing the implications of video replay application in football from the perspective of players, coaches, television professionals, referees, amateurs, sports lawyers, media and educators.

Manuel Armenteros is a visiting professor at the Carlos III University of Madrid. For over 10 years, he coordinated the training project between the FIFA Refereeing Department and the Carlos III University, Madrid, designing and implementing new technologies for referees and technical instructors. He is also a technology consultant and regularly collaborates with FIFA, UEFA, AFC and IFAB on a variety of projects.

Anto J. Benítez holds a degree in Information Science and a PhD in Audio-visual Communication from the Complutense University of Madrid. His thesis was titled "Espectáculo futbolístico y comunicación televisiva". Main lines of his research have to do with sports and media, new technologies of audio-visual production and its application in education. He is the author of the sole original volume in Spanish about producing and directing sports for television. He has been a television director for more than 15 years, some of which he was in charge of sports and broadcasting in Telemadrid. Thanks to that, he has planned and directed a huge number of sports events, especially in La Liga and La Liga 2. As an athlete, he competed at the highest level for 17 years in different rugby teams in Spanish Leagues.

Miguel Ángel Betancor is a professor of Theory and History of Education at the University of Las Palmas de Gran Canaria. He received his PhD in Psychopedagogy from the University of Las Palmas de Gran Canaria in 1988. He is the director of the Center for Studies on Sports Arbitration at the University of Las Palmas de Gran Canaria. He was also the coordinator of the online training project between FIFA and FIBA Europe and the University Las Palmas de Gran Canaria from 2000 to 2010, and the head of operations at FIBA Europe, where he worked on the design of the digital basketball record. He also introduced technology in referee training, providing an online platform to referees and instructions from 51 countries. Nowadays, he is the head of the European project "Erasmus PLUS SPORT", a new model of socio-healthy volunteering. He is also the head of the regional project "eSport Talent Canarias", which is aimed at the development of electronic sports and its recognition in the Sports Law and Gamification Life.

Routledge Research in Sports Technology and Engineering

The *Routledge Research in Sports Technology and Engineering* series showcases cutting-edge research from this increasingly significant discipline. Featuring the work of established and emerging scientists and practitioners from around the world, this series is an important channel for new and groundbreaking research in technology and engineering for sport.

Available in this series:

The Science and Engineering of Sport Surfaces
Edited by Sharon Dixon, Paul Fleming, Iain James and Matt Carré

The Use of Video Technologies in Refereeing Football and Other Sports
Edited by Manuel Armenteros, Anto J. Benítez and Miguel Ángel Betancor

The Use of Video Technologies in Refereeing Football and Other Sports

Edited by Manuel Armenteros, Anto J. Benítez, and Miguel Ángel Betancor

Routledge
Taylor & Francis Group
NEW YORK AND LONDON

First published 2020
by Routledge
52 Vanderbilt Avenue, New York, NY 10017

and by Routledge
2 Park Square, Milton Park, Abingdon, Oxon, OX14 4RN

Routledge is an imprint of the Taylor & Francis Group, an informa business

First issued in paperback 2021

Library of Congress Cataloging-in-Publication Data
A catalog record for this title has been requested

ISBN: 978-1-138-31204-3 (hbk)
ISBN: 978-1-03-208385-8 (pbk)
ISBN: 978-0-429-45555-1 (ebk)

Typeset in Bembo
by Swales & Willis, Exeter, Devon, UK

To our families:
To Patricia, Cris, and María Luisa,
For the time that we have taken from you;
For your understanding;
For your patient and resolute support.
To Erik, Gael, Gian and Gabriel; to Claudia and Ángela; and to Miguel and Ana.
For the strength that thinking of you gave us when we faltered.
To our parents.

Contents

Figures

Tables

Contributors

Manuel Armenteros is a visiting professor at the Carlos III University of Madrid. For over 10 years, he coordinated the training project between the FIFA Refereeing Department and the Carlos III University, Madrid, designing and implementing new technologies for referees and technical instructors. He is also a technology consultant and regularly collaborates with FIFA, UEFA, AFC and IFAB on a variety of projects.

He is the author of two software patents and is also the co-founder of LabHipermedia, a laboratory of technology that is a spin-off of the UC3M. His research interests include digital technologies and new methodologies in education of instructors and referees.

He is a Referee and Match Official Research Network member, and a reviewer in several sports and communication journals. He is also the co-director of two international workshops in new technologies in sports with high performance. Nowadays, he is involved in sport programs in order to promote fair play, sport, and media.

Carlos Bacigalupe was a football referee for more than ten years in the referee committee of the Balearic Islands, and since 2012 he has played a trainer role in the technical staff of the Madrid Football Referees Committee (CAFM), basically leading on-field technification sessions and helping to design teaching materials. Carlos shares this function with his task as Referee Observer in regional and national categories. He has also taken part in the training of referees through the publication of texts and the production of successful video tutorials related to both technical and match managerial aspects.

Currently, he is an associated member of the CAFM board of directors, responsible for the recruitment and training of new referees, with an active role in the "Promises and Mentors" program focused on the close follow-up and early identification of young referees with great potential. He also takes part in different projects in the university environment aimed at improving the referee decision process, and in the instruction program of Referee Observers and Monitors.

Anto J. Benítez holds a degree in Information Science and a PhD in Audio-visual Communication from the Complutense University of Madrid. His thesis was titled "Espectáculo futbolístico y comunicación televisiva". Main lines of his research have to do with sports and media, new technologies of audio-visual production and its application in education. He is the author of the sole original volume in Spanish about producing and directing sports for television. He has been a television director for more than 15 years, some of which he was in charge of sports and broadcasting in Telemadrid. Thanks to that, he has planned and directed a huge number of sports events, especially in La Liga and La Liga 2. As an athlete, he competed at the highest level for 17 years in different rugby teams in Spanish Leagues.

Miguel Ángel Betancor is a professor of Theory and History of Education at the University of Las Palmas de Gran Canaria (Universidad de Las Palmas de Gran Canaria). He received his PhD in Psychopedagogy from the University of Las Palmas de Gran Canaria in 1988. He is also the director of the Centre for Studies on Sports Arbitration at the University of Las Palmas de Gran Canaria, a research centre dedicated to refereeing. He was also the coordinator of the online training project between FIFA and FIBA Europe and the University Las Palmas de Gran Canaria from 2000 to 2010, and was the head of operations at FIBA Europe, where he worked on the design of the digital basketball record. He also introduced technology in referee training, providing an online platform to referees and instructors from 51 countries.

Nowadays, he is the head of the European project "Erasmus PLUS SPORT", a new model of socio-healthy volunteering. He is also the head of the regional project "eSport Talent Canarias", which is aimed at the development of electronic sports and its recognition in the Sports Law and Gamification Life.

Juan de Dios Crespo has been a lawyer since 1985 and is the owner of the law firm "Ruiz Huerta & Crespo". He obtained a Master's in International Sports Law, LL.M. (H.C. ISDE) and Master in Arts from Rey Juan Carlos University of Madrid.

He was an external lawyer of the Spanish Football League from 2000 to 2010 and he was also a legal advisor of a large number of football clubs all over the world. In addition, he was a legal advisor of football players, agents, federations, and leagues.

He has practised as a lawyer in more than 300 courts of refereeing for sport cases. He was a professor in several masters in sports law around the world.

He has been referee on several occasions: referee of the European Court of Arbitration of the European Handball Federation (EHF); referee of the International Chamber of Commerce; referee of the

Valencian Chamber of Commerce; referee of ALADDE (Latin American Sports Law Association).

He has published more than 100 articles on sports law and he has written or co-written 14 sports law books.

Juan Pedro Fuentes holds a Bachelor's Degree in Sport Sciences from the Polytechnic University of Madrid and a Doctorate in Sport Sciences from the University of Extremadura with Extraordinary Award in 2001. He is a senior lecturer in high performance tennis, with an award for teaching excellence in 2014, and coordinator of the Laboratory of Control and Motor Learning at the University of Extremadura. He is National Referee of Tennis at the Spanish Tennis Federation and Director of the Centre of Olympic Studies of Extremadura University (Spanish Olympic Academy – Spanish Olympic Committee) since 2002. His research interests are control and motor learning, and psychophysiological response in sports.

Raquel Gallego is a doctoral candidate at the Carlos III University of Madrid. She studied a double degree of journalism and audio-visual communication. She holds two Master's Degrees in Sports Journalism and in Radio Journalism. She has more than ten years of experience as a journalist. She has practised in different fields such as press, radio, and web. She played football for over 15 years, first in Murcia and then in Madrid. In addition, she obtained the title of referee in the Federation of Soccer Region of Murcia, where she acted as a referee. Nowadays, she is immersed in her PhD, which is related to sports values.

Alberto Luis García García has been a lecturer of television technologies at the Faculty of Communication of the Complutense University of Madrid since 2006. For more than 15 years, he worked with sports broadcasts of La Liga as a replay operator. He has given different specialization courses on the techniques of live broadcasting in television. His research interests are focused on television production technology, UHD (Ultra High Definition), virtual reality (which was the subject of his doctoral thesis presented in 2000) and postproduction.

Ángel M. López is a television director specializing in sports with 12 years of experience in the department of sports realization of Telemadrid, participating in the TV live broadcasting of international events like the UEFA Champions League. He is also expert in video replay operation for the ACB Basketball League, performing these tasks in the 2015–16 and 2016–17 seasons.

He obtained a Master's in Applied Research to the Media and he is doctoral candidate in new technologies applied to the performance of live sports at the Carlos III University of Madrid.

Otto Kolbinger is a research associate at the Chair of Performance Analysis and Sport Informatics at the Technical University of Munich. He has published several articles about the application of rules in sports and the evolution of rule changes. As part of his PhD project, he developed a framework for the evaluation of technological officiating aids and performed evaluative research about goal line technology and the use of vanishing spray on behalf of the German Bundesliga (DFL).

Marta Fernández Ruiz holds a Bachelor's Degree in Audio-visual Communications and a Doctoral Degree in Interactive Media from Carlos III University of Madrid. She also has a Master's Degree in Teacher Training in Compulsory Secondary and Upper Secondary School Education, Vocational Training and Languages from Rey Juan Carlos University. She works as a project manager in the field of multimedia educational products at Labhipermedia, with strong experience in the development of tools for football referees and refereeing instructors. She is adjunct professor at Carlos III University of Madrid and her research interests are new media, interaction design, and user experience and learning experiences mediated by the use of technology.

Manuel Sánchez Cid holds a PhD in Audio-visual Communication from the Rey Juan Carlos University. He is expert in audio and audio-visual media. For more than 27 years, he has performed a double job as a university teacher and a sound expert in audio-visual companies. He is a co-author of different patents as well as technologies applied in sports, apart from broadcasting. He has carried out numerous investigations and developments.

Juan Antonio Simón is a professor of history and sociology of sport at the European University of Madrid. He is also the director of the Department of Sports Sciences. He obtained his PhD in humanities (history of sport) at the Carlos III University of Madrid. For more than ten years, he has been working in the study of football and its influence in contemporary society from perspectives ranging from studies on national and regional identity, media, gender, and the use of football as politics. At the same time, he has played as a professional footballer for 15 years in different Spanish football clubs.

Esteban Stepanian is a lecturer at the Carlos III University and occasionally at Antonio de Nebrija University, teaching media in different schools of art and design. He had a long professional career in the television production field. His lines of research are related to the possibilities of images, the techniques of expression, audio-visual production, and, currently, focus on experimentation about using an academic audio-visual language for research and teaching history. He is the author of various audio-visual productions on this and other matters.

Jacob Tingle obtained an Ed.D. in Educational Leadership, is chair of the Sport Management program at Trinity, and serves as the Director of Experiential Learning. For twenty years, Jacob was a basketball official working at many levels of the game in the United States. His research focuses on organizational factors which impact the recruitment and retention of sports officials.

Ricardo de la Vega holds a Bachelor's Degree in Psychology, and obtained a PhD in Sport Sciences from the Autonomous University of Madrid. He is an associate professor of sport and exercise psychology at the Autonomous University of Madrid, and also a professor responsible for sport psychology at the Olympic Centre for Advanced Studies (Spanish Olympic Committee). In 2016, he was awarded the silver medal for merits in research and by the Spanish Armed Forces. His research interests are: control and motor learning, and cognitive, emotional, and psychophysiological response in soccer referees, elite athletes, and combatants.

Clinton Warren is a lecturer and affiliated scholar with the Global Institute for Responsible Sport Organizations at the University of Minnesota, USA. His research focuses on sport business strategy with an emphasis on revenue growth and sport service innovation. He has specific interests in ticket sales, spectator attendance, fundraising for non-profit sport, and sport service development. His work examines these and other topics within the context of soccer in the United States in an effort to better explain American soccer's position in the business of global football. He has spent a total of nine years working in the sport industry as a ticket sales executive and a sales and marketing strategy consultant. He has worked with organizations in MLS, the NBA, the NHL, minor league hockey, and lower division soccer in the United States.

Tom Webb (PhD, University of Portsmouth) is a senior lecturer in the Department of Sport and Exercise Science in Sports Management and Development at the University of Portsmouth, UK. He has conducted consultancy research for the Premier League (Professional Game Match Officials Limited) and published extensively in the subject area of match officiating in sport. His latest book, *Elite Soccer Referees: Officiating in the Premier League, La Liga and Serie A* was published in 2017.

Preface

Football is important. It is part of a large amount of people's day-to-day motivation and, in many cases, has a social function that has to do with belonging to groups and with shaping people's identity. At the same time, it is undeniable that football is of utmost relevance as a sport and as a business. Due to its global nature, controversies generated by football-related events have a great impact. Because of its characteristics, the actions that decide the result of a game are infrequent and thus have a correspondingly high value. In this respect, referees have always been regarded as being responsible for mistakes and, more basically, for supporters' frustrations.

Over many years, FIFA has looked for solutions in the human side of refereeing, mainly by trying to increase the size of the team of refereeing officials.

Growth in the media industry has meant that all sports are increasingly global – and there is none more so than football. At the same time, audio-visual technology has, since the era of digitalization, undergone huge progress, and this has been applied to football-related broadcasts. More and more frequently, images taken from new angles, with greater spatial resolution and filmed at higher speeds, are placing refereeing decisions in doubt. Football has thus suffered a sense of immobility while other sports have applied technical progress to assist referees' decision-making.

During this time, social agents asked themselves why this technology was not employed to help referees in football. Football supporters felt ignored. Calls for change were finally heeded, and coinciding with a change in FIFA management (cause or coincidence?), plans and a schedule were drawn up to apply technology to assist referees' decision-making. In this schedule, from March 2016 onwards, steps were also taken to use video. Shifting away from an event-based focus, it can be seen that the motion to apply VAR means that there is a concrete point of support – and not the less transparent methods previously used – in the analysis of a new variable: from now on, it will be possible to measure the accuracy of refereeing decisions.

The idea behind this book, while untested, has a logic every bit as unstoppable as a FIFA ruling: the following lines are an attempt to take an in-depth look, from interdisciplinary points of view, at the latest issues surrounding the use of video as assistant to referees' decision-making, to carefully observe case studies in different stages of application, and to try to learn from this analysis and improve further steps.

A wide range of the social agents involved has been examined and allowance is made for the learning period that referees need for the new situations and new skills required.

In addition, the work employs an equally broad selection of perspectives via the shared research and experiences of specialists and academics that were invited to participate. This has enabled a diverse overview to be established and in this regard the editors wish to thank all those who collaborated for their efforts and generosity.

The book is divided into five parts.

Part I deals with the *philosophical* need to approve and release a version of VAR. In the chapter, 'Between Scepticism and the Longed-Awaited Quest for "Football Justice"', Juan Antonio Simón conducts a survey of a large number of the most famous errors in refereeing and describes, in the words of the players, technical assistants, managers, and referees in football, some of the most important arguments for and against, providing a complete overview of current opinion.

Next, Juan de Dios Crespo, in 'The Contribution of VARs to Fairness in Sport', writes on the need for an organization that supervises the development and application of rules, the IFAB. He reviews the evolution of football rules chronologically and then poses a key question: innovation or tradition? Offering data from the first experiences of using VAR, he arrives at conclusions related to predictable changes in the area of fairness in sport in which this system will be involved.

Part II refers to the new relationships that referee decision-making in football will have to deal with: with audio-visual professionals' approach to production and with new channels and forms of communication that have to use the VAR system equipment so that their opinions are immediately and unmistakably accessible.

In 'Is it the Same for the TV Screen as for the VAR? Planning the Use of Cameras and Replays to Solve the Controversy', Anto Benítez conducts an in-depth study of the technique that television broadcast professionals developed over the decades to plan the visual construction and carry it over onto the screen to achieve a show with an attractive narrative and drama. The conclusion is that VAR, also based on cameras and repetitions, must use the same tools and skills but must employ them towards a different end, one that is normative. In this way, VAR must discover its own planning and its own way of implementing it and, although it should act more or less independently, it may base its technical development upon already existing methods.

Manuel Sánchez Cid and Alberto García García make a careful analysis of the technological elements that comprise a communication circuit such as that needed for VAR. Based on extensive experience as professionals and academics, their study, *The Interaction Between Audio and the Video Assistant Referee in Football*, sets out the characteristics, the threats, and the possibilities of a system that must be at once reliable, inviolable, and robust, and which has highly accurate protocols and a structure based on efficiency. The authors emphasize that this has not been granted the same importance as those areas related to sound and video. In addition, they provide ideas for the use of information registered by the audio channels and they consider whether transparency in its treatment would contribute to good results in refereeing and, beyond that, to the sporting aspect of matches.

Alberto García García and Manuel Sánchez Cid undertake an analysis of the procedure regarding repetitions in football broadcasts, both in its professional aspect and its reception by the audience and specialized media. The chapter 'Replays in Football: A Technical Resource That Serves as a Narrative and Post-Refereeing System' provides an interesting insight into the journalistic aspect of the narrative of replays, not only in its instantaneous nature but also in the possibility of its later use for entering into in-depth debate on past events. Indeed, such a process may continue until it becomes a means of conducting a type of *post-refereeing*, one that also prolongs the spectacle far beyond the final whistle.

Part III is dedicated to reviewing the well-established experiences of video refereeing in other sports.

The chapter 'Instant Replays in the Spanish Basketball League: ACB' thus illustrates the experience of the early stages in the implementation of the Instant Replay (IR) system in the so-called *second most important basketball league in the world*: the ACB Spanish League. Ángel M. López, an expert live sports broadcasting director, holds the post of certified IR operator out of scientific curiosity and is now preparing a doctorate by conducting research into the application of technology to the media-based construction of televised sports. This adds even greater value to his analysis of workflows and the projection of future possibilities.

In 'Instant Replay in the National Basketball Association', Jacob Tingle and Manuel Armenteros analyse the workings of the system of video assisted refereeing in the NBA. One standout feature, which the governing bodies of football would do well to learn from, is the transparency with which the league treats its replay decisions. The league, in fact, regularly provides a collection of game plays with commentary about the rulings, in a remarkable effort to make even the most controversial refereeing decisions understandable.

Perhaps two of the best known examples of the progress made by video-assisted decision-making for sports referees are those dealt with in the chapter 'Video Technology for Refereeing in Other Sports: Tennis and Rugby'. Authors Anto J. Benítez, Esteban Stepanian, and Ángel M. López, specialists in audio-visual technology applied to televised sport, focus on showing that while this technology is capable of improving the diagnostic capacities of the refereeing officials, it is still not valid in itself. Exploring the implications of progress in spatial and temporal resolution, the authors not only set out a promising direction for further study but also define limits of the technology's use that should be taken into consideration and serve as an example for other attempts at establishing norms. In the analysis of two very different sports, it can be seen that video in tennis is the highest form of substitution of the decision-making figure, whereas in rugby the limits of technology are very familiar to players, technicians, and supporters, and they deal with it with a sporting attitude.

Probably one of the longest traditions related to referee video assistance may be observed in the National Football League (NFL). While there have been ups and downs and substantial modifications throughout the history of its use (even to the point of abandoning the system and then recovering it), the relationship between its referees with video began when technology was still analogic. In 'Instant Replay in the National Football League', Jacob Tingle looks at a story that involves the adaptation of available technology to achieve better entertainment.

Part IV of the book defines case studies in different stages of the VAR system's application in football. In addition to providing valuable data and conclusions for future comparisons, it is very interesting to observe how some aspects are more controversial in some leagues than in others.

In first place, Carlos Bacigalupe, football referee and trainer on the technical staff of the Madrid Football Referees Committee (CAFM), analyses the evolution of the VAR protocol and regards it, not as a point of conflict, but rather as a tool to help refereeing and which thus offers greater guarantees of fair play in football while the match is taking place. Until a more or less definitive protocol is established, 'The Video Assistant Referee (VAR) Protocol' complies with the function of describing and commenting upon the state of rules from within the system and, as such, acts as an essential reference for the whole of this text.

In 'VAR Experiments in International Matches', Manuel Armenteros reviews its implementation during two intense years and almost a thousand matches. The points of view held by IFAB, FIFA, and some of the most important international sporting media are taken into account. Comment is thus made upon the main problems that have not only been detected in its implementation but also in its design and configuration; for example, stoppage time while decisions are made or how to homogenize those cases when the system is used. Experiments with "offline" and "live" stages are reviewed chronologically and a description is given of FIFA's efforts and the roadmap that has been drawn up to describe possible weaknesses in the system. This deals with the ability to resolve doubts from football's great institutions as to how to increase fairness in decision-making without threatening the essence of the sport.

Otto Kolbinger, in his 'VAR in the Bundesliga', describes his research into a European league which has pioneered the use of the VAR system. He creates a chronological overview of the evidence and describes the most important incidents as well as the changes carried out as a result of these. According to his own previously established framework, the author shows how this type of process has a high degree of complexity and warns of the difficulty of correctly evaluating the results.

In Italy, authorities responsible for refereeing began the implementation of VAR with the idea of making a pedagogical effort related to the stakeholders, above all, technicians, players, and the press. Juan Antonio Simón has a detailed account in 'VAR Experiments in the Italian Serie A League'. Results from his research show that the controversy subsided as the protocol was followed with increasing efficiency. Special note is made of the importance of introducing the referees to the theory and practice of assistance technology. Subjects of debate also appear, such as the possibility of centralizing VAR rooms or the importance of correctly communicating the results to supporters.

If there has been one place that has always needed some form of technology to assist refereeing in football, that place has been in the MLS. In 'VAR Experiments in Major League Soccer (MLS)', Clinton Warren analyses the initial working period in chronological order, revealing the main incidents and the modifications that occurred until the end of the season of 2018. He ends on a positive note, taking care to record supporters' impressions.

The chapter that presents the research of Raquel Gallego is especially relevant for the times in which we live, as it relates how the implementation of VAR has been experienced on social media. 'How Twitter Has Portrayed VAR' focuses on how institutions, main media, clubs, and other stakeholders have made use of Twitter in relation to VAR. Opinions of players and ex-players and feedback from media and supporters are also analysed. Raquel reveals that messages favourable to VAR evolved with the training strategy adopted by FIFA, IFAB, and other leagues, as well as its use, described as successful, in the 2018 World Cup in Russia.

Part V of the book deals with the inevitability of VAR and focuses on its consequences on the psychology and future training processes of referees.

In 'Educating International Football Referees: The Importance of Uniformity', Manuel Armenteros and Tom Webb diagnose the different interpretations that have been shown by referees and video assistant referees on some incidents. A detailed review is made about the technical training of elite football referees and referees' instructors in FIFA and its confederations.

Another perspective about the training of referees in the use and the consequences of the new digital environment is also examined by Marta Fernández Ruiz, Miguel Ángel Betancor and Manuel Armenteros, in 'Didactic Teaching Strategies for VAR: An Approach to Gamification and 21st-Century Skills'. This chapter analyses the needs and elements of a didactic model that covers the digital skills training requirements of VAR, which is integrated in the framework of new technologies. Objectives and learning strategies are described and proposal is even made at the level of didactic design for activities based on gamification or on the flipped classroom.

Starting with the idea that the introduction of VAR is affecting how refereeing decisions are made, Ricardo de la Vega and Juan Pedro Fuentes conduct a detailed study from the perspective of referees' psychology in 'Psychological Response of the Referee to the Introduction of VAR'. They consider the basic endogenous and exogenous sources of information and concentrates his research on the impact upon refereeing decisions in terms of cognitive, psychophysiological, and emotional processes.

It should be generally noted that the bulk of research has been conducted with the time limit of the 2018 World Cup finals in Russia. As a result, subsequent modifications of the protocols or events beyond this date will only have been briefly considered.

After this event, various leagues that had applied to implement VAR are now using it; others have formally applied for it via the IFAB. Football's regulatory authorities have acknowledged that they have been overwhelmed by the applications. At the same time, the use of the protocol in different leagues is producing new subjects for analysis and their results can be compared with those of the Russian World Cup. However, the media and human resources assigned for use in VAR in different leagues are very different from those used in the World Cup. There is no doubt that new research will be necessary on the implementation of VAR after the 2018 World Cup in Russia.

The establishment of new avenues of gathering data and comparatives will be important in order to test the evolution of expected results. Thus, only experience will be able to demonstrate the effectiveness of the "check list" (see Chapter 10), which is expected to be especially important in controlling incidents that occur far from where the ball is in play (in general, situations of violent behaviour that lead to the direct expulsion of the player), and whether this additional surveillance leads to any improvement in the behaviour of certain players.

The editors hope that this volume is highly useful, especially for those interested in refereeing and football, for the academic groups in communication and sport sciences, and, precisely because of the varied nature of the work, for the public that is curious about the media and sports in general.

Acknowledgements

We would like to thank the following people/companies for providing complete translations of all the words and phrases used in the book: Sebastian Salt (Preface), Harvey Holtom (Chapters 1, 2, 3, 8, 10, 11, and 18), Mark Sefton (Chapters 4, 5, 6, and 8), and Joaquín Marcos (Chapters 3 and 8).

We are grateful to Hailey Wilson, an undergraduate student research assistant at Trinity University, San Antonio, TX. Her contribution to Chapter 7 was invaluable.

We would like also to express our gratitude to Miguel Moscoso, Bruno Lázaro, Javier Fernández, Javier Benítez, Fernando Tresaco, Enrique Ortiz, Steve Bennett, Subkhiddin Salleh, Alan Snobby, Héctor Vergara, Krystina Szokolai, Peter Prendergast, and Alain Lim.

Part I

The Need for Video Assistance in Football

1 Between Scepticism and the Long-Awaited Quest for "Football Justice"

Juan Antonio Simón

Introduction: The Advent of Video-Assisted Refereeing: Changing to Improve?

In recent years there have been several studies from areas of research such as social sciences and humanities that have analysed the role of football referees, improvements in their training, and the relevance of the inclusion of new technologies in this process (Niandu, 2012; Ryall, 2012; Webb, 2014, 2016, 2017). However, so far there have been no studies of proven scientific quality that have addressed the impact that the implementation of VAR has had on the world of football, and how some of its main players are assimilating this revolutionary process. The main critics of the application of this new technology emphasize that its implementation causes multiple stoppages during the game, directly affecting its flow. At the same time, some actors who are clearly against VAR have had an impact, presenting errors as something that is part and parcel of the game of football, with both the errors made by the player and the referee being part of the essence of this sport. The true essence of football would appear to be in danger with the inclusion of new technologies, since the passionate relationship the fans have with the goal may be dampened when you have to wait several minutes to celebrate it being awarded. In the same way, it has also been pointed out that this technology is still subject to human interpretation, or that football is simply different from other sports such as American football, basketball, rugby, or tennis, where technology is already being used to review the most controversial decisions or the most decisive points.

But, together with the fiercest opponents of the inclusion of VAR technology in football, every day we are finding more and more defenders who believe that the application of these new technologies will help prevent many of the big mistakes that, as we will see later, have happened throughout the history of this sport. Equally, VAR allows the decision taken by the referee previously not only to be ratified, but also confirmed and strengthened. For those who defend this technology, the power to avoid obvious errors in referees' decision-making is a reason that more than justifies its implementation, even if to a greater or lesser degree it interrupts the flow of the game or transforms the traditional celebrations of players and fans after a goal.

This chapter begins with a historical journey through some of the most flagrant mistakes in the history of football, those refereeing errors that have marked the memory of this sport and have generated an important polemic regarding central issues such as the role of the referee, sporting "justice/injustice" in football, and the consequent need, or not, for the application of technology to reduce human error. Next, it is analysed how the assimilation process of this technology is taking place among its main actors: the players themselves, team managers, referees, and coaching staff. This will be done by means of analysing the information published since 2016 on this topic in the international press and online media, both those offering general information and those specifically dedicated to sports news; the impact of this transformation on the actors most directly involved; the problems generated by the implementation of such innovative technology in a sport that has historically been most reluctant to change; the criticism and praise of VAR that has arisen in each of the national and international competitions in which it has been implemented; and how this has been expressed through the main newspapers.

The History of Football Seen through Refereeing Mistakes

It would be impossible to list all the refereeing errors that have changed the history of this sport: decisions taken by referees in highly stressful contexts that have subsequently been proved wrong and that have directly influenced the outcome of the match, which have decided national and international titles and competitions, and that have made a large number of fans jump for joy or cry with frustration. A large segment of the world of football understood that these mistakes were part of the game, they were part of the very essence of this sport and should be assumed as such. Meanwhile, over recent decades, and especially since the development of the information society and the inclusion of innovative technologies in other sports such as tennis, basketball, or rugby, there has been an increasing number of voices that consider it unfair and inadmissible that we keep turning our backs on the use of new technologies in a sport such as football that generates passions and economic benefits in equal measure. FIFA, since Gianni Infantino took over the presidency in 2016, has backed the inclusion of VAR as a tool to avoid flagrant and unquestionable errors in football matches, ones which can decide the outcome.

In the memory of all football fans is the 1966 World Cup final in England, and what would go down in the history of this sport as "the ghost goal" by Geoff Hurst. The final was between the hosts and West Germany, and, with the score at 2–2 after reaching the end of normal playing time, it went into extra time. In the 101st minute, Hurst struck a shot from inside the penalty area that beat the goalkeeper and hit the underside of the crossbar before bouncing on the grass and being headed away by Weber. Gottfried Dienst, the Swiss referee, gave the goal, but even today it is impossible to be 100 per cent sure that the ball really went beyond the goal line. That shot made the score 3–2, a result that Hurst himself would later extend by scoring the fourth and final goal.

The 1978 Argentina World Cup was also marked by refereeing errors. In the final, between Argentina and The Netherlands, the referee did not sanction an elbow by Pasarella on Poortvliet nor a foul by Larrosa that would have earned him his second yellow card of the game. Spain 1982 was also not free of controversy in a semi-final that brought West Germany and France face to face, and in which the German goalkeeper Toni Schumacher came out to challenge Patrick Battiston and clattered him, causing a spinal injury and the loss of two teeth while the match referee did not even blow for a foul. Another mistake that has marked the history of this sport was the so-called "hand of God" by Maradona in the quarter-final of the 1986 World Cup in Mexico between Argentina and England. As the English goalkeeper, Peter Shilton, came off his line to try to catch a loose ball in the area, the Argentinian captain jumped up and hit the ball with his fist past Shilton and into the net. He ran to the side of the pitch to celebrate the goal while looking sideways at the Tunisian referee, Ali Bennaceur, and the linesman, neither of whom had seen the handball. No less controversial was the refereeing decision that meant that France got their ticket to the 2010 World Cup, by drawing with the Republic of Ireland (1–1) thanks to a goal that was preceded by a clear handball by Thierry Henry. The French player himself told the press at the end of the match that

> The ball hit my hand, I will be honest. It was a handball, you can clearly see it. (Sebastien) Squillaci went to jump with two Irish players, I was behind him and the next thing I know the ball hit my hand. It was a handball, but I'm not the ref. I told (the referee) but he said to me the same: "You are not the ref."
>
> (White, 2009)

Italian fans still remember the controversial decisions of the Ecuadorian referee Byron Moreno, during the Round of 16 match between South Korea and Italy in the 2002 World Cup that led to their elimination. He awarded a dubious penalty in favour of the hosts in the first half, sent off the Italian striker Francesco Totti in extra time by showing him the second yellow card for diving when in reality he had been tripped over by a rival, and disallowed a completely legal "golden goal" by Damiano Tommassi for an inexistent offside. The elimination of the Italian national team is still remembered bitterly in the country.

In that same World Cup, Spain saw how in the quarter-final tie against South Korea, the referee Al Ghandour disallowed a goal by the Spanish striker Fernando Morientes, considering that the ball had left the field before Joaquin centred to him. The images, however, showed that the ball really had not crossed the goal line. This happened in extra time, and the goal would have given the victory to Spain and put them through to the semi-final because it was a "golden goal".

But the discussions about the refereeing decisions do not end here. The mistake made by Graham Poll during the 2006 World Cup, when he showed three yellow cards to the Croatian player Simunic before sending him off in the group stage match against Australia, or the performance of the referees in the knockout rounds of the 2010 World Cup in South Africa, raised once again in the minds of the public the idea that the time for technology to be applied in football had arrived. Regarding what happened in the 2010 World Cup, it is interesting to remember that in the Germany–England clash in the Round of 16, the referee Jorge Larrionda and his assistant did not give a goal by Lampard, whose looping, long-range shot, after hitting the underside of the crossbar, bounced almost a metre inside the German goal and would have levelled the scores at 2–2. "The referee's big mistake was what prevented us from advancing in the competition, we are talking about a match in which, if the referee had given the goal, everything would have been different," said Fabio Capello, coach of the English national team. Lampard himself acknowledged his disappointment and his "support [for] the use of technology for the goal line, it was a huge failure and a big disappointment" (Carlin, 2010).

In the Argentina–Mexico match, the Argentinian player Carlos Tevez scored to make it 1–0, but he was clearly offside when Messi played the pass to him. The Italian referee Roberto Rossetti, after consulting with his assistant, decided to give the goal even though the whole stadium at that time was watching the repetition of the play on the electronic screen, proving that Tevez was in the wrong position. During the World Cup in South Korea and Japan, in 2002, the repetition of the images on stadium screens triggered the old dispute over the clash between refereeing and technology. FIFA studied the case and decided to ban repeats in the 2006 Germany World Cup. However, in the World Cup in South Africa, footage was again shown in the stadiums, giving rise to great controversy. From that moment FIFA prohibited replays (Torres, 2010). These refereeing decisions generated great controversy in public opinion. A large number of media outlets took up positions in favour of the inclusion of new technologies to avoid such situations once and for all. Among them *La Gazzetta dello Sport*, who addressed FIFA President Joseph Blatter and the members of the International Board to complain, among other things, about how

> the serious, and unfortunately unavoidable, refereeing oversights put a strain on the credibility of football, which has remained almost isolated, compared to other great sports, in not opening up to new times: various technologies help, for example, the "reading" of basketball, rugby, tennis, American football, etc.[1]

> [...] "We believe that football and your competence have the possibility to get out of this serious situation, both by increasing the number of referees and by studying in depth the possible contribution of television images or other forms of technology".[2]
>
> ("Al presidente della Fifa", 2010)

FIFA's response continued to be contrary to the application of new technologies, declaring that "as regards the goal-line technology, FIFA's position remains in accordance with the decision taken by the International Board in March", in which it was decided to discard the idea of a chip embedded in the ball and the help of the video images (Fest, 2010). On the other hand, FIFA continued to bank on increasing the number of officials on the pitch to minimize errors. Since the 2009–10 season, the Europa League has used two assistants and two area judges to accompany the referee in each game. They were incorporated in the Champions League the following season and subsequently included in all competitions ("Los cinco árbitros", 2010). Each area judge would be placed next to the goal, just behind the goal line, and would be a key element of consultation for the referee at the time of clarifying a possible penalty or if a ball has completely crossed the goal line. FIFA was opposed to the use of technology to correct human errors, resorting to the argument of the game's universality, since it should be possible for this sport to be played in the same way in a small town as at a professional level, and stressed its simplicity, the human factor, the passion of the discussion generated by the incidents, the economic cost of implementing the technology, the risk of replacing the referee with a machine, and the dynamism of the game.

The Opinion about VAR of the Main Actors in International Football and its Impact on the Media

Players and Former Players Give Their Thoughts on VAR

It is necessary to remind readers that before using VAR in the 2018 FIFA World Cup in Russia, this technology had already been trialled in different leagues and international competitions in its two modes, offline and live (see Chapter 11). During the FIFA Club World Cup held in Japan in 2016, the Video Assistant Referee (VAR) could not avoid taking centre stage from the first qualifiers. In the semi-final between Kashima Antler and Atlético Nacional on 14 December 2016, referee Viktor Kassai awarded the first penalty in an official FIFA match with the help of VAR, taking about two minutes to make the final decision, which clearly indicated that there was a need to improve the existing communication between the referee and the video assistant to reduce the excessive waiting time. In the other semi-final between Real Madrid and Club América, there were again communication problems between referees, which provoked harsh criticism in some media. These first problems of coordination in the decision-making generated the first criticisms and doubts about the improvements this new technology would really bring.

Real Madrid footballer Luka Modric was very critical of VAR experimentation in declaring that "I do not like this video-arbitration because it creates a lot of confusion. I hope this rule does not continue, for me it is not football." The same footballer recognized that in the meeting that the referees had held with the players before the competition to explain what the characteristics of this technology were and how it would be applied in the field, he had not paid much attention "[...] because I hope that this rule does not continue." Other teammates, like Lucas Vázquez, recognized how strange it was for the players themselves to have to wait until a referee's decision was confirmed: "it's a little weird to wait for someone to decide from the outside. Perhaps the essence of football, which are the decisions taken on the spot, is lost a little." Even the manager himself, Zinedine Zidane, was very cautious and reserved his personal opinion regarding the new technology, but pointed out that "things should be clearer for everyone, more so if the technology wants to improve them" (Giovio, 2016).

A year later, in the same competition, the FIFA Club World Cup 2017, in the semi-final between Al-Jazeera and Real Madrid there was once again controversy when a goal by Casemiro was disallowed for offside after more than three minutes of waiting. Gareth Bale declared after the game to Spanish television that "I do not like VAR to be honest. I believe that football is better without it," while Zidane himself admitted that it was not pleasant to be waiting so long for a referee's decision, although he understood that "you have to take the positives from these things and try to improve the negative things" (Jones, 2017) (for more details see Chapter 11).

Both in the Italian and Portuguese national championships, which had been among the first leagues to introduce VAR, and after the discussions that emerged in the early days and which led to the recognition that it was necessary to make certain improvements in specific aspects of the protocol, the use of VAR began to be perceived positively from large sectors, demonstrating that it is significantly reducing the number of errors that occur in matches, although there is no doubt that the controversies and discussions regarding certain refereeing decisions have continued (see Chapter 13).

On the other hand, the introduction of VAR in German football raised serious doubts in public opinion regarding the true effectiveness of this technology and its proper application. After the controversy that was aroused in the media after the first fixtures in which it was used, the Fußball-Bundesliga decided not to apply VAR in offside decisions, and in November 2017 the project manager and supervisor of VAR-Centre in Cologne, Hellmut Krugg, was sacked (see Chapter 12). The German magazine *Kicker* reported in early 2018 that 47 per cent of a total of 219 players in this country were against VAR; while only 42 per cent accepted it should continue and the remaining 11 per cent did not position themselves either in favour or against (Pérez, 2018).

One of the great legends of football, Diego Armando Maradona, has defended the application of VAR from the first moment by understanding that football could not be left behind in this sense: "if technology advances as it advances, if all sports use it, how are we not going to consider it for football?" One of the aspects that has been criticized from the beginning is the loss of the rhythm of the game and the excessive time it takes to make a final decision. Maradona is very clear in pointing out that

> people get annoyed when you call something that is not, when they do not give you a goal that was, for example, technology brings transparency and quality; gives a positive response to the team that decides to attack and take risks.
>
> ("Maradona: 'Con tecnología'", 2017)

It was not the only time that Maradona has defended the need for VAR. In 2018 he made a statement defending the application of the new technology in football and criticized some of the great names who had directed the sport in recent years: "the referee has one more weapon to tell the public that this new FIFA is another way for the transparency we asked for when football was in the hands of that thief Blatter and that thief Grondona" ("La razón de peso", 2018).

Michel Platini, former UEFA president and a famous ex-player, argues against video refereeing in football, offering two reasons: "football has also based its popularity on injustice. You can remember them and talk about them in the bars. You can talk about 1982, France–Germany, it was an injustice like the hand of Maradona, or that of Henry." Another of the aspects that Platini wields to position himself against the application of VAR are the stoppages involved in reviewing each of the moves or actions: "it would mean ever-more-frequent stoppages to check infractions of various kinds and that this would be 'the death of football,' since football is a game that ideally flows on and on" (Early, 2018). Platini came to declare that it was "against the spirit of football", and that many times the footage can deceive the referees in their assessment: "Since the beginning, I am against VAR because, when I was a player, I realised that the televisions did not always tell the truth of the ground" (Perform, 2019).

In Italy, the errors that were noticed during the first days of implementation of VAR, and that in most cases were caused by the need to improve the coordination between the on-field referee and the VAR, aroused harsh complaints among the players themselves. Regarding the above, the goalkeeper and captain of Juventus, Gianluigi Buffon, after the match against Genoa in which VAR had taken the lead role for the media, by modifying two decisions that became separate penalties, told the press that

> I do not like VAR like that, we are going over the top. This seems more like water polo, not every contact in the area should be sanctioned. And I say it to my regret, because Juve attack, and with this system we would get 55 penalties. But this is no longer football, it is laboratory football.

In spite of everything, Buffon stressed the importance of changing the excessive tension with which, in his judgement, refereeing decisions were experienced:

> If we were all calmer, more objective, more accepting of errors, we would take any result with more humanity. Video refereeing, in moderation, can give great results and improve this sport, but not like that, I do not like it, I repeat it.
>
> (Calemme, 2017)

In other less prestigious international competitions, like the Malaysia Super League, most of their coaches seem to be in favour of this technology despite the fact that it will probably still be several years before it can be implemented in the country. Australian coach and former player Durakovic said, "Everyone makes mistakes and if the decision is right, then it's right. […] Referees don't always see everything so I'm in support of video technology as long as it does not hold up the game too much" (Singh, 2018). Malaysian football coach and a former footballer Sathianathan said football is now too fast for the human eye to follow. "The pace of modern football has increased, and referees are only human, and they struggle to cope with it." Even he thinks that "because now the play is very fast, the breaks in play make the game interesting" (Singh, 2018).

In England, and after its use for the first time in the international friendly held in November 2017 at Wembley between the national teams of England and Germany, VAR was presented at club level in January 2018, specifically in the match between Brighton and Crystal Palace (2–1) in the third round of the FA Cup. In the middle of that same month, in the match in the same competition between Leicester City and Fleetwood Town, the first modification of an initial decision of the referee was seen. After initially ruling out a goal by the Leicester player Kelechi Iheanacho for offside, the official was notified by VAR and finally awarded it (Miller, 2018).

The Real Madrid defender and captain, Sergio Ramos, expressed his doubts about how VAR was being applied after the defeat to Real Sociedad in a league match. The referee did not interpret a play involving Vinicius in the Real Sociedad area to be worth a penalty, and did not resort to VAR. Ramos called for the technology to be used in this type of play: "… the one on Vinicius is scandalous. I've always supported VAR, but this leaves you with your head in your hands. Why wasn't the opportunity taken [for the referee] to have another look at it?" (De Juan, 2019). Weeks later, the Real Madrid captain commented to the media after the match against Atlético de Madrid that "VAR is fair enough; I am very much in favour" (Torres, 2019).

Luis García, a former player for, among other clubs, FC Barcelona and Liverpool, also showed his doubts about video-assisted refereeing on his personal Twitter account during the recent Russia World Cup:

> VAR makes players lose the focus of the game, waiting for a possible change of the referee decision. Argentina had a corner and players were looking at the referee instead of winning the ball in the air!! Just an opinion.
>
> (García, 2018)

It was obvious that some players were largely unaware of how VAR protocol worked before the start of the World Cup, especially in the case of players who played in national competitions in which this technology had not yet been introduced. Such is the case of the English forward of Leicester, Jamie Vardy, who after the friendly played against the Italian team felt that the system still had to improve because "most of it is down to players and supporters not knowing what is going on" ("VAR in football", 2018).

The Ambivalent Opinion of Team Managers Regarding VAR

Like players, team managers are another of the key groups in the world of football and a determining element for VAR to be applied successfully. During the first months of the implementation of video arbitration, various positions could be observed among the managers, ranging from those who were radically opposed to those who defended at all costs the inclusion of new technologies in football as the only way to make the game fairer.

France coach Didier Deschamps has been a defender of this technology from the beginning. In March 2017 he stated that the TV replays will prove the "evolution of football" after his team had lost against Spain, a game in which VAR had overruled a goal for France ("France coach Didier Deschamps", 2017). During the last World Cup in Russia, the French national coach once again celebrated the use of VAR when the video arbitration gave his team a penalty in the match against Australia (2–1), and said that the technology "allowed an error to be corrected" and that

> at the moment an error is corrected it is positive, I cannot say it is easier, since we suffered it in a friendly match and the decision was against us, that's the way things are and we have to get used to it. We know that in certain situations VAR could be used.
>
> (Deschamps, 2018)

An international reference of the prestige of Alex Ferguson (former manager of Manchester United) reflected on the application of this technology at the end of August 2017 in the UEFA Elite Coaches Forum, held in Nyon. In his opinion, and as ambassador of the coaches' group in UEFA, he commented that it was necessary to give VAR a chance, although he was not sure that it was being implemented in the correct way: "VAR, I don't think anyone in the room is completely convinced, but we're prepared to wait and see some of the examples" (Greaves, 2017). Fabio Capello, former manager of the English team and clubs as prestigious as Real Madrid and Juventus, among many others, has also recognized that he has always been favourable to technology, but considers it necessary to be "carefully evaluated in some episodes, because there are shadows that it is necessary to understand" (Capello, 2017).

In the final of the 2017 UEFA Super Cup, Manchester United lost 2–1 to Real Madrid. In the post-match press conference, José Mourinho declared that the result would not have been the same if there had been video-assisted refereeing, claiming that this would "explain Madrid's dominance in the first half because their first goal was offside. With VAR it would have been ruled out and we would have had another game at 0–0"(Melgar i Anguera, 2017). Mourinho has been one of the coaches who has most defended the application of this technology, as demonstrated in the statements made to FIFA.com in 2017:

> We all need it. Professionals cannot lose or win matches and titles because of a refusal of this evolution. Sponsors, owners and investors must feel that technology is there. Also, referees especially need and deserve protection. They need the technology to help them, protect them and to support them.
>
> ("11 questions for", 2017)

In contrast, Iran manager Carlos Queiroz clearly showed his displeasure with VAR, when in the match against Portugal during the 2018 World Cup, a player from his team went down after being elbowed by Cristiano Ronaldo who, in his opinion, had committed a clear red-card offence but for which he only received a yellow card. Queiroz, ostensibly angry, vented his opinion in the press room: "My suggestion, Mr Infantino, is you put up your hand and say: 'Sorry, VAR is not working so we have to stop it,' or the communication must be clear like rugby, when everybody knows what is going on." On the other hand, his Portuguese colleague Fernando Santos had seen the incident in that match in a completely different way: "I'd have to watch it again to talk about it but I think VAR did its job and that's what we have to accept" (Walker, 2018). What seems clear is that VAR will not be able to put an end to controversy in football any time soon.

Jorge Jesús, manager of Sporting Clube, one of the most prestigious clubs in Portugal, was emphatic when he welcomed the introduction of VAR into his national competition in May 2017:

> Both Sporting CP and the Portuguese Football Federation deserve to be congratulated for this decision. The video assistant referee is really important for Portuguese football. We have great coaches and clubs in this country, but we don't have the financial resources others do and it is important that we are on the cutting edge of technology. It won't fix everything, but it will certainly help.
>
> (Journal Sporting, 2017)

Some Italian coaches also expressed their opinions after the first five months of VAR's application. In this regard, the Inter Milan manager, Luciano Spalletti, recalled that it was necessary to adjust some of the aspects of VAR, but that the numbers were saying it was working correctly and that "it would be devastating at this moment to go back".[3] One of the critical points made by this manager referred to the moment when offsides are flagged, which in some cases are delayed too long while waiting for the response of the video: "Our linesmen are good, they know how to choose, they have a percentage of correct that means they should be given more responsibility than the VAR system allows them."[4] Stefano Poli, the Fiorentina manager, also showed his support for VAR: "I believe VAR is an element of greater justice. There are always mistakes made, but now the number is much lower. I think that in the future, some of VAR decisions will be screened in the stadium, making them public."[5] Regarding the possibility that the coaches themselves could request the intervention of VAR during the match, he mentioned that "the rules don't allow it and the referees have stressed this situation, but the referees are improving and are now better than they were at first, thanks to the efforts that have been made, so things are working out well" ("Il Var e i suoi limiti", 2018).[6]

Luis Diego López, the Cagliari manager, also gave his opinion, which was in favour of the VAR system, and pointing out how important it was to keep having meetings of this type in which they had the opportunity to discuss the most controversial plays that had occurred over the previous months: "[…] obviously there have been mistakes, but the important thing is to keep growing, and this meeting has served exactly that purpose". Even one of the managers who had been most critical of the technology over the initial period, Simone Inzaghi, of Lazio, recognized that, despite his opposition to the VAR system, the analysis of the early results showed that the process was going well: "it is helping football and it is only fair to carry on ahead". Inzaghi praised the decision to hold meetings of this type, in which both referees and managers were able to analyse the controversies that had arisen in recent games ("Serie A, Spalletti", 2018).

Marcelo Lippi, one of the most prestigious coaches and with a considerable amount of success in this sport, speaking from his position of manager of the China national team, voiced his scepticism regarding VAR, although he was not against the application of technology *per se*, but reduced it to very specific situations such as "phantom goals": "Ok the 'goal line technology', but otherwise I would leave everything as before" (Cina, 2017).[7] In contrast, Mauricio Pochettino, manager of Tottenham Hotspur, believes that mistakes are part of the game and that it is necessary to assume them: "Football is also about mistakes, like us, players can make mistakes, referees make mistakes. It has always been like that and we understand football in that way" (Pinochet, 2018).

What Do the Referees Think of the New VAR Technology?

One of the most interesting aspects of the study of the implementation of VAR in football is to analyse how the inclusion of this technology is affecting the role that referees have traditionally played on the pitch, and how their authority in the development of the game can be modified. From the first moment, FIFA has been very clear in emphasizing that, despite the use of VAR, final decisions are still taken by referees. For that reason, it was decided that before changing a decision in a game, the referee should be able to see the videos of the play at the edge of the pitch. In this way, FIFA wanted to make clear that the referees still have the final responsibility for all decisions during the match (see Chapter 10).

On this subject, the former Premier League referee, Graham Poll, has clearly defined the two positions in which the referees currently stand regarding the inclusion of VAR:

> From people I have spoken to, opinions are split over the system being trialled. Some are happy to have their authority diluted and, like goal-line technology, would be happy for the final decision to be taken away from them. However, others do not want to have their authority undermined. In terms of credibility, it can also be difficult for a referee to admit they are wrong. However, referees are generally happy for a VAR experiment. The game has changed. If everybody is judging referees' decisions on replays – and as long as players continue to try to deceive referees – then it makes sense that officials are given the same tools.
>
> (Poll, 2017)

It is interesting to note that in January 2018, the members of the Italian Refereeing Association (AIA) and the Serie A coaches met to analyse the results of VAR after the first phase of implementation in Italian football. One of the first conclusions they reached is that there is no doubt that errors had existed, but at the same time there was no possibility of retracing the path taken. The meeting was led by the highest representative of the AIA, Marcello Nicchi, as well as by the person responsible for appointing the referees, Nicola Rizzoli, and Roberto Rosetti, the supervisor of the VAR project. Marcello Nicchi, president of the AIA, has defended that "the VAR system is a product that works. There have been mistakes, and there will be more, but if we think that in just five months we have got to this point, then I am not at all troubled" ("Il Var e i suoi limiti", 2018) (see Chapter 13).

The chairman of the FIFA Referees Committee, Pierluigi Collina, declared after the conclusion of the Russian FIFA Confederation Cup that he wanted to share the message of one of the referees of the tournament regarding VAR, in which he recognized that he was enjoying the matches because he felt less pressure. Collina argued that the referees

> know that sometimes they can make the wrong decisions simply because they do not have the angle of vision that one of the more than 30 cameras can offer. VAR is a very positive instrument for sport, because it helps referees avoid mistakes.
>
> ("Elogios a Rusia", 2017)

Collina also informed the media that the decision on whether VAR would finally be used in the next World Cup in Russia would be taken in March 2018. Similarly, he did not want the fans to forget that VAR "is an experiment at an advanced stage, the results achieved so far will be evaluated at the next IFAB meeting". Although a final decision has not yet been made, FIFA is already working on the preparation of the referees. Gianni Infantino himself, as president of FIFA, supported in January 2017 at the Globe Soccer meeting held in Dubai the use of technologies to help referees in the taking of decisions. After admitting his initial scepticism about VAR, the positive results made him change his mind radically and he then understood that "new technologies can make refereeing decisions much simpler when there are doubtful episodes which could impact single games or whole tournaments" (Amici, 2017).

The IFAB secretary, Lukas Brud, also acknowledged in December 2017 that it was still "not very clear when the video assistant referee should intervene", and that it was necessary to work towards an "international harmonization, it is a question of training and experience". Brud wanted to make it clear that he was trialling "something totally new and, as no referee wants to make mistakes, they prefer to resort to VAR one time too many rather than one too few" (AFP, 2017).

As happened in Italy, at the beginning of January 2018, José Fontelas Gomes and João Ferreira, respectively president and vice president of the Conselho de Arbitragem (CA) of the Federação Portuguesa de Futebol (FPF), organized a technical meeting together with 16 of the 18 representatives of the country's top-flight teams. In that meeting, the basic principles of the use of VAR were recalled and some of the most controversial moves that had taken place in the international competitions and some of the most common difficulties encountered in the application and interpretation of this technology were all analysed. The CA also informed in this meeting about the role that David Elleray, the technical coordinator of the IFAB, was to have, one in which he would work together with the FPF in training referees to use the VAR system. In 17 days, in the Portuguese league, VAR had had 991 checks, with a total of 41 plays being reviewed, 28 of which led to the referee modifying his initial decision (Federação Portuguesa de Futebol, 2018).

But the implementation of VAR in Portugal was also accompanied by controversy. In December 2017, the CA sent the clubs a dossier in which they were reminded of VAR's intervention mechanisms and the indications that had been given since the beginning of the season were clarified so as to reinforce them. Over the previous months, the CA had been accused of duality in the application criteria of video refereeing and some clubs such as Benfica had even requested that audio communications between the on-field referees and the assistants in VAR be disseminated. Regarding this aspect, the CA was adamant:

> [We will] never divulge anything at the request of any club and take a dim view of clubs making demands of this type, creating noise and trying to confuse the fans, in a clear show of contempt for the pedagogical action.
> ("CA reforça lição de VAR", 2017)

José Fontelas Gomes had been very clear from the outset of VAR's application in Portuguese football, stressing that the main objective of this technology was to "end some serious errors, those that we all see on television. However, the error will always exist in football. We are human beings. Video arbitration seeks to assist the referee, but the error will continue to exist."[8] What was seen was the important interest the CA had in carrying out instructive action with all the actors involved in football to help them understand in depth how VAR works:

> We have trained everyone, We have also had an impact on social communication, we opened a Twitter account to reach the fans, we tried to give all the information. As in all projects, there is a maturation phase. There is always distrust at the beginning of each process.
> ("Fontelas Gomes", 2017)[9]

In Spain, former referees were also critical of how some of their serving colleagues were applying VAR protocol. In particular, Arturo Daudén Ibáñez, who was an international referee between 1998 and 2005, expressed his opinion regarding the controversial play that occurred in the league match between Real Madrid and Real Sociedad in January 2019. The referee was criticized for the decision he took then as well as for not referring to VAR, and Daudén Ibañez, in his Twitter account, commented that "refereeing is about 'taking decisions' under pressure. Little use is being given to the pitch-side monitor. The correct decision of #VAR would have been 'go back to see the action of #Vinicius'" (Daudén Ibáñez, 2019).[10] Similarly, former Spanish referee Eduardo Iturralde González would also give the press his opinion regarding VAR and the need to extend the protocol, advising that it would be opportune to be able to "give the referee the opportunity to review the television replays", or "to be able to show the images at the end of the game, especially the offsides". There are some measures that Iturralde González believes could improve the application of this technology, though he has always positioned himself as a convinced defender of its benefits: "I do not know what interest is behind the attempt to discredit a tool that has prevented 59 errors in 20 days" (Iturralde González, 2019).

Conclusions

In January 2018, FIFA's Chief Commercial Officer, Philippe Le Floc'h, told the press what he thought: "It is fantastic to use technology in football because it ensures justice, avoiding clear and obvious mistakes." He also recognized that FIFA was looking for sponsors to take advantage of VAR replays during the World Cup in Russia and include in these spaces advertising of technological products: "There are several companies in the sector that have shown interest in the project" ("Fifa, svolta storica", 2018). These statements confirm that the implementation of VAR in international football is an unstoppable process, but at the same time it is necessary to manage it in an appropriate manner, so that from other segments of the football world it is not understood that its integration has as a final objective to try to generate greater economic benefits through advertising or distort the essence of this sport.

In a testing phase like the one we are currently experiencing, it is important that the detailed knowledge of the operating process of this new technology be disseminated to the masses in an appropriate manner. FIFA and IFAB together with the national federations and the referees' committees of each country must understand that VAR will not succeed if suitable pedagogical action is not taken. Many of the controversies that have arisen in the countries where this technology is currently being developed are due to the lack of knowledge of VAR's protocol of action. It is necessary that the big television platforms and the most important press head offices work together with the referees' committees of each country, to shape communication strategies adapted to each national championship. In this way, a didactic function must be carried out, both among coaches and players, as well as media professionals and, of course, among the fans.

At the same time, it is convenient to deepen the development of new methodologies for the training of referees, and in matters that until now were completely alien to their function, such as the use of new technologies adapted to audio-visual environments and production and television performance. It would be very important for each country to create a centre specialized in the training of referees in VAR, in the same way that it is also essential that a reference centre for the training of referees in this type of technology be developed internationally, allowing everyone to access training of the same quality and with the same "standards", carrying out refresher courses and updating throughout the year.

It has been more than a century since athletics tests used the "photo finish", in the NBA and in European basketball the "instant replay" is applied to review doubtful plays, in the same way that video has been used for some time in sports such as rugby and hockey, and since 2006 the "Hawk-Eye System" has been used on tennis courts to correct the possible errors of officials. For much of its history football has been characterized by its radical opposition to any technological innovation, having to wait until quite recently for FIFA to decide to back the development of video refereeing. Nevertheless, VAR, unlike "Hawk-Eye", still has an interpretation component, so it will still be impossible to eliminate completely human error. The march towards greater fairness in football by means of the use of new technologies now seems unstoppable, but it is still too early to say how this will change the way the sport is practised and the experience of the fans, both those present in the stadiums and those watching in their own homes.

Notes

1 Original: "Le gravi, e purtroppo inevitabili, sviste arbitrali mettono a dura prova la credibilità del calcio, che è rimasto quasi isolato, al confronto di altri grandi sport, nel non aprirsi ai tempi nuovi: varie tecnologie aiutano, per esempio, la 'lettura' del basket come del rugby, del tennis come del football americano, etc."

2 Original: "Riteniamo che il calcio e la vostra competenza abbiano in sé la possibilità di uscire da questa seria impasse. Sia aumentando il numero degli arbitri sia studiando con profondità il possibile apporto delle immagini televisive o di altre forme di tecnologia"

3 Original: "sarebbe devastante in questo momento tornare indietro".

4 Original: "I nostri guardalinee sono di qualità, sanno scegliere – dice – hanno una percentuale di scelte corrette da potersi prendere più responsabilità di quelle che la Var consente"

5 Original: "Il Var credo sia un elemento di maggior giustizia. Gli errori si commettono sempre, ma in numero di gran lunga minore. Penso che in futuro si arriverà anche a trasmettere nello stadio certe decisioni del Var rendendole pubbliche".

6 Original: "Il regolamento lo vieta e gli arbitri hanno sottolineato ancora questa situazione, ma gli arbitri stanno migliorando rispetto all'inizio grazie ai grandi sforzi, quindi le cose stanno procedendo bene"

7 Original: "Ok la 'goal line technology', ma per il resto lascerei tutto come prima".

8 Original: "Acabar com alguns erros grosseiros, aqueles que todos vemos na televisão. Mas o erro vai existir sempre no futebol".

9 Original: "Fizemos formação a todos os clubes da I Liga. Temos feito também na comunicação social, abrimos uma conta de Twitter para elucidar os adeptos. Tentámos dar toda a informação. Como em todos os projetos há uma fase de maturação. Há sempre desconfiança no início de cada processo".

10 Original: "Arbitrar es 'tomar decisiones' bajo presión. Se está dando poco uso a la pantalla de pie de campo. La decisión correcta del #VAR hubiese sido 'vuelve a ver la acción de #Vinicius'".

References

11 Questions for Jose Mourinho. (2017, January 6). *FIFA.com*. Retrieved from www.fifa.com/the-best-fifa-football-awards/news/y=2017/m=1/news=11-questions-for-jose-mourinho-2862355.html.

AFP. (2017, December 14). Arbitrage vidéo: L'IFAB reconnaît des difficultés par l'intermédiaire de son secrétaire, Lukas Brud. *L'Equipe*. Retrieved from www.lequipe.fr/Football/Actualites/Arbitrage-video-l-ifab-reconnait-des-difficultes-par-l-intermediaire-de-son-secretaire-lukas-brud/859326.

Al presidente della Fifa, Joseph Blatter, e ai membri dell'International Board. (2010, June 29). *La Gazzetta dello Sport*. Retrieved from www.gazzetta.it/appellogazzetta/.

Amici, Federica. (2017, January 7). Alongside FIFA's no. 1, Gianni Infantino, the Dubai International Sports Conference and the Globe Soccer Awards light up Dubai. *Globe Soccer*. Retrieved from www.globesoccer.com/blog/alongside-fifas-no-1-gianni-infantino-dubai-international-sports-conference-globe-soccer-awards-light-dubai/.

CA reforça lição de VAR em dossiê enviado aos clubes. (2017, December 9). *O Jogo*. Retrieved from www.ojogo.pt/futebol/noticias/interior/ca-reforca-licao-de-var-em-dossie-enviado-aos-clubes-8974485.html.

Calemme, Mirko. (2017, August 26). Buffon: "El VAR así no me gusta, esto parece waterpolo". *Diario As*. Retrieved from https://as.com/futbol/2017/08/26/internacional/1503780142_463289.html.

Capello: "Io al Milan? No, già dato! Che spettacolo Napoli-Inter". (2017, October 23). *Sport Mediaset*. Retrieved from www.sportmediaset.mediaset.it/calcio/calcio/capello-io-al-milan-no-gi%C3%A0-dato-che-spettacolo-napoli-inter-_1178864-201702a.shtml.

Carlin, John. (2010, June 28). "Es inexplicable que no se use la tecnología". *El País*. Retrieved from https://elpais.com/diario/2010/06/28/deportes/1277676002_850215.html.

Cina, Lippi: "Non torno in Italia da allenatore. Var? Tecnologia solo per i goal". (2017, May 1). *ITAsportPress*. Retrieved from www.itasportpress.it/calcio/cina-lippi-non-torno-in-italia-da-allenatore-arbitri-ok-la-tecnologia-ma-solo-per-i-goal/.

Daudén Ibáñez, Arturo [@daudenibanez]. (2019, January 6). Arbitrar es "tomar decisiones" bajo presión. Se está dando poco uso a la pantalla de pie de campo. La decisión correcta del #VAR hubiese sido "vuelve a ver la acción de #Vinicius". Tuit] Retrieved from https://twitter.com/daudenibanez/status/1081991527008796673?ref_src=twsrc%5Etfw%7Ctwcamp%5Etweetembed%7Ctwterm%5E1081991527008796673&ref_url=https%3A%2F%2Fonefootball.com%2Fmx%2Fnoticias%2Fcriticas-al-var-por-todos-los-lados-es-23575973%3Fvariable%3D20190406.

De Juan, Manu. (2019, January 6). Real Madrid's Ramos slams VAR after Vinicius denied penalty. *Diario As*. Retrieved from https://en.as.com/en/2019/01/06/football/1546813199_772362.html.

Deschamps, Didier. "El VAR ha permitido corregir un error". (2018, June 16). *El País*. Retrieved from www.elpais.com.co/mundial-rusia-2018/didier-deschamps-el-var-ha-permitido-corregir-un-error.html.

Early, K. (2018, January 29). Ken Early: Give Coaches a Limited Number of VAR Appeals. NFL System Would Prevent the New Practice from Being Too Finicky and Overbearing. *The Irish Times*. Retrieved from www.irishtimes.com/sport/soccer/ken-early-give-coaches-a-limited-number-of-var-appeals-1.3371625.

Elogios a Rusia 2017 y al VAR en la rueda de prensa de clausura. (2017, July 1). *FIFA.com*. Retrieved from http://es.fifa.com/confederationscup/news/y=2017/m=7/news=elogios-a-rusia-2017-y-al-var-en-la-rueda-de-prensa-de-clausura-2899626.html.

Federação Portuguesa de Futebol. (2018, January 11). CA reúne com clubes sobre vídeo-árbitro. Retrieved from www.fpf.pt/pt/News/Todas-as-not%C3%ADcias/Not%C3%ADcia/news/14473

Fest, Sebastián. (2010, June 28). Una negativa insostenible. *Página 12*. Retrieved from www.pagina12.com.ar/diario/deportes/subnotas/148424-47677-2010-06-28.html.

Fifa, svolta storica: "A Russia 2018 ci sarà il VAR". (2018, January 22). *Sport Mediaset*. Retrieved from www.sportmediaset.mediaset.it/calcio/calcio/fifa-svolta-storica-a-russia-2018-ci-sar%C3%A0-il-var-_1193246-201802a.shtml.

Fontelas Gomes e o videoárbitro: "Vai existir sempre erro no futebol". (2017, August 7). *O Jogo*. Retrieved from www.ojogo.pt/futebol/noticias/interior/fontelas-gomes-e-o-videoarbitro-erro-vai-existir-sempre-no-futebol-8690189.html.

France Coach Didier Deschamps: Video Technology Is Future, Even Though It Cost Us Goal against Spain. (2017, March 29). *The Telegraph*. Retrieved from www.telegraph.co.uk/football/2017/03/29/france-coach-didier-deschamps-video-technology-future-even-though/.

García, Luis [@LuchoGarcia14]. (2018, June 16). VAR Makes Players Lose the Focus of the Game, Waiting for a Possible Change of the Referee Decision. Argentina had a Corner and Players Were Looking at the Referee Instead of Winning the Ball in the Air!! Just an Opinion. [Tuit] Retrieved from https://twitter.com/LuchoGarcia14/status/1007982910228582400.

Giovio, Eleonora. (2016, December 17). El vídeo siembra el caos. *El País*. Retrieved from https://elpais.com/deportes/2016/12/15/actualidad/1481819838_456390.html?rel=mas.

Greaves, Russell. (2017, August 31). We're in limbo – Ferguson yet to be convinced by VAR. *Goal*. www.goal.com/en-za/news/alex-ferguson-video-assistant-referee/y5fkfgzouxvx1v1mrm6oa1yv2.

Il Var e i suoi limiti, arbitri e tecnici a confronto. Spalletti: "Ma tornare indietro sarebbe devastante". (2018, January 15). *Repubblica*. Retrieved from www.repubblica.it/sport/calcio/2018/01/15/news/var_errori_allenatori_arbitri_spalletti_inzaghi-186537837/?ref=search.

Iturralde González, Eduardo. (2019, January 23). Uno de los puntos de mejora del VAR consistirá en abrir el protocolo. *Diario As*. Retrieved from https://as.com/opinion/2019/01/23/portada/1548200429_495918.html

Jones, Mark. (2017, December 13). Everyone Was Left Confused by VAR in the Al Jazira vs Real Madrid Club World Cup Semi-Final. *Mirror*. Retrieved from www.mirror.co.uk/sport/football/news/al-jazira-real-madrid-var-11689952.

Journal Sporting. (2017, May 6). Jorge Jesus on the Early Kick-Off, His Possible Starting XI and the Video Assistant Referee. *Official Website of Sporting Clube de Portugal*. Retrieved from www.sporting.pt/en/tags/video-assistant-referee.

La razón de peso por la que Maradona está en contra del VAR. (2018, March 27). *Sport*. Retrieved from www.sport.es/es/noticias/futbol-america/razon-peso-por-que-maradona-esta-contra-del-var-video-6718367.

Los cinco árbitros llegan a la Champions. (2010, August 27). *Marca*. Retrieved from www.marca.com/2010/08/27/futbol/liga_campeones/1282916109.html

Maradona: "Con tecnología, mi gol a Inglaterra no hubiese contado". (2017, July 25). *FIFA*. Retrieved from http://es.fifa.com/about-fifa/news/y=2017/m=7/news=maradona-con-tecnologia-mi-gol-a-inglaterra-no-habria-contado-2901436.html.

Melgar i Anguera, Blai. (2017, August 9). Mourinho: "Con el VAR el 1-0 no habría subido al marcador". *Mundo Deportivo*. Retrieved from www.mundodeportivo.com/futbol/internacional/20170809/43436776760/mourinho-con-el-var-el-1-0-no-habria-subido-al-marcador.html.

Miller, Nick. (2018, January 16). Kelechi Iheanacho and VAR Help Leicester to FA Cup Win Over Fleetwood. *The Guardian*. www.theguardian.com/football/2018/jan/16/leicester-city-fleetwood-town-fa-cup-replay-match-report.

Niandu, T. (2012). The Fallacies of the Assumptions Behind the Arguments for Goal-Line Technology in Soccer. *Sport, Ethics and Philosophy*, 6(4), 451–466.

Pérez, Gorka. (2018, January 8). "La mitad de la Bundesliga rechaza el VAR". *El País*. Retrieved from https://elpais.com/deportes/2018/01/08/actualidad/1515414870_424275.html.

Perform. (2019, March 10). Michel Platini Slams VAR as "against the Spirit of Football". *Diario As*. Retrieved from https://en.as.com/en/2019/03/10/football/1552217724_197839.html.

Pinochet, J.M. (2018, March 1). Por qué estoy en contra del VAR, el sistema de videoarbitraje en el fútbol. *BBC*. Retrieved from www.bbc.com/mundo/deportes-43243126.

Poll, Graham. (2017, November 8). England and Germany Will Do Battle at Wembley with the Use of Video Assistant Referee Technology, but There Are Still Major Problems with It. *Daily Mail Online*. Retrieved from www.dailymail.co.uk/sport/football/article-5063757/England-vs-Germany-use-Video-Assistant-Referee-technology.html.

Ryall, E. (2012). Are there any Good Arguments Against Goal-Line Technology? *Sport, Ethics and Philosophy*, 6(4), 439–450.

Serie A. Spalletti: "Var? Impossibile tornare indietro". Inzaghi: "Sono contrario, ma giusto andare avanti". (2018, January 15). *Sky Sport*. Retrieved from https://sport.sky.it/calcio/serie-a/2018/01/15/incontro-allenatori-serie-a-var.html.

Singh, Devinder. (2018, March 4). M-League Coaches in Favour of VAR. *New Straits Times*. Retrieved from www.nst.com.my/sports/football/2018/03/341321/m-league-coaches-favour-var.

Torres, Diego. (2010, June 28). "Bochorno en pantalla gigante". *El País*. Retrieved from https://elpais.com/diario/2010/06/28/deportes/1277676006_850215.html

Torres, Diego. (2019, February 10). Sergio Ramos: "El VAR es justo, yo estoy muy a favor". *El País*. Retrieved from https://elpais.com/deportes/2019/02/09/actualidad/1549721709_299253.html

VAR in Football Leaving "Players and Fans in the Dark" – Vardy. (2018, March 28). *The South African*. Retrieved from www.thesouthafrican.com/var-in-football-players-and-fans-vardy/.

Walker, S. (2018, June 25). Iran's Carlos Queiroz berates Fifa Over Cristiano Ronaldo VAR Escape. *The Guardian*. Retrieved from www.theguardian.com/football/2018/jun/25/iran-carlos-queiroz-fifa-cristiano-ronaldo-var-escape.

Webb, T. (2014). Elite Refereeing Structures in England: A Perfect Model or a Challenging Invention? *Soccer & Society*, 18(1), 47–62.

Webb, T. (2016). Referees and the Media: A Difficult Relationship but an Unavoidable Necessity. *Soccer & Society*. DOI: 10.1080/14660970.2015.1133414.

Webb, T. (2017). *Elite Soccer Referees: Officiating in the Premier League, La Liga and Serie A*. London: Routledge.

White, Julia. (2009, November 19). Thierry Henry Admits to Handball. *Daily Star Sunday*. Retrieved from www.dailystar.co.uk/sport/football/109047/Thierry-Henry-admits-to-handball.

2 The Contribution of VARs to Fairness in Sport

Juan de Dios Crespo

Introduction

There are different ways of looking at what effect substitutions have on the world of football and, like everything else in life, there are those who criticise them and others who support them. When substitutions were first allowed in football, there were those who cried out that, with them, the essence of the game was lost, since it was a sport "for men", and if a player was injured, he should either play on heroically or go off, and his remaining team-mates would have to "man up" and make up for his loss.

Later, more substitutions would be allowed and now they are even allowed in extra time. It is true that football has changed, and has become more physical, and nowadays nobody disputes the need to change players during a match. To argue anything different would be to go against the health of the players and should, therefore, not be considered.

There were also arguments against the application of the system of yellow and red cards, a novelty initially, used to punish certain actions either lightly or more severely. Nowadays, it not only seems antiquated to dispute the validity of this, but there are calls for the introduction of an orange or green card, an intermediary measure, which would mean the offender would be sent off for a limited period, as happens in handball (International Football Association Board [IFAB], 2017a).

The arrival of the fourth official, a direct auxiliary for the match referee, was also not to everyone's liking at the time. Nowadays, their role is that of intermediary between the coaches on the touchline and the principal referee on the field, and it is a role that everyone has become so accustomed to that it is almost as if it had always been there. Now, the Additional Assistant Referees (AARs), who oversee play in the area, are just part of the scenery and younger spectators may not recognise football as such when looking at footage and photographs from not so long ago.

Human beings adapt to change, and more so when that change is logical. All of the above-mentioned changes are that and, apart from a certain unwillingness to accept them in the early stages, have ended up becoming the norm, disputed by few, while some even find it hard to believe they have not always been a part of the game.

The addition of another "referee" to review actions *a posteriori*, as is the case with VARs, is not only something that is not negative, but showed in the 2018 Russia World Cup that it is needed. While some of those who are most resistant to change might still grumble about the "loss" of a kind of original footballing paradise, one in which errors were almost wished for in order to be able to argue about them afterwards, the immense majority of fans, both those at the stadia and those watching on TV, have benefitted from the justice VARs have imparted and, deep down, have admitted that what everybody really wants, whether it works in favour of their preferred colours or a national team, is for the final result to be real, not one that is influenced by poor refereeing decisions or the guile of a particular player.

VARs, then, are on the way to becoming part and parcel of the game for football fans, even though there are many competitions that are yet to adopt it. As of now, what is surely the greatest club competition in the world, the Champions League, does not employ the system, although UEFA, under a lot of pressure from all quarters, has announced it will be introduced in this season's competition, from the Round of 16 in February 2019, and from the beginning of the 2019–20 edition. It will also be employed in the UEFA Nations League Finals in 2019 and in the final stages of the UEFA European Under-21 Championship (Union of European Football Associations [UEFA], 2018). Many countries, however, are still reluctant to put it into practice. It should also be borne in mind that not every country has the technology or, let us not forget, the money, to make VARs available for all, as has been in the case of the Brazilian Football Confederation (CBF), where clubs refused to take on the expense, which amounts to $6.2m annually (Fernández, 2018).

This will be, without any doubt, the next step to be taken because, if the other rule changes that have changed football are universal, and are followed from the World Cup right down to the lowest-level competition in the poorest country on earth, the VAR system is not only not applied because of people's reluctance to put it into practice, but because there is not the economic capacity for everyone to benefit from it.

History of the Laws of the Game and Disciplinary Rules of Football: the International Football Association Board (IFAB)

The International Football Association Board, which was set up in 1886, is the guardian of the unified laws of the game of football. It is made up of the four British football associations (the English, Scottish, Welsh, and Northern Irish FAs), who each have a vote, and FIFA (which became a member in 1913) which, with four votes, covers the 211 remaining (at the current time) national associations.[1]

The mission of the IFAB is to serve the world of football as the independent guardian of the Laws of the Game, meeting regularly to manage, debate, define, and amend those Laws and to discuss their rightness, with the possibility of trialling new applications in order to see whether they are finally introduced as new rules.

Football flourished in the British Isles from the eighth to the nineteenth centuries and featured a considerable variety of local and regional versions (Fédération Internationale de Football Association [FIFA], n.d.), long before IFAB was founded. However, with the growth in popularity of the game, the need to formalise the rules became urgent (IFAB, n.d.-a). Nowadays, football is regulated by the "Laws of the Game 2017/2018" (IFAB, 2017b), put out by the IFAB and which gathers together and explains in detail the famous 17 Laws of this sport, and which are the commandments that rule it.

The timeline of the Laws of the Game can be divided into six periods, fundamentally (IFAB, 2018a):

- 1863–1970: before football became the most popular sport in the world, it lacked a uniform set of rules; therefore, in its first years of existence, the IFAB established the first fundamental rules of football.
- 1970–1980: in this decade there were significant modifications with reference to fouls and violent conduct. The introduction of red and yellow cards in 1970 was a milestone decision that helped shape football as we know it today.
- 1980–1990: the changes in the rules of the game introduced in the period from 1980 to 1990, among other things, focused on offsides, the awarding of penalties, and time-wasting.
- 1990–2000: in these years there were notable changes to the rules of the game, such as the introduction of the concept of the "obvious goal-scoring chance", "diving", "backpasses to goalkeepers", and new adjustments regarding fouls and unsportsmanlike conduct.
- 2000–2016: further changes were made not only to make football safer, more enjoyable and fairer, but also to harmonise it with the modern world, making the most of available technology, which is particularly evident in recent changes with reference to, for example, the pitch and "Goal Line Technology".
- 2016 to present day: the 130th Annual General Meeting of the International Football Association Board that took place in Cardiff on 5 March 2016, approved a period of two years of live experiments with video assistance in order to correct clear errors in determining situations that directly affected the outcome of matches. For this, referees are involved, considered as match officials, who take advantage of this videographic support to improve the quality of the game, reducing the number of refereeing errors. The so-called Video Assistant Referee (VAR) was introduced (IFAB, n.d.-b). The 132nd IFAB Annual General Meeting was held in Zurich, Switzerland, on 3 March 2018, and it was decided unanimously that the VAR system

would be applicable at the 2018 FIFA World Cup in Russia. This decision caused a confrontation with the previous FIFA president, Sepp Blatter, who was totally against its general application (Cooke, 2018). Also, UEFA is still reluctant to implement it in its competitions (Cooper, 2018), and this has also generated much controversy, as it has been understood that it was turning its back on modernity. In fact, it was called for at the time, energetically, by some clubs who felt they had been deprived of justice in the 2017–18 Champions League (let us recall only the case of Juventus of Turin, and the famous penalty that was awarded against them in the quarter-finals of that competition and that allowed Real Madrid to qualify for the semi-finals, although perhaps a VAR might not have resolved the situation in a different way from how the referee did at the time (Minder, 2018)).

Security versus Doubts: Which Is Better for Football?

On the one hand, we have innovation and progress that guarantee greater security by reducing the risk of errors, while, on the other, is the history of a sport that has a hundred-year-old tradition.

This, therefore, raises the question: what is better for football: innovation or tradition?

It should be pointed out that the aim of the VAR system is not to achieve 100 per cent accuracy for all decisions, since there is no desire to impede the essential flow of the game or the excitement of football that stems from the almost uninterrupted action that is so characteristic of this sport.

Referees take many decisions during a football match. It would be impossible, without changing the essence of football radically, to review every decision, and that is not the function of the VAR system (see Chapter 10). As well as this, it must be employed in moderation because, if not, it would be impossible for the game to continue in a flowing way as it would be constantly interrupted. It has, therefore, been decided that the use of VARs should be limited to the following four situations (IFAB, 2018) (see Chapter 10 this volume):

- goal/no goal
- penalty/no penalty
- direct red card (not second caution)
- mistaken identity when the referee cautions or sends off the wrong player of the offending team.

In all of these situations, VARs are only called upon after the referee has taken a decision (including that of allowing play to continue) or if there has been a serious refereeing error, that is to say, an unseen action. The

VAR, however, must be considered as a back-up mechanism for the referee (as are the assistant referees who, at times, indicate mistakes that they have seen, verbally, by means of head-set communication), who must still be the principal actor in the taking of decisions, being the only one of the officials who is a "participant in the game" (with the exception of the help the auxiliaries may provide at given moments).

The VAR system, though, does represent a milestone in the history of football, in which technology takes on a crucial role in the way matches unfold and the results, and, therefore, is without doubt likely to change footballing habits. Other sports have introduced it earlier, as is the case of tennis or American football (see Chapters 8 and 14) to name just two, and not just for poetic justice, but because the sport itself has not only become a different game, but a specific market, in which the individual (in the case of tennis) or collective (in the case of football, etc.) athletes, or clubs, play not only for the honour of winning, but also have a lot of money at stake.

As was seen at the beginning of this chapter, football has, throughout its existence, been able to take on new rules, and these changes have always meant certain repercussions. The challenge is to keep these repercussions to a minimum, and to alter the essence and customs of football as little as possible but, at the same time, reduce the errors that may influence the outcome of matches. This is because, even though the post-match arguments are something fans enjoy engaging in, the anger felt on knowing that something unjust has happened is not good for sport and, as mentioned above, the consequences for players and clubs, in terms of money, are nowadays of much greater importance.

The flow of money that has come into the game through the television companies and other means of broadcasting, as well as from sponsors, has meant that for some time now there has been a demand for greater transparency in the game and true justice based on the checking and rectifying of refereeing mistakes.

There are some who say that opposing VARs is to deny evolution and that we should apply progress to society with the aim of, in the case in hand here, improving the game of football. On the other hand, there are those who claim that the use of VARs will inevitably and radically affect football in the sense that it will lose that natural uncertainty that comes from its high pace and the lack of interruptions, and will become ever more like American sports in which the instant replay reduces the risk of error to a minimum, eliminating any kind of doubt.

With a VAR, the first goal Maradona scored, by means of the "hand of God", in the 1986 World Cup quarter-final between Argentina and England would have been disallowed, while the goal that did not count but which should have given Spain the lead against Brazil, in the same tournament, would have been awarded. Similarly, France would not have eliminated Ireland in the qualifying game for the 2010 World Cup as Thierry Henry controlling the ball with his hand would have been spotted. Much has been written about these events and there have even been philosophical essays discussing whether "that hand of God" was about Argentina settling old scores with England.

As can be seen, for football, with all the passions it arouses, the most important refereeing error ever in a World Cup was seen as something unassailable. Some think that the trick the Argentinian footballer pulled off not only did not go against the ideals of football, but rather it gave the game a human quality …

The "history" of football, though, could have changed so many times with a possible VAR system that, obviously, some advocate its immediate introduction while others believe that the romanticism of post-match arguments in the bar would be rendered impossible if it were to be applied. Even still, recent events have shown that some decisions that have been taken using a VAR are not so clear cut and the argument as to whether it was applied correctly or not goes on.

What, therefore, is best for football? Security or doubts? Since 1986, football has evolved a lot and has become a true worldwide business in which clubs have become, in a great many countries, trading companies, above all where there is a strong economy. In this context, is it in the common interest of the parties concerned to guarantee the security of match results, and therefore that of the whole system, at the risk of partially giving up the romantic and traditional look and feel of the game?

At the World Sport Congress held in Barcelona on 2–3 March 2018, there was a heated discussion on the good and bad things that VAR could bring to football, and a former chairman of FC Barcelona came out absolutely opposed to it because, he said, it would not contribute everything that one believes it should. One of the responses he was given at the round table was that some fairness is better than none.

As well as this, there was discussion about as from what moment the VAR should be used, because while there might be no error at the moment in which a goal is scored, and awarded, what should happen if the move leading to the goal had started from a foul that was not blown by the referee? Could the VAR be applied as far back as the initial moment in which the error occurred?

We shall not philosophise here about the principle of causality and the fact that every event (goal) has a cause (a good move but preceded by a refereeing mistake), because that is not the aim of this chapter. Some day, when the VAR system has been implemented completely, perhaps the next step would be to have an even greater technological control of football. That, though, is a different story.

Fans and VARs

VARs are being accepted this season, 2018–19, their first, in several countries and compeitions among which we can point to the German

Bundesliga, Italy's Serie A, la Liga in Spain and the English Premier League, though for only certain fixtures in the latter. In order to understand fans' opinions regarding the application of the VAR system, we must refer to those who are directly experiencing the technology.

In this sense, a survey carried out by the SWG company ("Gli italiani applaudono", 2017) shows how 79 per cent of fans in the Italian Serie A are in favour of VARs (37 per cent very much in favour, and 42 per cent quite in favour), while 13 per cent are against and 8 per cent abstained.

Some 40 per cent say VARs are especially useful for referees, while 34 per cent believe it is a revolution that could benefit football as a whole because it would increase the credibility and competitiveness of the game (above all, bearing in mind that 84 per cent of fans of small and medium-sized clubs are in favour of VARs).

The most controversial aspect concerns the way in which VARs are used. In fact, only 45 per cent of fans are satisfied with how it is applied, while 55 per cent would like it to be applied less (26 per cent), more rigorously (18 per cent), or not at all (11 per cent).

Consequently, Italian fans have, by and large, accepted the introduction of VARs into football, as they guarantee greater security and fairness, without interfering with the traditions and distinctive features of this sport. However, the technology is not yet perfect and needs to be studied more and improved if it is to achieve its aim.

It is equally certain that the Russia World Cup has also had some influence in terms of acceptance, and the percentages after the tournament are without any doubt greater than they might have been. On every occasion in this top tournament that the VAR was used, it was successful, which explains its greater acceptance.

The fact is that FIFA had the courage, it should be said, to use its greatest competition, one watched by thousands of millions worldwide, as the first in which VARs would be employed. This could have turned out badly should there have been a big mistake, let's imagine, during a game in the knock-out stage. Quite the opposite occurred, however, and their bravery earned the reward of a greater awareness of this technology and a wider acceptance of it.

The Transformation of Football through VARs

Table 2.1 summarises the results since VARs were first employed as an experiment in March 2016. These results are related to the "live" use of VARs in professional football matches (IFAB, 2018).

The usefulness/uselessness of VARs is evaluated by means of the statistics quoted in Table 2.1 and the results obtained since it has been applied. As can be seen in Table 2.1, the impact of VARs on the 804 matches analysed has been to contribute greater security (with accuracy of 98.9 per cent, making a decisive impact on 8 per cent of the matches) without upsetting the nature of play (time "lost" by VARs represents less than 1 per cent of the total).

Table 2.1 Summary of results of "live" VAR experiments from March 2016 to March 2018

Result	*Explanation*
20+ national associations and competitions	Australia, Belgium, Brazil, China, CONMEBOL, Czech Republic, England, FIFA, France, Germany, Italy, The Netherlands, Poland, Portugal, Qatar, Saudi Arabia, Korea Republic, Turkey, and the USA.
804 competitive matches	A further 700 friendly/training/exhibition matches have used VARs, but the results are not included in the analysis as they are not comparable with "live" VAR use in competitive professional matches.
56.9% of checks were for penalty incidents and goals	Majority of match-changing situations relate to penalty/no penalty incidents and goals; red card incidents – 42.3%; mistaken identity negligible.
56.9% of checks were for penalty incidents and goals	In the 804 matches there were 3,947 checks for possible reviewable incidents – vast majority were "background" checks which did not interfere with the game.
Median check time of the VAR is 20 seconds	Most checks take place quickly whilst play continues or during the "normal time" of a stoppage, e.g., during the goal celebration, so have no impact on the flow of the game.
Initial accuracy for reviewable categories of decision is 93.0%	Without VARs the quality of decision-making in the match-changing reviewable categories is already extremely high.
68.8% of matches had no review	The flow of many games has not been interrupted by a review, as 533 matches did not have a review. Only 42 matches (5.2%) had more than one review.
Average of one clear and obvious error every three matches	Before and during the experiment, statistics clearly show that only one decision in three matches is a "clear and obvious error" in the reviewable categories of decision.
Decision accuracy with VAR is 98.9%	The accuracy of decisions in the reviewable categories has increased by 5.9% to 98.9% (100% accuracy impossible due to human perception and subjectivity in decision-making).
Decisive impact of the VAR in 8% of all matches	In 8% of all matches the VAR had a decisive impact on the outcome of the game. The experiment also showed that 24% of all

Result	*Explanation*
	matches were positively affected by the involvement of the VAR (changing an initial incorrect decision by the referee).
The median duration of a review is 60 seconds	For decisions based on head-set communication from the VAR, the median time = 39 seconds. When there is an on-field review, the median time = 70 seconds.
The average time "lost" due to the VAR represents under 1% of overall playing time	Use of VARs has a very small impact on the overall playing time "lost" compared with all other situations where playing time is "lost" (typically: free kicks (9.5%), throw-ins (8%), goal kicks (6%), corner kicks (4.5%), substitutions (3.5%) etc.).
A "clear and obvious error" was not corrected in 1 in 20 matches (5%)	This figure (5%) is very encouraging considering the short testing period and inevitable human errors in perception, decision-making, unfamiliarity with technology, communication, etc.; it should improve over time with more experience, training, etc.

Source: IFAB.

As well as this, a study carried out in Italy of the first 210 matches of the 2017–18[2] season compared the results obtained through the application of VARs with regard to the previous season, when this new technology had yet to be employed.

In the 210 Serie A and TIM Cup matches there were 1,078 plays (an average of 5.1 per match) that were looked at, with 60 corrections made and 49 mistakes avoided, while only 11 errors were committed (of which 7 influenced the final result) (Ceniti, 2016). Table 2.2 sums up the results obtained ("Var, gli arbitri", 2018).

It can, therefore, be seen that the actions contributing to cut the effective playing time were reduced. With the introduction of VARs, the number of yellow and red cards shown has gone down, and as well as that, there has still been no sending-off as a result of a protest (there were five in the previous season) and that is because the players have more trust in the technology when decisions are taken. This also leads to players showing more respect for VARs, meaning there is a clear drop in the number of protests and the amount of diving or violent and unsporting conduct.

In total, the number of decisions that were changed by the VARs was 60 (approximately three per match) and the time necessary to take the

Table 2.2 Results without and with VARs in the Serie A and TIM Cup in the 2016–18 seasons

	2016–17 (*without VARs*)	*2017–18* (*with VARs*)	*Difference*	*Variation %*
Yellow cards	931	756	−175	−19%
Protests	71	61	−10	−14%
Diving	13	10	−3	−23%
Unsporting conduct	59	50	−9	−15%
Red cards	51	40	−9	−22%
Violent conduct	17	13	−4	−24%
Fouls per match	26.2	24	−2.2	−8%
Penalties	73	77	+4	+5%
Offsides	887	866	−21	−2%

Source: Marco Bonfiglio (Bonfiglio, 2018).

decision in question has been shortened. In this sense, in the earlier matches the referee needed 2 minutes and 35 seconds to review a given piece of action, while now the average time necessary is 1 minute and 15 seconds. The time needed to review a decision remotely has gone down from 1 minute 22 seconds to 29 seconds, which increases the effective playing time of games by 40 seconds (and by 10 seconds in stoppage time).

There is, however, no total security nor uniformity of opinion about the utilisation of VARs, above all because it is left to the discretion of the referee to determine when to consult them. The principal problem to be dealt with in the future is to get all the interested parties, including team managers, to agree that the benefits outweigh the disadvantages.

The following step might then consist of reviewing the balance of power between the VAR and the referee. As well as this, in order to involve the fans in attendance at the stadium more, the video images that the referee and the VAR are reviewing could also be shown on the screens.

That is to say, the VAR has achieved what was argued it would not: a better use of playing time and reduced conflict. Like any review procedure, it was thought that the time it would take up would be time not played, and this is indeed the case. It is compensated, on the other hand, by less time spent in arguments between players, and between players and the referee, with the consequence being that there is more effective playing time and, in addition, a reduction in disciplinary problems on account of waiting for the result, during which time the players calm down.

This is something that had not been foreseen and has been a little surprising, even perhaps for the advocates of VARs themselves. A player who, in a moment of tension involving a goal, an offside, or a play that he considers unfair, has to calm down as soon as the referee decides to consult the VAR, and his aggressiveness towards the official or his opponents is therefore put to one side.

This is another argument in favour of VARs and one which, without any doubt, will be of greater benefit to football than was thought: more effective playing time and improved discipline.

Conclusions

It is possible to see, in broad outlines, what VARs can contribute in terms of fairness in sports, both on the playing field and subsequently. On the pitch there is less chaos and the outcomes are more objective and, definitively, fairer. Off the pitch as well, there are fewer disciplinary cases to be dealt with, especially those involving clashes and conflicts stemming from arguments between players and between players and referees.

As well as this, and over time it will no doubt be seen, other aspects of sports discipline are going to change, such as another possible "hand of God" or a penalty that has been provoked. It will not be easy, nor will it happen straight away, because many footballers have it in their genes still, but the new generations will see that these tricks do not have positive consequences and they will, if not disappear, largely diminish. But for that to happen, we will have to wait still, because we are in the early stages and changes cannot happen so quickly. It is, however, more than likely to happen and youngsters will have to change their way of playing.

Innovations take time to prevail, that is nothing new, and all advances that humans have made have had their detractors who have stood in the way or slowed them down. However, just like 30 years ago we could not even imagine mobiles such as they exist now, and our children think it odd to see rotary telephones, or telephones fixed to the wall, and social networks mean we are constantly connected (for better and for worse), now it is football's turn to move into the modern age it has always been reluctant to embrace. The need to be *à la page*, both owing to the obligation of ensuring "fairer" results and to satisfy the demand that the investments made in football not be unduly affected by injustices brought about by players' tricks or refereeing mistakes, all point to VARs becoming unstoppable.

Other sports, such as American football, have their own VAR system for reviewing plays; in basketball minutes can be lost while video replays check whether the ball went through the hoop in time, and in tennis everybody awaits the Hawk-Eye line technology's decision. Football, however, still has that old, constant, historic patina about it, which actually is not true given that it has undergone changes, as seen above. The only thing missing now is for the system to be applied worldwide, which is the only way it will get the necessary recognition. FIFA has kicked off by employing the system in the 2018 Russia World Cup and now, perhaps, they will have to spend money to allow many member countries to adopt it for their competitions, given how hard it is to use it if the economic means are lacking.

Another decisive point, apart from the technological factors and the equipment needed for VAR to be used, will be the preparation of the "operators" of the system, former referees in the main, but it may be necessary to train up a whole new generation of VAR specialists who have possibly never trodden the playing field. This is surely one further step towards its total and definitive implementation.

This change, however, seems to contain the fire of Zeus, which he did not want to give to humans, and which Prometheus eventually gave us. Here, the VAR-fire exists and the IFAB (and FIFA) have been our Prometheus. It is now up to us, though, to use it wisely, and not to let ourselves get burnt by it…

Notes

1 All decisions regarding a change in the Laws require a three-quarters majority.
2 The 19 matchdays of the first half of the season and the first game of the second half, to which were added 12 Coppa Italia games to make a total of 210 games (minus the Sampdoria-Roma and Lazio-Udinese fixtures).

References

Bonfiglio, M. (2018, January 1). Var, le statistiche dopo il girone di andata: meno cartellini e più rigori rispetto al passato. Retrieved from www.infobetting.com/blog/statistiche-var-2017-18-girone-di-andata/.

Ceniti, F. (2016, January 16). Var che numeri, ma sul mani si cambia. *La Gazzetta dello Sport*, p. 4.

Cooke, R. (2018, March 1). Former FIFA president Sepp Blatter against using VAR at World Cup. Retrieved from www.skysports.com/football/news/11095/11271684/former-fifa-president-sepp-blatter-against-using-var-at-world-cup.

Cooper, D. (2018, February 28). UEFA explain decision not to use VAR in next season's Champions League [video file]. Retrieved from https://sport360.com/video/football/269800/uefa-explain-decision-not-to-use-var-in-next-seasons-champions-league.

Fédération Internationale de Football Association. (n.d.). History of Football – Britain, the home of Football. Retrieved from www.fifa.com/about-fifa/who-we-are/the-game/britain-home-of-football.html.

Fernández, M. (2018, February 5). Clubes vetam árbitro de vídeo no Brasileirão-2018; veja como cada um votou. Retrieved from https://globoesporte.globo.com/sp/futebol/brasileirao-serie-a/noticia/brasileirao-de-2018-tera-venda-de-mando-de-campo-e-grama-sintetica.ghtml.

Gli italiani applaudono la Var. (2017, September 9). *La Gazzetta dello Sport*. Retrieved from www.fcinter1908.it/ultimora/gds-gli-italiani-applaudono-il-var-il-79-e-favorevole-i-tifosi-della-juve-invece/.

International Football Association Board. (2017a). *The IFAB strategy and "play fair!" initiative* (Circular 10). Retrieved from www.play-fair.com/data/Strategy_Paper_EN_150dpi_Doppelseiten.pdf.

International Football Association Board. (2017b). Reglas de juego 2017/18. Retrieved from http://static-3eb8.kxcdn.com/documents/275/092922_180517_LotG_17_18_ES_150dpi_DoublePage.pdf.

International Football Association Board. (2018, January 22). 132nd Annual Business Meeting. 22 January, 2018. Media Package. Information on the Video Assistant Referee (VAR) experiment, incl. provisional results. Retrieved from http://static-3eb8.kxcdn.com/documents/639/165902_220118_IFAB_Media_Package_ABM2017_all_media_FINAL.pdf.

International Football Association Board. (n.d.-a). History of IFAB. Retrieved from www.theifab.com/history/ifab.

International Football Association Board. (n.d.-b). Video Assistant Referees experiments. Retrieved from www.theifab.com/projects/video-assistant-referees-vars-experiment/background-scope.

Minder, R. (2018, April 11). Juventus' Near Miracle Against Real Madrid Ends in Controversy. *The New York Times*. Retrieved from www.nytimes.com/2018/04/11/sports/real-madrid-juventus-champions-league.html.

Union of European Football Associations. (2018, December 3). VAR to be used in UEFA Champions League knockout phase. Retrieved from www.uefa.com/insideuefa/about-uefa/news/newsid=2586230.html.

Var, gli arbitri: «Finora 1.078 check, solo l'1% di errori. (2018, January 15). Corriere dello Sport. Retrieved from www.corrieredellosport.it/news/calcio/serie-a/2018/01/15-36983591/nicchi_positiva_la_sperimentazione_del_var/.

Part II

Video Broadcasting Limitations

3 Is it the Same for the TV Screen as for the VAR? Planning the Use of Cameras and Replays to Solve the Controversy

Anto J. Benítez

Introduction

Live football broadcasts are one of the most watched events on TV, so much so that they could be said to be universal in character. FIFA estimates that more than 3.25 billion people watched some matches in the FIFA World Cup (FWC) Russia 2018, and that at least 1.12 billion watched the final live. The total audience reached was more than half of the world's population over the age of four (FIFA, 2018).

A TV viewer anywhere on this planet can understand what is happening on the screen and be carried away by the emotions the game arouses. Football is one of those daily activities that has a great deal of meaning for humanity. The view of a match on the screen is basically the same everywhere, save for the natural differences in location, lighting, architecture or playing surface, and availability of media. Also, there are cultural preferences in the way of playing; there is room for diversity in a sociological view: "football's diffusion across the world has enabled different cultures and nations to construct particular forms of identity through their interpretation and practice of the game" (Giulianotti, 1999).

The global vision about mega events (Maurice Roche, cited in Horne & Manzenreiter, 2002), on the other hand, builds a strong current of feedback. Major sporting events on TV, football matches especially, shape the taste of the audience, their competence in the way that the TV viewers demand certain well-known patterns of coverage.

These formulas for bringing major football events to the screen have been studied and discussed fully. British investigators compared the 1974 World Cup (WC) in Germany, produced by Deutsches Olympia Zentrum, a consortium of ZDF and ARD to act as host broadcaster (HB) of the competition (Buscombe, 1975: 1), with what was offered by British TV. The results were published in a British Film Institute Monographic. Some of them applied shot-by-shot film strategy to HB footage analysis (Barr, 1975; Ryall, 1975).

Facing an environment that had technical limitations, some of their conclusions showed that *national styles* appeared on the screens for football matches, attached to a superior comprehension of cultural differences. For example, the article by Charles Barr (1975: 53) concluded that German TV showed on screen a *team-centred* proposal, while the British one was *personality-centred*. A similar dichotomy had already been dealt with by Umberto Eco in "The Open Work", whose first edition is from 1962 (Eco, 1990).

By observing their way of doing TV, other studies have tried to find differences between individual or national styles in the way the HB expresses itself when broadcasting football or other sports activities, generally to support certain cultural distinctions (Chandler, 1988). Coming forward to present-day technology, a similar analytical approach compares the way the most heated rivalries in the English Premier League, Spanish La Liga and Saudi Professional leagues are broadcast (Benítez & López, 2015).

In perspective, an early conclusion of these investigations is that there is an evident diversity in the way football reaches the TV screen, but there are more similarities than differences. Today, if they exist, the cause can be applied more to the availability of coverage facilities than to structural differences. In fact, when facing a global, and therefore singular, audience, it is only natural that multinational companies, with similar broadcasting experiences, obtain the exploitation and production rights for major sports events, not only those for football. Though referring specifically to track and field coverage, this sentence could be used for sports in general: "a viewer anywhere in the world must be able to comprehend both the competition itself and the story line" (Kalevi Uusivuori and Tapani Parm, quoted in Owens, 2007: 10).

Accepting as football *mise en scéne* (MES) anything that has to do with what happens on the playing field in front of the fans or whatever concerns the show if there were no cameras to broadcast it – which is inconceivable nowadays – FIFA dedicates great efforts to homogenizing it, with 211 different federations all focused on the same objective. It is the intent that whatever happens on the field has to look the same and will be the same anywhere it is played by keeping an alert eye on, and providing education about, the application of the Laws of the Game. The FIFA FUTURO III training programmes (Armenteros et al., 2019) for referees and assistants work to make sure the rules are applied homogeneously so that anything happening on the pitch will be responded to in the same way by the referees. Finally, to make sure that the objective, which consists of applying the rules equally to all matches everywhere, so they look very similar, is achieved, the organization uses a large database with related videos (Armenteros et al., 2016) that keeps growing after every international event – in all age categories – with comments and pre-agreed designs by panels of instructors of the utmost experience and authority (Armenteros et al., 2019).

Important questions spring up: what is the goal for referees in a football match? What is the objective of FIFA, the organization that keeps and maintains them? Provisionally, they are about ensuring the match is played within the terms of fair play and does not allow for any doubts or mistakes in the application of the rules. It must be taken into account that for officials there is always room for interpretation – but this must be immediate and involves actions that have happened quickly – and referees' decisions should not leave any room for complaints. It is important to consider that football is a game that is intended to flow and also that the ideal of fair play is a cultural construct, one that has no owner – with the permission of the COI (International Olympic Committee) – and is in a constant state of evolution and redefinition.

In opposition to the concept of MES, in the sense of Kulechov (1956: 161) or Bordwell and Thompson (2002: 156), and in connection with this text, a different dimension, given the name of visual construction (VC) procedures (VCP), is considered. This is the result of the processes made to transfer the MES of a given football match to audio-visual form, by means of a device that captures images and sounds and builds a linear programme with immediate editing techniques (Benítez, 2006). These procedures have their own entity: "take, editing and projection, three phases that in cinema production are totally different and have their own physiognomy, here (in live TV) are identified" (Eco, 1990: 229).[1] Then, *two universes of analysis* can be established, corresponding to different dimensions of symbolic action that coexist in the definition of sports: presentational and performance (Reeves, 1989: 209), both of which, in this proposal, corresponds to VCP and MES.

In the same way, for the authors of VCP, what will their objective be? And those of the organization who contracts them? For some researchers, the directors of sport transmissions are "auteurs" who employ television's technical magic to create unique and dramatic spectacles (Gruneau, 1989: 135, using Morris & Nydahl, 1985). The director of a match documents the event while building a television show, one that is not lacking in drama, the expression of which he has planned and directs.

For networks the main goal is to catch the attention of the TV viewers in a wide sense and keep it alive throughout the whole duration: "the main concern of television producers is to prevent viewer boredom" (Morris & Nydahl, 1985: 109). So, TV sports broadcasting is, firstly, a show and then sport. As long as the viewer is interested in the match, "the network is interested in the viewer; to the network the game is to catch the viewer" (ibid). In the TV ecosystem and under the menace of channel-surfing, this implies that spectacularity takes priority as a main goal even above the intention of providing information, or even when it is intended as reporting or news (Iozzia & Minerva, 1986: 61). The idea is to use an efficient and irresistible narrative; "ultimately the main responsibility of the sportcaster was not that of a reporter – to accurately inform the audience – but that of an entertainer to keep the fans glued to the television set" (Rader, 1984: 132).

The TV Show and Good Television

For networks and directors, the ideal consists in getting what is understood as *good television*. This term is very recurrent in the analyses of sports broadcasting. Professionals believe that reaching perfection in this field will finally bring its rewards in terms of audience and critiques. This term evolves; what was done successfully in a given season may or may not be to the audience's liking a short time afterwards. The development of audio-visual technologies offers possibilities that are worth exploring as they may be passing fads. While FIFA watches over the good application of the rules – and therefore that of the MES of football – no organization agrees with others as to what *good television* will be.

Several scholars, with examples not necessarily about football, claim that to appreciate the qualitative aspects of an audio-visual text, it is necessary to consider professional practices.[2] To understand what can be seen on the screen it is as important to know what the professionals and creators understand as good television (Gruneau, 1989; Whannel, 1992; Silk et al., 2004) as it is the way in which they imagine a given sport (Gruneau, 1989; Whannel, 1992: 100). According to the idea launched by Tom Ryall, the singular vision a broadcaster has of football is reflected on the screen during the transmission of a match (Ryall, 1975: 36). What is inside a frame and which of the cameras is offered to the viewer at each moment is a proposed construction influenced by a set of informal and conventional rules shared by the members of the production team, and which operates as an "underlying grammar that structures the selective construction of the images and the narrative" (Gruneau, 1989: 146).

The production team adapts to the requirements of the audience and in some way evolves with its commodity (Silk et al., 2004: 155, quoting Jhally), while, although "fully aware of the normative industry values and procedures required for producing good television" (Silk et al., 2004: 158), in some ways it lets them be influenced by the actual tastes of every director in a kind of "on the job socialization" (ibid). Knowledge of this type of "industrial wisdom" (Silk et al., 2004: 146) is necessary to understand that "the decisions about what to cut; whether to move the camera in or out, or what action replay to show need to be taken at speed and must therefore be made with reference to a pre-existing system of codes and values" (Silk et al., 2004: 149).

Garry Whannel explained some of the characteristics for a good sport television: "absorbing narratives, exciting action and the presence of characters and stars" (Whannel, 1992: 100). This concept, although slippery, is prone to *crossover*. As an example, the adaptation of different camera positions from American football for taking offside shots by TVE in the 1960s (Santamaría, 2002: 65); the seminaries to adopt the western concept in the FWC Argentina 1978 (Whannel, 1992: 62–3); or the formation by German and Italian directors so that the Canadian Sport

Network could face the FWC USA 1994 – the implemented style was used in the transmissions of the Canada Cup of Soccer (Silk et al., 2004: 156–7).

VC Is Always a Construction from the Director's Point of View: Total Objectivity Does Not Exist

Good practices do not mean VC is waterproofed against ideological influences. In an early stage of reviewing television programmes, Umberto Eco, referring to live takes and sport on television, used to say: "a live take is never a speculative result of the moment but always an interpretation – even if, in certain cases, in an infinitesimal portion – of that moment" (Eco, 1990: 229).

Later still, the debate was about the difference in expressiveness between cinema and TV, and if TV transformed the reality of what was happening on the pitch. In 1975 Collin McArthur wrote that what was expected was that the event would be "passed 'raw' to the audience through the eye of the camera" (McArthur, 1975: 8), but there should be no doubt that the television image was but a construction with reference to added value of sense in every characteristic of the appearance – be these colour broadcasting or black and white, wide shot or close-up, normal speed or slow motion, all correspond to a decision (Buscombe, 1975: 4). In the following decade, the image on TV was considered a "representation" and therefore different from the event represented (Novel-Smith, 1981: 160); it was also said that "what is 'shown' on TV is always the result of a complex process of selection" (Gruneau, 1989: 134).

Michael R. Real had given a different value to the VCP of the Super Bowl, although it is convenient to clarify that he was referring to the total television *packaging*, including the pre- and post-match periods, publicity, etc.: "rather than mere diversionary entertainment, it can be seen to function as a 'propaganda' vehicle strengthening and developing the larger social structure" (Real, 1975: 42). According to Lawrence A. Wenner, "audience experience must be thought of as a cultural indicator rather than proof of hegemonic control" (Wenner, 1989: 41). The ideological charge is, in the end, unavoidable; Brien R. Williams in 1977 established that the networks imposed their ideological vision in the "medium event" above the one of the stadium (Wenner, 1989: 32). In the conclusions of Williams' article, "the television coverage of a live, unscripted event such as a football match imposes its own structures and provides its own ideological viewpoints" (Williams, 1977: 139). The term unscripted will be discussed later. In the early 1990s, Whannel fixed the analysis of sport television as a conflict, a system of tensions between realism and entertainment (Whannel, 1992: 94).

At the beginning it was stated that football on the screen is quite recognizable in all latitudes. In spite of that, it is accepted that certain style elements, as transmitters of ideologies or simply of different concepts of *good television*, are detectable at local levels from one network to another. Scholars have noticed and remarked these differences with reference to American football (Williams, 1977), skiing (Gruneau, 1989), sports (Whannel, 1992), or football (Silk et al., 2004; Benítez, 2006, 2013a). Whannel proposes a scheme to characterize the position of the HB starting from a triangle whose three vertexes are information, entertainment, and drama (Whannel, 1992: 67), and this scheme has been applied in detail to football in the Spanish case, concluding that in the 1990s different operators acting simultaneously in the Spanish League clearly showed what type of coverage they aimed to provide, for information solely – TVE; for drama, playing up the partisan emotion – FORTA, public regional broadcasters; and exalting the pure show – CANAL+, the only pay channel – proving that the same football match could be broadcast in different ways, depending on the *style* of the HB (Benítez, 2013b).

Starting from the moment that FIFA has begun to implement the VAR, several questions come up that can only be answered by knowing the process of live football MES well. Who decides, and with what intentions, the contents of the footage the VAR uses? Or who makes the spectacular VC system available for use as a system for helping referees take important decisions during the match? What criteria are used to plan framing and recording? Are they optimized for use in the application of the rules of the game? Is it the ideal procedure to make it so that video can help refereeing?

This text is not going to consider the dimensions of narration and comments by announcers or specialists, even being aware of their relevance for a good number of interesting analyses (i.e., Wenner, 1989; Barnfield, 2013). The reason is evident: the VAR should be impervious to narrations, comments, and opinions while viewing the match from the point of view of applying the rules. Likewise, examples will continue to be given, or comments will be made, on different sports, which are considered valid for comparison with football.

Creative Figures in the VCP of Football Transmissions: Producer, Director, Realizador, Réalisateur

Starting from casual or not quite detailed approaches, a number of stereotypes have been spread about live football VCP: as matches are unpredictable – unscripted events – the director improvises continuously; spatially the spectator is placed in the position of the main camera; football audio-visual planning consists of placing the largest number of cameras possible on the same side of a virtual line that crosses the field from goal to goal and the execution on the fly, as the game unfolds; a dramatic construction parallel to the narration of the match cannot be given (adapted from Benítez, 2006: 5).

Beginning with the understanding of the available technology and of its evolution (see Chapter 8), and a deep and detailed knowledge of football, directors develop the technical procedures of the VC. In an ideal scenario, they acquaint themselves with the event, imagine how they could use the screen to please their public, determine the technology to use, conceive a plan to do it, choose the human team, instruct the team in their own planning specifications, create workflows for the professionals involved, and direct their execution. All these skills demand of them the skills expected of audio-visual creatives, technological and logistical consultants, CEOs, leading executives, scouting managers, coaching or systems analysts.

In the British television tradition and its heirs, the professionals in charge of VCP are the producer and the director. According to Jim Owens, "the producer is the responsible person for planning and logistics before the event. The producer works with the director to implement the overall production plan. Together they are responsible for overseeing all elements of the production of an event" (Owens, 2007: 11). He says of the director:

> under the guidance of the producer, responsible for the creative and aesthetic portions of the production, directors must know what they want and how they are going to accomplish it. To coordinate the various aspects of the broadcast, they have to understand the capabilities of the equipment and the crew they have been assigned. The director sets the tone and pace of the production and serves as crew motivator. The director must be able to visually tell the story, to document the event in a way that allows the viewers to feel as if they are part of the action.
>
> (ibid)

Without lessening the importance of the producer's role, who apart from supervising the work of the cast and directing the commercials and video material prepared beforehand, this text will give a closer view of the figure of the director. In countries with different production practices, such as France and Spain, the *réalisateur* or *realizador* – considered the equivalent of the director – also usually assumes the responsibilities of guiding the talent and handling the material recorded during the broadcast. For Benjamin Rader, "by quickly shifting images the director tries to lace the events with a continual series of visual surprises. Close ups and replay shots could add to the fan's appreciation of beautifully executed plays" (Rader, 1984: 196).

More formally, the directors of a live football broadcast lead a device that gets its prime materials from the sports stage and returns an audio-visual narration prepared for their public according to a preconceived plan, in such a way that it is not based on random or capricious elements, but on a determined plan and carried through by aware and capable professionals (Benítez, Armenteros & Sánchez, 2011: 36–7).

The Football Script

This plan is made possible mainly because a football match does have a scripted structure and therefore the director may foresee and prepare the events that will happen. Despite the fact that many authors defend that all sports action is unforeseeable and the answer on the screen is purely improvisation, matches on the screen can be analysed and structurally reduced to the categories of characters, action, space, and time (CAST) in two layers of analysis: those of MES and VCP (based on Benítez, 2006, 2013b) – a construction process that the director conceives, prepares, and later executes. VC has been considered a transition (Morse, 1983: 48), translation (Chandler, 1988: 4, 68), or transformation (Whannel, 1992: 96), meditated and immediate from what happens in the MES layer to the narration offered to TV viewers.

According to Barnfield, "sport on television is comprehended through the convention of narrative drama" (2013: 327). The spectacular keys of football, whose function is narrative, are identical to those of fiction stories. They develop conflicts through the actions of their characters who are trying to reach their objectives while opposing those of their antagonists. Simple and close story plots that speak of individual qualities and human groups, of their skills to predominate over the rest of the species and escape extinction, through force, death-denying resistance, speed, skills, but also through trickiness, tactics, intelligence, adaptation, and overcoming adverse circumstances, coordination, or teamwork (Benítez, 2013b: 106).

In football, a number of characters are seen by a community as their representatives who fight for a common goal: to win. Contrary to the fiction plots, the TV viewer can choose both leading and antagonist roles. The MES requires two contenders in competitive style and, although the natural state of football is balance (Gréhaigne, 2001: 97), the conflict comes from trying to break the initial deadlock. If victory is not achieved immediately, revenge can come in the medium or long term, as in sport there are usually further opportunities. The balance can be altered in just one play (or sequence), in one day in particular (or chapter), over a whole season, in such a way that may be points of re-enlistment with the narrative (Benítez, 2013b: 106–7).

The director, following this structural logic, may foresee events, and possible schemes, to build his story. According to Kalevi Uusivuori and Tapami Parm (producer/directors, YLE television network, Finland), talking about the sports director's role,

> directors must understand the way in which viewers watch television: how they think, feel and want to be entertained. This means that the director must find ways to capture the viewer's interest. This is best

> done by doing thorough homework and proper storytelling: the two keys to better coverage.
>
> (interviewed by Owens, 2007: 10)

From a more operative point of view, more in relation to the action every protagonist may communicate through his or her expressive activities, and with respect to what the director may be expecting (and therefore foresee) of the MES within a frame, a footballer may be doing a whole repertory, a more or less limited *series of actions*, that will allow him to express himself on the pitch, according to different criteria (based on Benítez, 2006):

- *athletic series*, individual performance: running, accelerating, slowing down or stopping, walking, jumping, falling to the floor, changing direction or rhythm;
- *technical series*, individual: striking the ball, with the foot or the head or other body parts, stopping it, shooting, passing, dribbling it, throwing it in, catching it, tackling an opposing player;
- *tactical series*, involving another player: getting back to the defensive position, maintaining the position, pressing an opponent, awaiting another player's initiative, moving the defensive lines forward or backward, passing or running, trying to get away from the opponent's marking, looking for numerical superiority, combining with teammates, carrying out set plays;
- *normative series*: normative valid actions, non-valid, normative subject to interpretation;
- *gestural series*: any action or reaction related to body or spoken language.

The main objectives of the director are to document the football event in such a way that no relevant fact is left behind in any of its possible story plots, and to produce a competent TV show with a narrative that reaches the public and keeps them hooked, as far as dynamism, aesthetics, and drama are concerned:

> but [the] director's goal and responsibility is to make the production interesting and entertaining for the audience, so there is a very basic objective to be met in the use of increased production values so to achieve this end, and it must be determined before the production begins.
>
> (Owens, 2007: 105)

Principles of Planning

The term *planning* can be used to denominate

> a set of instructions that are imagined by the director and that articulates the story in its *instantaneous* facets – over which subjects are followed by a camera in every game situation, with every individual framing, what

> shot size is to be used, what kind of movements are allowed, what happens if … – and the *linear* facets – what to bring to the screen, or for how long, which actions may remain latent, which ones ebb away, what characters must be shown in isolation, etc.
>
> (Benítez, 2006: 190)

The way the MES unfolds cannot be determined, but it can be foreseen. A reply can be given in VCP for each action in the MES: that is, the idea of planning. And, like a procedure, planning must pay attention to two layers of selection: spatial – the framing – and temporal – editing – and one spare of bringing the plan up to date depending on the modifications that occur every instant. If two layers of selection exist, there will also be two corresponding levels of omissions and ellipsis. After that, the mechanisms of the narrative construction in the VCP will be developed.

The director *guides the viewer's gaze*, that is, he determines the point of interest at each moment of the coverage:

> In viewing the contest at the stadium, the fan takes in a large view – the players, the playing field, the other fans – and must decide what to focus on. In the mediated version, such 'focus' is decided for the fan by the television director who is making a decision about what is visually important.
>
> (Wenner, 1989: 31)

The television directors have amassed seven decades of experience on how to plan and direct teams to bring football matches to the screen. The introduction of the VAR and what changes it will imply for a live broadcast will be discussed later, but now is the moment to ask what principles the planning of a football match will be based on for it to be an expression of good television.

The principle *of knowledge of the script*. Let this brief CAST analysis of the MES of football matches serve as an example. Football can be characterized as a sport of collective confrontation – with numerous participants – which is positional and strategic, played with a ball on a flat surface of short grass, which is rectangular with symmetrical, marked-out playing zones, in which scoring is low, and, as a result, all action that potentially leads to scoring is of great value, which flows and has few interruptions, is of a more or less prefixed duration, over two halves, among other determining features. MES organizes the play as a kind of Roman circus (González, 1992: 68–9), with the public totally surrounding the field of play and expressing themselves throughout the match.

The *principle of the alternating point of interest*. Some sports have a clear point of interest: the ball while it is in play, what spectators want to watch is how it is moved around (Eco, 1990: 230). There is, therefore, an obligation to show it. Visually, from the analyses of any given match, a tension between showing the action or showing the players arises, it being understood that if the gesture is shown, the action is not seen and vice versa. From this point a cutting pattern structure can be created, to be discussed later, showing the ball while in play and using the pauses for the close-ups (Whannel, 1992: 100).

The *principle of balance*. This means avoiding taking the side of either of the contenders, but at the same time trying to ensure that whatever happens is documented with equity, no matter who is involved, wherever and whenever it happens.

The *principle of the all-present eye*. This is to allow viewers to be able to appreciate the action without obstacles. This principle coexists with the evolution of the technology and the competence of the viewer, who demands greater visual variety, depth of analysis, precision, coherence, and pertinence into the cutting pattern. In the 1940s and 1950s, care was taken to seat the spectator in the best place possible in the stands and to move him as little as possible from there. Over time, another philosophy prevailed: *the TV viewer must have the best experience of the event possible*, although some rules of construction of the visual space are still maintained, but that will not be discussed here.[3]

The aesthetic principle. This bears in mind the objective of spectacularity, when selecting the point of view, especially in recordings. For example, repetitions in the axis (that is, moving from or to the camera) may be preferred for several reasons: they keep the main characters in frame and approach subjective points of view of the goalkeeper and the one who has taken the shot or set up the play, and it is easier for the camera operator not to lose sight of the ball – although keeping correct focus may be more difficult – they function to give the idea of the technical difficulty, as a suitable scale can be maintained and there is less uncertainty with the trajectories, framing can be more harmonious and therefore better composed aesthetically.

The principle of ellipsis. As in any narration, the dialectic between showing and hiding. There are two different dimensions for ellipsis, at the sequential level and that of the frame. Considering that football coverage is always multi-camera, in the first one there is a selective tension between the cameras that are being shown on the screen and those that are omitted; this will lead to replay sequences being prepared. In the second one, the question is how to delimit each frame from each one of the cameras. Then guidelines must be created to frame (delimiting which subject to follow, width and height), to focus, and to keep lead and head room.

The principle of action-reaction. Adapting from physics, it could be explained as "any action in a football pitch may create a reaction on the part of any one of the characters in the scene". Not only that of the players or officials, but also the fans or important persons in the stands, of the players and the coaches on the benches, etc. In this respect Sandy Grossman, director of many Super Bowls and NBA finals, said:

> in football there are at least twenty seconds in between every play, and there are lots of things that you can show: the coaches, the huddles, the players; there is room for a great variety of things besides the game, and everyone has a different reaction to what you cut away to.
>
> (interviewed by Kuney, 1990: 18)

In football, those 20 seconds do not always exist, but the idea is applicable.

Multi-camera configuration was, provisionally, accepted. In each play, a large number of characters may act in a way that will give them a prominent role, located in different places at the same time in the scene:

> events may take place simultaneously. The totality of the action may be distributed over a large area or too fast to be covered from one angle, and thus demand a combination of coverage from selected vantage points and close-up cameras fluidly following the action.
>
> (Engström et al., 2008: 116)

It is necessary to pick up relevant actions thus placed and for them also to be taken from different points of view and with visual variety.

If there are different points of view for an action, a *principle of construction in sequence* will be necessary for continuity to happen (Eco, 1990: 229). Bearing out what André Bazin says about the results of cutting and editing in cinema: "when the essence of a scene demands the simultaneous presence of two or more factors in the action, montage is ruled out" (Bazin, 1990: 77). Theorists have described a zero-degree style that has dominated cinema for many years, according to which a non-modal writing is possible, without style, with an apparently absent author and an invisible and transparent (Bürch, 1983: 20) technique. In televised football, realistic construction in montage (Eco, 1990: 246) functions, implicit while the ball is in play but, to the contrary, it does not work or remains very diffuse in the sequences of analysis built of replays. However, in law courts, evidence coming from images in movement, by their own essence and narrative, are watched with much attention: "Silbey convincingly debunks three enduring myths about film as evidence, (1) that film is objective and unbiased (2) that its meaning is unambiguous and obvious, and (3) that film transforms a viewer into an eyewitness" (Jessica Silbey cited in Dickson, 2013: 13).

The principle of division of the MES. This looks at continuous play in manageable units. In the case of football, that division comes from a rules-based action, indicated by the referee. When the ball goes out of play by leaving the field or crossing the goal line, or "play has been stopped by the referee" (IFAB, 2018), according to Law 9, the action that was taking place is considered to have finalized. This segment of time, variable due to many circumstances, is what we know as play. The director turns the unit of action of the MES, play, into one narrative unit of the VC, sequence (Benítez, 2006: 191). This way the TV director, without time for reflection, must identify logical phases while they are still chronological phases (Eco, 1990: 237). The alternation and the time of possession of the ball by each team are variable, no shifts nor determined use of the ball exist. There is always a preparation time, however minimum, before the game restarts, and the director usually will use this time to punctuate his VCP story, closing the present sequence and anticipating the next one.

The fact that play stops with greater or lesser frequency determines two modes in the MES of football: when the ball is in play or when play is stopped. These two modes mutually exclude one another, even if we must pay attention to the indications of the referee, as at times they could be confusing: "the production activities during in-play action and the transition between in and out-of-play are very complex, mobile and unpredictable, and these features generate challenges both for the coordination of the production team and the maintenance of meaningful broadcast footage" (Engström et al., 2008: 118).

Most studies have estimated the average time the ball is in play to be between 55 per cent and 65 per cent over the duration of the match (Castellano, 2008). According to valid measurements, in the present decade this percentage would be between 57 per cent and 63 per cent (see Chapter 10). It is evident that the director must develop a consistent plan for the time that the ball is not in play (at least 40–45 per cent of the overall time) that could create similar interest to the time when the ball is in play.

Operations Plan and Narrative

The director devises a plan, in which he will involve all his operators, to obtain several takes explaining each play and will make it into a sequence in the context of the match. Also he designs how to convert every action when the ball is in play in the context of every play, distinguishing the scale of shots according to a kind of *narrative depth* (Benítez, 2006: 211): general or wide shots for the tactical series; whole body shots or *full* – in the jargon – for the technical ones; and closed shots for the gestural series.

Harmony and balance between this diversity of takes, which is fundamental to show the game fully, resume, explain, or comment on the important parts, taking into account that any camera would be part of a sequence of repetitions; understandable enough so the viewer can form an opinion about what has been going on or of originating purely dramatic and/or spectacular sequences.

To obtain good results on the screen the director must have a great command of the technology and logic. The plan has these two layers and in the technological dimension must determine at least the location – the study of the architecture of the venue, the weather, the placement of each camera, etc. – the type of camera mount, lens, and the temporal resolution of every camera. These decisions have much to do with what risks the HB is prepared to take regarding the event, and what he wishes to portray. In the logical layer, decisions have to be taken with regard to the use factors, such as instructions for the exposure, shutter speed, and depth of field, colour design, type, and cadence of the camera movements to use, etc. (see Chapter 8), and with respect to the operations plan for following up the point of interest, focus, framing, and keeping head and lead room.

The Canonic Structure

As has already been seen, each play of the MES seen is translated to a different narrative sequence in the VCP. But these have a common structure. Brien R. Williams presented in an article in 1975 a sort of *canonic sequence* (with the ball in play) studying the different VC of every one of the big American television chains, when broadcasting American football. In spite of the different ways of handling it, he found that "the typical sequence of shots depicting play" consisted of a few shots of preparation, with wide short or medium shots of the characters who look like they are going to be the protagonists, the normal coverage shot which showed that tactical figure, the play-action camera zoomed in on the ball carrier, and "a close up as the ball carrier extricated himself from the pile-up and moved back to a re-forming huddle" (Williams, 1977: 137).

Plays, reactions, and repositioning for the start of the next play can define in a structured form any succession of actions in the MES in a football field. VCP also have their own structure, although it is very dependent on the periods when the ball is in play. A typical block of shots could be the following: *live action, shots of protagonists, replays, shots of the next play being prepared* (based on Benítez, 2006: 192; 2013b: 269).

Roles and Assignments

The director defines each possible action and distributes follow-up assignments and responsibilities to each one of his camera operators (i.e., Owens, 2007: 122–3) when the ball is in play. To optimize his means, he also assigns different missions in this mode when the ball is not in play, when play is stopped. To do that he distinguishes roles of action. In a header that goes wide or over, the passer, the shooter, or the goalkeeper may have singular reactions. When one player challenges another in midfield, and there is a foul, more often than not one of the players involved will end up on the ground, but there will be one offender and one who is offended against. After an offside is flagged, the player in the offside position or the one who passed to him can be shown. The name or number is not so important at that moment, only the role the player was performing.

While foreseeable actions happen, those plans are executed by means of procedures, which are similar to algorithms (Benítez, 2006: 189, based on Gréhaigne, 2001: 154–6), that give the sets of instructions that respond to various states of the game. If an operator has been assigned a given mission and what was foreseen actually occurs, he will act according to the received orders. Referring to ice hockey, although also applicable: "predictable situations that reoccur throughout the production may have predefined patterns that support editing decisions" (Engström et al., 2008: 116). In many cases this operator must decide on possibilities that, although foreseen, fall outside the normal. In that case, both the director and the camera operators will decide as they go along, applying heuristics instead of algorithms. They discard the absurd solutions and take rapid decisions that give value to what is viable and more pertinent, trying to adjust their decision to the generalities of the plan and then returning as soon as possible to the algorithm flow. They base their actions on their knowledge and experience.

To summarize, for camera operators, the sets of instructions that form the algorithm contain certain parameters: mode of operation (ball in play, ball out of play), target (role or roles that they must frame or follow), narrative depth (scale), following instructions, focus, framing (lead room and head room), reaction to contacts, reaction to tally, preventive behaviour in each play. Also, for those in charge of making the adjustments, decisions have been taken over the technology such as spatial and time resolution, mount types, lenses power, etc., and about its placements and way of applying in the using factors: exposition, shutter speed, depth of field, and colour design.

Replay Planning

The director conceives the plan that details how and when to take advantage of a pause in the play to insert a period of analyses in replay sequences or *playlists* and the procedure for showing them; he also designs tools and processes of visual effects such as formulas for displaying match stats, plays, and other details about football.

The replay planning also has its technological and logical layers. The number of devices for recording and their capacities are not always a decision of the director but the result of the kind of hardware available. The task of the director consists of securing the simultaneous recording of as many camera signals as there are available and the plan to explore the takes can then be efficiently applied at great speed. It is a question of guaranteeing the possibility of replaying some of the best takes that present the informative function – documentation showing what is of importance of the things occurring in the match – spectacular – extoling the sports aesthetics or purely photographic – and dramatic – managing the intrigue and the emotion of the TV viewers (see Chapter 5). He can take the decision to distribute the different signals among the replay operators (ROs) so that similar takes can be reviewed simultaneously.

The director has to make sure there is a possible explanation for each play and also be ready to offer an analysis and possibly an opinion to his audience. Now, he also has to help and support the referee's decision, so he takes on a new function that will be seen in more detail: the normative function.

It is convenient to quickly look over how the workflow of the execution of the replay plan has been until now. The director assigns to his ROs the cameras that have to record in their machines, and explains the sequence construction algorithm: the type of play, the number of replays, the ideal order of the takes, and the reproduction speed. Referring to American football, Sandy Grossman's collaborators have instructions about the routine tasks: what to do in situations at the goal line; kicks after touchdowns, kickoffs, punts, etc. (Kuney, 1990: 18). In ice hockey a cameraman explains his assignments: "patterns are scripted and commonly understood by all of the participants involved" (Engström et al., 2008: 117).

At the time of executing the plan, the director awaits the opportunity when the ball is out of play to give some shots of the players involved in the play and open a replay sequence. While that sequence is getting to the screen following his instructions, directly controlled by his assistants, he follows the MES details very attentively in case it is necessary to return to the game immediately or to prepare the transition shot to the live VCP. Thus, he can adapt the duration of the different sequences, or clear the way for any of the playlists he had waiting and ready to use.

The director also must foresee and give indications about the possibility of heuristic variations in replay sequences. Furthermore, he explains a plan with general ideas about the urgency, pertinence, and periods of validity of the takes that are waiting to be replayed (Benítez, 2006: 222.). It is important that the ROs know and participate in the camera plan and the cutting plan, as that way they will be more able to take decisions better and more rapidly.

The work of the ROs is based on the time code (TC). If they foresee that an action may become a replay, they'll start their procedure: mark the take (register of TC of a frame over all the cameras that the operator records simultaneously), select the take according to the director's plan, return to the mark, quickly previewing to see if the take is correct and may be replayed, place themselves in the right frame to start (the definitive mark); say he is ready with the take to broadcast; eventually prepare more takes; in the case that the sequence is going ahead, start the reproduction at the indicated speed; as soon as possible put the take or takes in the pertinent playlists, tagging (type of action, reason for the validity of the take, characters concerned, etc.) to be able to access faster later on.

Plan Modifications

During the execution phase of the plan, the director omits many more camera feeds than he puts on the screen; for this reason a lot of takes will be candidates to form part of replay sequences or playlists. In the selection of the ones waiting to be emitted, factors appear which depend directly on the design and planning of the VCP, while others depend on the skill and practice in the operation or on the various different modifying situations of the plan. In this respect the director is prisoner of his own plan (Benítez, 2013b: 320); enclosed in his van he can only appreciate what his camera operators keep in frame at every moment.

John Goldlust explained that "the broadcast of any sport event is 'constructed of three embedded' events that occur simultaneously" (cited by Barnfield, 2013: 338): the game event, the stadium event, and the medium event. This starting point serves to outline that, if time is stopped at any moment of a television live broadcasting of a football match, variable simultaneous modifiers and singularly at least three layers can be analysed:[4]

- the *sport snapshot*, referring to what is happening on the pitch but also in the competition as a whole, including the standings and results elsewhere;
- the *scenic snapshot*, which embraces anything related to the MES in the sense of the arrangement of characters and other stage elements such as meteorological conditions, unforeseen actions on the part of the fans, etc.;
- the *media snapshot*, having to do with the placement of the hardware, on the one hand, and the points of interest contributed by public opinion, on the other.

The order of actions for ROs can be outlined thus: *mark-selection-checking-preparation-start playing*. The more capable and expert ROs need between 10 and 20 seconds to prepare each take when the actions are clear enough and the game does not indicate an important play is about to occur. In the case of working with four ROs, four takes can be prepared and some also tagged in 20 seconds. Unforeseen things occurring in different situations may lengthen these periods: for example, a player crosses the line of the camera view – scenic snapshot – or the sweep panning of a camera makes the ball appear blurred in the instant it looks like the hand was going to touch it – media snapshot – etc. In these cases, each take may need over 30 seconds' preparation time.

It is evident that putting a sequence of repetitions on the screen demands a range of skills and know-how, apart from a polished teamwork methodology. The director and his team are used to responding to the stimulus of a play that is happening in the MES with foresight of what can happen and anticipation to take the adequate decisions. It is not only necessary to put into play a series of techniques with great skill, but also to do it in the proper order by means of an important individual contribution from each person in the team. The ideal state is the *flow* in the sense of Csikszentmihalyi (1997), in which no matter how complicated the challenge of the MES, the answer in the VCP makes the technique not an impediment or obstacle but a perfectly oiled and calibrated vehicle. It is equally necessary to know the weaknesses of every derivative from the plan.

Discussion: The Commitment of the Director and the Arrival of the VAR

Football referees are required to make efforts that go beyond their capacity as human beings (see Chapter 8), which they solve with experience, physical fitness, precise interpretation, good positioning, and other skills. To achieve flowing play, referees are expected to take rules-based decisions on the spot. It is difficult to justify why football has taken so much time to start using technology to improve results in refereeing. In times of standard definition TV, video was far from being infallible (Benítez et al., 2008). Even though it is not perfect yet, it has been inevitable that an attempt would be made to use video assistance.

The VAR work starts with the review of the video record of all actions that happen in the match. It has already been seen that the video record contains the feed from each camera. Each camera operator or every system to register images must have instructions to ensure his work achieves the desired objectives and does it with the available means and time. This type of activity is included in the procedure that has been called VC here and it is easy to see that the concept of VAR shares this nature. So, it is convenient to observe the differences that may exist between a conventional HB planning and the one the VAR would need, or better, to ask what the VAR needs to plan its VCP, so that the steps taken before are the right ones.

In an interesting review of models that study *situation awareness* applied to the performance of sports officials, Timothy Neville and Paul Salmon establish that

> the regulation of the behaviour of players, highlights the inherently different goals that the official has within the game system. The goal of the official is to judge and administer the rules of the game; while the goal of the game itself is to determine a result between two competing teams (or players).
>
> (Neville & Salmon, 2015: 3)

This principle can be applied to directors and officials. The purpose of the following lines is to point out the differences of their different concepts regarding plans and execution of the VCP.

The objective of the referee consists in keeping the match within the Rules of the Game and in an atmosphere of correctness and fair play (according to the question asked to Carlos Bacigalupe, Head of Training at the Refereeing Committee in Madrid, Spain). Football is a sport that has to obtain maximum continuity in the play; the capacity of the referee to achieve this will be one of the most-valued aspects of his work.

Therefore, the intention of creating an entertaining show means it is very clear that the objectives of the director and those of the refereeing team are different. The director only considers the normative layer as a reference to clear up controversies. In a previous section it was argued that to reach determined audio-visual targets/goals it was necessary to start from a technological level and also from a logical one, and to do so with the principles already outlined. With regard to the foreseeable actions to which it is applied – those seen above – it is clear that the main interests of the VAR are the normative series and the gestural-reaction.

FIFA's starting point is the motto *minimum interference-maximum benefit* (IFAB, 2017). It is necessary to point out at this juncture that neither the technology nor the services provided by the HB should create obstacles for the objective of the VAR to be achieved.

Technology Helping Refereeing

The VAR system, in principle and as we deduct from its name, reviews videos as in instant replay (see Chapters 6, 7, 8, and 9), reinforced occasionally by a system using graphic tools that helps for diagnosis and representation, in the style of Hawk-Eye (see Chapter 8). As well as this, football employs goal-line technology (GLT) to clarify if the ball has crossed the goal line or not, but this device, based on positioning and not so much on video use, falls outside the scope of this review.

In the chapter that is focused on technology (Chapter 8) some crucial advances for the objectives of the VAR are mentioned, especially in terms of spatial and temporal resolutions. With all their pros and cons, the first one allows points of view to be established at a tactical scale, and increased, if necessary, into technical scale or closer, and the second allows movement to be analysed in detail without blurring or imprecision, all within parameters of perfect sharpness and image quality. However, the director of the HB determines framing and subject in accordance with a point of interest that has a lot to do with his vision of football as an audio-visual spectacle, and to achieve it he uses narrative principles like ellipsis and the construction of live-editing sequences. Thus, at times, his planning includes camera shots from suitable spots, but with spatial restrictions: a shot of the lower part of a body while the ball is being played, or a close up when the referee stops the match due to a foul that leaves two players shoving each other out of the frame.

In another example, an operator who has been assigned a technical shot following the ball, must know what to do if two players fight for a ball in the air, whether to stay with them even if the ball has left the frame, or follow the ball, thus leaving the players out of frame and possibly omitting something that could have been against the rules a moment later. This simple decision is of crucial importance for live narrative, at which it is aimed, namely explaining actions in a sequence of analyses. In cases like this example, if there is not an alternative take of the same action, from more than one camera, the video would not register these infringements. The provisional conclusion is that, by using the same VC system, different objectives may or may not be obtained. Since the director does not have to focus on the rules being respected, there should be an independent device to register footage that is not at the service of the HB. As has been said about American football, "furthermore, because the replay official relied exclusively on camera angles provided by the television broadcast, network executives felt that the NFL was compromising the independence of the broadcast" (Oldfather & Fernholz, 2009: 50).

On the other hand, neither would it be right that the HB should find images showing an infringement and the VAR not have them available. In version 8 of the VAR handbook, it can be read that "the integrity of the VAR system would be undermined if the broadcaster could show footage not available to the VAR/referee which contradicts the VAR/referee decision" (IFAB, 2017: 6) (see Chapter 10). Then, the VAR would have to analyse, as well as its own footage, that coming from the HB. The description of the procedure for repetitions shows that the greater the number of cameras recording, the more complicated it is to select takes and the more complex the coordination work becomes. And, it must also be taken into account that some actions will have to be visualized in slow motion.

A group of sportsmen acting at the same time, on an extensive stage with a prepared multi-camera device to record their activities, means the director's team needs to be mentally prepared for a sort of polyhedral reading in multiple space[5] and time perspective. Referring, once again, to American football, "on an extra point conversion, 20 of the 22 players on the field participate in such physical contact" (Real, 1975: 38). In a corner kick in soccer the problem is not as bad with regard to the number of players, but more complex in its spatial distribution.

For decades, the television teams have been developing the abilities to master these complications. They conceive the MES as a sort of audio-visual cubism, being able to locate the best angles and takes instantly to review any play and reduce them to a camera number or video channel. These skills require learning and experience; they cannot be acquired magically. Today is not a question of will or automatism; it is not enough to simplify or to try to imitate the present procedures: "much of this research seems to assume that we can produce simple tools that ape the professional ones. However, in reality, professional live broadcast is highly skilled and intense" (Engström et al., 2010: 1503).

Video, although very evolved technologically and a very valuable tool, is not infallible (Dickson, 2013: 20) and it is not valid just by itself (Nlandu, 2012). Somewhere else this affirmation has been sufficiently discussed and argued (Benítez et al., 2008) (see Chapter 8), but there are recent examples of important occasions when it has not been able to eliminate all doubts. The key play of the Champions League semi-final of the 2017–18 season, Juventus vs. Real Madrid (who went on to win the tournament) in April, was a contact between Benatia and Lucas Vázquez close to the Juve goal area, in minute 92; the match had almost finished, with a result that would have required extra time. A penalty, which classified the Spaniards for the final, was awarded. This play may be seen from all possible angles and different solutions can be given, but never with indisputable evidence (see Chapter 9). As a technology, video cannot give a solution by itself in plays of this kind, thus requiring a human to interpret and take responsibility for the final decision.

Occasionally video has been used in an improper way from its technological implications. Cameras construct images from a single point of view, in a projection that theory calls pin-hole geometry, with rules that assign to each point of the tri-dimensional space, by means of projective transformations, a point in the bi-dimensional plane of the image. With this concept it is intended to give a solution to a problem of dynamic position of a minimum of four mobile elements – two attackers, the one with the ball and the one who moves into position to receive it and play it, the last defender, and the ball – with respect to a line that runs parallel to the goal line – by a camera which constantly moves its focal centre around the tripod's axis in combined and fluent movements, tilt, pan, and zoom, to follow play. In these circumstances it is necessary to proceed from epipolar geometry, which always considers two instantaneous points of view to be able to determine relative mobile positions. It is a difficult challenge, but it has been proved to be possible (Tamaki & Saito, 2015) even without the use of synchronized cameras or dedicated systems, just like the ones used for football coverage, but with all the elements in sight and without obstacles. A greater number of points of view should be used to circumvent the problem of some points being hidden in the scenic snapshot. Dedicated multi-camera systems have already been installed in numerous stadiums, but in the medium term, work should be done to make solutions available in open systems, without commercial dependencies.

Already integrated into the VAR system, the validity of video may continue to be discussed. In the FWC Russia 2018, on the last day of the Spain vs. Morocco Group Stage match, the result was 2–1 to Morocco and the game was about to finish. Spain took a free-kick from the right side of the attack, the ball swung in close to the penalty spot, and Iago Aspas scored. Initially the AR indicated offside, but the VAR reviewed the play and determined it was a goal. The takes that were offered only convinced the Spanish fans.

The take from the camera in line with the defence and located at low altitude cannot show the moment when Aspas' foot hits the ball and therefore is not situated in a place from which to show whether he is in a correct position or not, as his body and his head are shown clearly to be in front of every part of the Moroccan defenders' bodies. In the majority of shots that pick up a right-winger crossing the ball into the area situated to the left of the master camera, the ball can hide the boot, and this is applicable to similar situations that, counting them up, means a high number of plays over the course of a whole match. Afterwards a virtual offside line is displayed over the elevated shot – the camera situated almost on the edge of the roof – but in the frame the ball cannot be seen, being shown instead in an additional take by picture in picture effect. The moment when the ball is touched cannot be seen sharply, and besides, the take from above shows the figures on the pitch at an angle, making a precise view of the relative positions difficult. The final result saw Spain go through to the next round as group leader; Morocco were eliminated.

VAR Principles

The plans of HB and VAR are analysed comparatively according to the established planning principles. In the *principle of knowledge of the script*, in the case of the VAR, it would be the rules-oriented script of a given match: the rules of the game and a lot of interpretation. One of the most important consequences would be the maintenance of the rules, which implies impartiality, restraint, and justice (similar to the *principle of balance* given for the HB). Applying the *principle of guiding the viewer's gaze*, the refereeing team cannot discriminate as infringements could happen at any moment over the whole course of the match, at any place on the field or in several places at a time. While the play is going, the ball is a point of interest but not the only one.

While a slight distraction of a member of the television crew may cause a dull or incorrect sequence, but one that is avoidable, for an official to lose concentration at a key moment could have serious consequences, including influencing the match result or, worse, that of a competition. The officials must, working as a team, help the referee take the right decisions, applying the rules with two characteristics: assuming *responsibility* for the decision and its *immediacy*, to keep play flowing. So, this principle of *decision-taking* is in tension with the principle of *division of the MES* explained before; the director tries to compartmentalize the action and the referee, in opposition to this, tries to ensure that the flow continues.

While the ball is not in play, the director tries to build sequences of close-ups and replays, while the refereeing team takes care to organize the sport layer, and more and more the scenic one: for instance they mark the order in the position of the players with a spray when there is a foul near the areas. When stopping the action, the referee indirectly centres the gaze on himself, completing the MES. Also, the referee takes part in the organization of the media layer with the explanation given of what the VAR says; this is the first time that the VCP has become involved with and modified the MES of football (for the GLT detection system this is not necessary). As a result of the VAR planning, three phases or modes can be found affecting the director: the ball in play, the ball not in play, and the phase of the application of the VAR in which – for the time being – the screen is left to the VAR.

Going against the premise that directors work on, aesthetics is not a usable criterion for the VAR. The greatest variety of camera angles and scales are needed to be able to evaluate at every moment that the rules are being adhered to and trying to avoid any undesirable incident in the other layers of modification – the sport snapshot, the scenic, or the media. For example, in the penalty areas, any contact might give rise to a penalty, so nobody inside it or very near it can be ignored by the referee. The *principle of ellipsis* would not be valid, because it would be a critical error to hide anything. However, the *principle of action-reaction* must be taken into account by the referee, if only from the point of view of the rules, and therefore without zooming in to close-up scale (which would hide a great part of the action).

The *principle of construction in sequence* requires a mention all on its own. This means the VAR has to receive footage that does not always help in the taking of decisions as it is only faithful to the criteria of a formula of cinematic expression. Traditionally a sequence that has been built by cutting and pasting several takes together instead of a one-off shot means the viewer has to suspend disbelief and, to avoid the sensation of manipulation, measures are forced and not especially effective. The example of Iago Aspas' goal in the Spain vs. Morocco game, which has just been cited, used this method as there was no other single shot that records all the movement of the ball, the position of the attacker, and that of the defensive line with clarity. The VAR should look for its own combination of angles, resolutions, and editing.

VAR Narrative

Television has been creating a narrative with a dramatic facet: "television has invented, in effect, an original form of drama, an audience experience that could not have been conceived prior to the existence of technology now available for recording, storing and retrieving live action" (Morris & Nydahl, 1985: 102), but keeping a link with reality: "[to] create a compelling 'story line' which originates from and is related to, but which is not actually itself, 'real life'" (ibid).

To other authors, the mere fact of the combination of different times in the same story already constitutes an opportunity to create new narratives and not only for television uses: "furthermore we suggest that our focus on the juxtaposition of media with different temporal trajectories, or what we have termed as temporally hybrid media, might inspire new kind of narratives both in video productions and games development" (Engström et al., 2010: 1503).

Like all VCP, the VAR needs a narrative, eliminating of course its spectacular and dramatic facets, but one that communicates its intervention to the main referee and explains it to the fans in the stadium and television viewers. It is very important to understand that, although its construction is more rushed – because of the urgency and immediacy required – than replay sequences the HB would put together, which would look similar, its meaning does not finish within the stadium, as it affects very important decisions whose impact will long outlive the refereeing on the day (see Chapter 5): each VAR intervention (because it extends doubts about the infallibility of the referee) creates consequences. It is clear that the fans at the stadium would be more than happy to take the decision to award a penalty or not, as if the place really were a Roman circus. Since they are not invited to form part of the decision-making process, the time they spend waiting should not be a moment for distrust or anger and perhaps football should consider in this sense the experience of other sports (see Chapter 8). Although the example does not refer directly to video it is valid:

> unless competition organisers decide otherwise, the only individuals in the soccer stadium that will see whether or not the ball has crossed the line are the match officials through the use of their watch. The technology would hence only be used to aid referees and not for entertainment purposes.
>
> (Winand & Fergusson, 2016: 10)

The *minimum interference-maximum benefit* principle that the IFAB wants to apply adds pressure and urgency to the process that VAR must follow. Not even the director of the HB has to be in such a hurry to offer the sequences of repetitions nor does he have an obligation to proceed with so much precision, since whatever the former concludes will determine, as has been said, how the game will continue, while the second will always be able to open a narrative parenthesis later to add other previously unseen takes or to analyse details discovered with posterity.

It has been stated that the sequence of plays in itself in the MES of a football match gives it a specific structure in its VC: *live action, shot of protagonists, replays, and preparation of the next play*. The VAR's intervention could be outlined in a completely different way, one that is much more complex: development of the play, definition of the play, stopping play to apply the rules, application of the advantage rule (if the VAR confirms the referee's decision in the *silent check* process (see Chapter 10)), or the application of the VAR, consisting of a time waiting for the definition, VC of the explanation, communication and application of the rules, and restarting of play.

In these pioneering times an interesting debate has opened about whether the HB, and more exactly the director, will be a part of the new narrative which is moving away from its initial objectives. If he is inevitably to take part in the game, he will have to assume a new role in the MES and participate, given that he has the experience, the knowledge, and the preparation to do so. Even if he is not integrated into the refereeing team, keeping him out while the VAR acts seems to be a waste of resources. As long as the production company that provides the service to the VAR and the HB coincide, there will be no dispute, as in the present circumstances a total collaboration will make interests converge: the VAR develops without obstacles and the HB increases its income, so everybody wins. But sooner or later VAR will have to take on a certain independence and it will be necessary to negotiate if that time on the screen is handed over and in what conditions.

Execution of the Replay Plan

It has been seen how the different objectives of the VAR could mean the planning of camera coverage in a given match will not meet its needs. With the execution of the replay plan (which is a part of VCP), the same thing could happen. FIFA's initial working hypothesis included a process with time enough to see a possible error, check it by identifying the clearer takes immediately, notify the referee, review it but passing on only the ones which can better help the main referee – together with his comments – and take a decision. At one time it was estimated that this lapse would take about ten seconds. It has already been demonstrated that this was wishful thinking (see Chapter 10). In the first place, deciding there has been an error means having observed the way the play has unfolded, looking at the indications (or lack thereof) of the referee, and comparing it with the takes in case of doubt. Here a series of factors intervene that could make a variation in the response time: the number of cameras; the skills to understand the plans of the director and to conceive a polyhedral space; the presence of high-speed cameras; the experience and the degree of autonomy and joint responsibility of the team; the existence of hierarchies in the decision-making (even if a hierarchical structure refines the results, it will also make the process slower); the oiling of the communication procedure (see Chapter 4); the reproduction speed that makes an efficient analysis of what has happened; the possible modifications owing to the various sport, scenic, media, and the new normative snapshots and the pressure that they can create.

The structure of the sequence of repetitions itself has a different make-up for the HB director from the one the VAR will have. In some questions it may easily coincide, for example with the preference of the repetitions in the axis because it may be important from the point of view of Laws that those involved in the play be kept in frame while the same play is being shown. In the case of a shot from outside the penalty area that ends up being a goal, the director could define well-known structures which are repeatable depending on the play, for example: a wide shot from a lateral camera; a wide shot from behind the goal; a full shot from behind the attacker in line with the shot; a shot from behind the goal; a close-up shot of the attacker celebrating.

But the points of interest are different for the VAR: possible offside positions; possible contacts between players, other times where observation of rules must be checked, etc. It is a question of observing attitudes for which the framing and coverage followings are not designed, and which are too elusive to be structured. This will trigger heuristic processes until the techniques to produce algorithms are refined. It does not seem simple at this moment for specialists in the handling of replay systems and the observation of the field of play in a polyhedral concept to have enough abilities to detect behaviour that breaks the rules nor for officials to be able to handle solvency plans and repetition devices. The framing also could be designed in a different way, one that does not correspond to the already-mentioned tactical, technical, or gestural actions.

The director and his crew can begin to foresee the immediate future of the events that are fit to form part of his VCP based on, for example, the principle of action-reaction analysis. According to his plan it is more or less simple to mark plays and be focused on reactions while looking for explanations of what has happened. For the referee, however, it is more complicated to foresee rules infringements, especially those that are done wilfully. It is true that there are moments in which contact can be supposed, but in a corner kick in which normally there will be a great number of potential points of contact, where to look? As well as this, he must remove from his decision-making criteria the reactions of certain players given to theatrics. He will have to apply a different kind of foresight in those situations.

Who decides what is to be recorded, and are they the right people to do it? Who announces what plays are suitable to be replayed to the VAR? What system of communication should be used? Who builds the technique that is to be used? How can the VAR's ROs be sure that one of them does not have the two best takes and, therefore, slows down the decision-making more?

Conclusions

The arrival of VAR to football refereeing means the application of Visual Construction Procedures (VCP) for the team of officials. These techniques have been developed and continuously refined since the 1950s for the objectives of the HB, trying to bring to their viewers' TV screens live programmes that document the match, but that also have a high dose of audio-visual thrill and an interesting and absorbing dramatic charge. Researchers and theoreticians have followed with growing interest not only the results on the screen but also the production process, at least from the 1960s and 1970s. It can also be seen that the VC brings with it an ideological charge, even if in an involuntary way.

The director assigned by the HB has the responsibility to plan and execute the VCP to achieve those objectives. The procedures of planning and execution, especially those of the replay sequences, have been seen in detail and analytical keys have been given in order to face all the steps of the process. The first phase of the TV crew's decision-taking, the technological one, must take into account *mise en scène* (MES) elements and becomes, among other consequences, a process regarding number, type, lenses, mounting type, and placement of the cameras, and number and characteristics of the recording devices. A second, logical, phase, presents the instructions for the use of technology, with precise indications for the operation of image adjusting, different assignments of the camera, and replay operators depending on whether the match is in ball-in-play or ball-not-in-play mode. From the application of the same type of analysis, the VAR needs independent planning applied to a normative script, since the objectives for which it uses its own VC are different from those of the HB, meaning a certain degree of autonomy is advisable in the use of the VAR, and this will be facilitated by the present level of the available technology.

As the objectives to be achieved in the respective VCP of HB and VAR do not coincide, a different technique is needed, one which can be foreseen with a different use of the media and the time and important variations in the workflow. To arrive at an approximation, an additional layer of analysis – the normative one – has been proposed, to the already-existing sport, scenic, and media ones. The observation in this new ambiance reveals, among other aspects, a new phase to bring to the screen, VAR mode, that intertwines with MES, determining this one on occasions and which, in this way, invalidates the idea that the MES can be carried out with or without the presence of the cameras.

In addition, it is underlined that the usual point of interest for the VAR is not limited to the ball when it is in play and that a number of decisions taken by the director of HB – for example the scale in the frames, or the composition of the shots that are registered to be replayed – are not valid for the objectives of helping the referee. Therefore, the same system of VC cannot have two different objectives – the normative and the narrative or spectacular – without producing incoherent results.

It is possible, and it should therefore be viable, to implement an additional circuit of high spatial and temporal resolution cameras planned by the VAR, to complement the coverage of the HB and which could be installed in the venues. Elimination of the HB takes is not advisable due, eventually, to different situations in which these could be more telling than those of the VAR system.

On the other hand, refereeing should not leave out, in its training programmes, the specialists in VC while they try to find an ideal application procedure.

Notes

1 Quotes from this book are in Spanish in the original.
2 Gary Whannel, in 1992 *Fields in vision: Television sport and cultural transformation*, draws a detailed approach to production practices of television organizations, especially in Chapter 3: Production practices and professional ideologies.
3 For a deep discussion of classical narrative audio-visual rules applied to sport VCP, see Benítez et al. 2011.
4 For a detailed explanation and examples about modification situations, see Benítez, 2013b: 81–85.
5 See interview with Nuno Pereira: www.youtube.com/watch?reload=9&v=B_BA1uXqpgo

References

Armenteros, M., Benítez, A. J., Fernández, M., De la Vega, R., Sillero, M. & Sánchez Cid, M. (2019). Collaborative learning methods and multimedia tools for the education and training of instructors. The case of FIFA referee technical instructors in *The International Journal of Information and Learning Technology*, *36*(5), 395–409.

Armenteros, M., Domínguez, A., Fernández, M., & Benítez, A. J. (2016). The video database for teaching and learning in football refereeing. *World Academy of Science, Engineering and Technology, International Journal of Social, Behavioral, Educational, Economic, Business and Industrial Engineering*, *10*(9), 3134–3139.

Barnfield, A. (2013). Soccer, broadcasting, and narrative: on televising a live soccer match. *Communication & Sport*, *1*(4), 326–341.

Barr, C. (1975). Comparing styles: England vs West Germany. In Buscombe, E. (ed.): *Football on television* (pp. 47–53). London: British Film Institute.

Bazin, A. (1990). *¿Qué es el cine?* (2nd ed.). Madrid: Rialp.

Benítez, A. J. (2006). Espectáculo futbolístico y comunicación televisiva. June 2, 2006 (Doctoral dissertation, PhD Thesis in Information Science–Universidad Complutense de Madrid, Available at: http://hdl. handle. net/10016/11731).

Benítez, A. J. (2013a). Deporte y televisión durante la Transición. Nuevos argumentos de programación deportiva utilizados en los inicios de las Televisiones Autonómicas. In Manuel Palacio (ed.): *Las imágenes del cambio. Medios audiovisuales en las transiciones a la democracia* (pp. 35–61). Madrid: Biblioteca Nueva.

Benítez, A. J. (2013b). *Realización de deportes en televisión*. Madrid: Instituto RTVE.

Benítez, A. J., Armenteros, M., & Sánchez Cid, M. (2011). Análisis de la realización de transmisiones de fútbol según variables cuantitativas. Indicadores evolutivos en tecnología y narración de 2000 a 2010. In Sierra, J. (ed.): *La tecnología audio-visual al servicio de la sociedad* (pp. 34–50). Madrid: Fragua.

Benítez, A. J., & López, Á. M. (2015). Comparing sports broadcasting: tracing the style of football telecast in Saudi Arabia, England and Spain. *Fonseca, Journal of Communication, 10*, 79–117.

Benítez, A. J., Sánchez-Cid, M., & Armenteros, M. (2008). La retransmisión televisiva y el arbitraje ¿Es posible que el vídeo ayude al fútbol? In Armenteros, M. (ed.): *TADAR Tecnologías Aplicadas al Deporte de Alto Rendimiento* (pp. 167–178). Madrid: CSD.

Bordwell, D., & Thompson, K. (2002). *Arte cinematográfico*. México: Mc Graw-Hill.

Bürch, N. (1983). *Praxis del cine* (1st ed., Gallimard, 1970, París). Madrid: Ed. Fundamentos.

Buscombe, E. (1975). Introduction. In Buscombe, E. (ed.): *Football on television* (pp. 1–7). London: British Film Institute.

Castellano, J. (2008). Análisis de las posesiones de balón en fútbol: frecuencia, duración y transición. *European Journal of Human Movement, 21*, 179–196.

Chandler, J. M. (1988). *Television and national sport: the United States and Britain (sport and society)*. Champaign, IL: University of Illinois Press.

Csikszentmihalyi, M. (1997). Finding flow. *Psychology Today*, 1–7.

Dickson, C. (2013). *Cela n'est pas une passe en avant (That is not a forward pass)*. Retrieved from: http://dx.doi.org/10.2139/ssrn.2577097

Eco, U. (1990). *Obra Abierta* (1a ed., 1962). Barcelona, España: Ariel.

Engström, A., Esbjornsson, M., Juhlin, O., & Perry, M. (2008). Producing collaborative video: developing an interactive user experience for mobile TV. In *Proceedings of the 1st International Conference on Designing Interactive User Experiences for TV and Video* (pp. 115–124). ACM Press.

Engström, A., Juhlin, O., Perry, M., & Broth, M. (2010). Temporal hybridity: footage with instant replay in real time. In *Proceedings of the SIGCHI Conference on Human Factors in Computing Systems* (pp. 1495–1504). Atlanta, GA: ACM Press.

FIFA (2018). *2018 broadcast and audience executive summary*. Retrieved from: https://resources.fifa.com/image/upload/2018-fifa-world-cup-russia-global-broadcast-and-audience-executive-summary.pdf?cloudid=njqsntrvdvqv8ho1dag5

Giulianotti, R. (1999). *Football. A sociology of the global game*. Cambridge: Polity.

González, Requena J. (1992). *El discurso televisivo: Espectáculo de la posmodernidad*. Madrid: Ediciones Cátedra.

Gréhaigne, J.-F. (2001). La organización del juego en el fútbol. Barcelona: INDE.

Gruneau, R. (1989). Making spectacle: a case study in television sports production. In Wenner, L. A. (ed.): *Media, sport & society* (pp. 134–156). Newbury Park, CA: Sage.

Horne, J., & Manzenreiter, W. (2002). The world cup and television football. In Horne, J. and Manzenreiter, W. (eds.): *Japan, Korea and the 2002 World Cup* (pp. 195–212). London: Routledge.

IFAB. (2017). Video assistant referees (VARs) implementation handbook for competitions.

IFAB. (2018). Laws of the game 2018–19.

Iozzia, G., & Minerva, L. (1986). *Sport e Televisione, un matrimonio di interesse.* Torino, Italia: Edizione RAI.

Kulechov, L. (1956). *Tratado de la realización cinematográfica.* Buenos Aires, Argentina: Futuro.

Kuney, J. (1990). *Take one. Television directors on directing.* New York: Praeger.

McArthur, C. (1975). Setting the scene. In Buscombe, E. (ed.): *Football on television* (pp. 8–15). London: British Film Institute.

Morris, B. S., & Nydahl, J. (1985). Sports spectacle as drama: image, language and technology. *Journal of Popular Culture, 18*(4), 101–110.

Morse, M. (1983). Sport on television: replay and display. In Kaplan, A. (ed.): *Regarding television* (pp. 44–66). Los Ángeles: AFI.

Neville, T. J., & Salmon, P. M. (2015). Never blame the umpire – a review of situation awareness models and methods for examining the performance of officials in sport. *Ergonomics, 59*(7), 962–975.

Nlandu, T. (2012). The fallacies of the assumptions behind the arguments for goal-line technology in soccer. *Sport, Ethics and Philosophy, 6*(4), 451–466.

Novel-Smith, G. (1981). Television – football – the world. In Bennett, T. et al. (ed.): *Popular television and film* (pp. 159–170). London: BFI Publishing / Open University Press.

Oldfather, C. M., & Fernholz, M. M. (2009). Comparative procedure on a Sunday afternoon: instant replay in the NFL as a process of appellate review. *Indiana Law Review, 43*, 1-45.

Owens, J. (2007). *Television sports production* (4th ed.). Oxford: Focal Press.

Rader, B. G. (1984). *In its own image: how television has transformed sports.* New York: Free Press.

Real, M. R. (1975). Super bowl: mythic spectacle. *Journal of Communication, 25*(1), 31–43.

Reeves, J. L. (1989). TV's world of sports: presenting and playing the game. In Burns and Thompson (ed.): *Television studies. Textual analysis* (pp. 205–222). New York: Preager.

Ryall, T. (1975). Visual style in Scotland versus Yugoslavia. In Buscombe, E. (ed.): *Football on television* (pp. 35–46). London: British Film Institute.

Santamaría, V. (2002). Planificación de eventos deportivos (I). *Shooting, 1*, 62–66.

Silk, M., Slack, T., & Amis, J. (2004). Bread, butter and gravy: an institutional approach to televised sport production. In Rowe, D. (ed.): *Critical readings: sport, culture and the media* (pp. 146–164). Maidenhead: Open University Press.

Tamaki, S., & Saito, H. (2015). Reconstructing the 3D trajectory of a ball with unsynchronized cameras. *International Journal of Computer Science in Sport, 14*(1), 51–68.

Wenner, L. A. (1989). Media, sport and society: the research agenda. In Wenner, L. A. (ed.): *Media, sports & society* (pp. 13–48). Newbury Park, CA: Sage Publications.

Whannel, G. (1992). *Fields in vision: television sport and cultural transformation*. London: Routledge.

Williams, B. R. (1977). The structure of televised football. *Journal of Communication, 27*(3), 133–138.

Winand, M., & Fergusson, C. (2016). More decision-aid technology in sport? An analysis of football supporters' perceptions on goal-line technology. *Soccer & Society*, 1–20.

4 The Interaction between Audio and the Video Assistant Referee in Football

Manuel Sánchez Cid and Alberto Luis García García

Introduction

According to Carlos Bacigalupe's definition of a VAR (see Chapter 10), this is "a surveillance system in which representative images and audio signals are analysed in order to resolve doubts and/or confirm decisions". However, despite being called Video Assistant Referee, audio communication is a fundamental functional element of this technological advance. Audio is the essence of the communication between all members of the team involved, the officials on the pitch, and the surveillance team in the back room. Moreover, it is important to remember that the audio signal from the pitch can also be assessed and analysed by the VAR, with the potential to re-call both the audio signal between the officials and the VAR, as well as the audio recorded for broadcast (mainly ambient sound). Despite the fact that, as FIFA representatives Pierluigi Collina, Massimo Busacca, and Roberto Rosseti agreed that in its current format VAR helped achieve near perfect refereeing during the World Cup in Russia 2018 (99.3 per cent success) (Moñino, 2018), structural problems in terms of communication have been reported in the past, as confirmed by Francisco José de la Fuente in the "Workshop on Audio-visual Technology and Innovation in High Performance Sport" (De la Fuente, 2017).[1]

As such, there appear to be two issues related to the audio communication in VAR that require improvement: on the one hand, clearer and more effective protocols for audio communication should be established; and on the other hand, decisions must be taken regarding the transparency of the audio content, given that according to news articles (Fuentes, 2018), FIFA itself still considers the content of refereeing communications during matches sensitive material. Indeed, in terms of the latter, FIFA is careful to avoid possible leaks and it restricts the use of such recording by third parties, as seen during the World Cup in Russia 2018, when they allegedly initially refused to release the audio signal recorded between the main official and VAR to the Brazilian Federation, ultimately releasing partially edited details at a later stage but not the entire recording (Kallás, 2018). In addition, to help clarify confusing situations, this audio signal may also be considered a useful training aid, as it helps illustrate the correct and incorrect decisions taken during matches.

In our opinion, the use of audio in the VAR process should be given more consideration by the organizations associated with football and the media than is currently the case. Both these parties systematically place stronger emphasis on the visual information, as well as how this material is structured and handled, without addressing the issues related to the audio signal in sufficient detail, or the communication protocols involved. This attitude illustrates a clear disregard for the potential of the audio signal when trying to clarify controversial situations, where it can be at least as informative as the video footage. Accordingly, we will address this aspect in the following section, describing the associated and related technological developments.

VAR Technology: A Necessary Evolution

Background

There has been much international debate about the need to implement systems to avoid unfair refereeing decisions due to human error. Indeed, sports like American football, rugby, field hockey, tennis, basketball, and cricket (see Chapters 7, 8, and 9) have been implementing systems that incorporate different technologies to improve referee communication and oversee decisions during the game for years (more than 20 in the case of the NFL). However, while football is the latest sport to implement a VAR system, this is not currently the case in all countries nor in all competitions. According to data published by EuropaPress in March 2018 ("¿Cómo funciona el VAR ...?", 2018), there are more than 20 national associations and tournaments that currently implement or are close to implementing a VAR, including: Saudi Arabia, Germany, Australia, Belgium, Brazil, China, CONMEBOL, FIFA, USA, France, UK, Italy, Japan, Mexico, Netherlands, Poland, Portugal, Qatar, Czech Republic, South Korea, and Turkey. Moreover, VAR has been tested in the U-20 World Cup, the 2018 Confederations Cup, the FIFA Club World Cup and the FA Cup. Nevertheless, some leagues as iconic as the Premier League, the Brazilian league, Europa League, and the Champions League have reservations and are sceptical about the benefits of its application (Pérez & Alcalá, 2018), delaying its implementation. Indeed, UEFA's president, Aleksander Ceferin, referred to certain confusion regarding the use of VAR in a statement made in February 2018.

In February 1992, one of the authors of this chapter (Sánchez Cid & Pueo, 2017)[2] proposed to the Head of Referees Committee of the Royal Spanish Football Federation (Real Federación Española de Fútbol – [RFEF], 2018) that a system allowing communication between the main referee and an external supervisor could be implanted to assist the former with more conflictive decisions, enabling audio-visual footage of the game to be reviewed. The committee's Chairman pointed out that at that time, they were aware of three people who had come up with this idea worldwide: Franz Beckenbauer, a Spanish security technician, and Manuel Sánchez Cid (co-author of this article). Given the relevance of this initiative, a study was commissioned on which an internal debate could be launched by the Committee, potentially taking this up with the relevant European authorities, if approved. The idea was passed on to the regional television channel, Telemadrid,[3] for endorsement and further development, but the necessary support was never attained, and the impetus behind this innovative proposal gradually diluted. At that time, and given that the project was in a preliminary conceptual phase, no credible patent application exists. Nevertheless, the members of the management team at Telemadrid who participated in the discussion of this idea are known, including Alejandro Gómez, Javier Reyero, and José Joaquín Brotons (director, presenter and head of sports programming at Telemadrid, respectively), as were the members of the aforementioned Referees Committee.

It is worth noting that the biggest concern of the Spanish Referees Committee at that time was to ensure the communications were secure, specifically in order to prevent potential illegal activity of third parties regarding the legitimate ownership of the broadcast. It is clear that 26 years on, this is still a priority on an international scale, as the audio recordings of the referees' exchanges continue to be to a large extent private, as witnessed by FIFA's response to the Brazilian Federation's request to access to the recordings of the referees' conversations. This is an issue that will be discussed later, as the authors believe that transparency is a key element in any process that claims to be explanatory.

Another issue raised by the Spanish Referees Committee was the need to generate a solid protocol that enables the system to be applied in the different national competitions, given that its implementation should be standardized in all top-class matches. Nevertheless, economic factors remain a significant obstacle to the more general implementation of VAR, whereby those countries that can commit more resources to football will be able to implement the system more easily and extensively, although even in those countries the minor leagues that generate less revenue could be left out. Spain is one such example, where VAR will be present in all the national league matches of the highest category (La Liga) during the 2018–19 season, as stipulated in the agreement signed between La Liga and the RFEF (2018), although it won't be in use in the second national category (currently known as "La Liga 123") and it appears that it will also not be used in the "Copa del Rey" (a knock-out tournament similar to the FA Cup in England, the "Copa Italia", the French League Cup, and the Portuguese League Cup, among others). However, its implementation in these lower leagues and in the cup has not been ruled out in the near future (Pérez & Alcalá, 2018). Spain is not an isolated case and similar decisions have been taken in most countries currently implementing VAR. Nevertheless, this highlights the degree of discrepancy in the planning and setting up of the system, since aside from disagreements between the different entities and governing bodies, its application should in theory be universal in the top categories in all countries.

Technological Innovations in Football

A wide variety of technical resources currently have FIFA approval, and these can all be consulted in an extensive database of certified products. Among the different support technologies developed there are some that are of particular interest, such as: automatic goal detection (a first iteration appearing in 2012); portable electronic performance and tracking system (EPTS); ball certification; artificial grass; and, lastly, the VAR system.

As such, FIFA established a Quality Control Program[4] to establish industrial standards for all the products and technology officially used or employed in football. The Department of Technological Innovation at FIFA carries out a global analysis of its needs, and it supervises the process and assesses the viability of potential innovations from start to finish, up to their standardization. Nevertheless, as far as VAR is concerned, a number of articles have appeared in the media reporting the acknowledgment and alleged breaches in the use of the patent covering VAR by FIFA. The invention was patented by several inventors, such as: the Spaniard Antonio Ibáñez de Alba, who claims to have filed a patent request on 1 September 1995 in the name of Spanish businessman Mario Conde (Rosado, 2018); the Spaniard Francisco López, who according to journalistic sources filed a patent application at the Ministry of Education and Culture in September 1999 (*El fútbol del siglo XXI – Tecnología de futuro para los equipos arbitrales – Football in the 21st Century – Future technologies for referees)* (García, 2018); and the Bolivian Fernando Méndez Rivero, who published his project in the media in 2005, presenting it to the Bolivian Ministry of Education and the National Service of Industrial Property (SENAPI, its acronym in Spanish) (Hurtado, 2018). This controversy is similar to the previous disagreement around the authorship of the patent for the referee's spray, which was incidentally banned during the World Cup in Russia by a Brazilian Tribunal (Chiappetta, 2018), although it was finally utilized, disregarding the magistrate's precautionary measures. Pending conclusive court decisions, these circumstances illustrate the importance of these innovations to international football organizations and their repercussions in the evolution of the sport.

Video and Audio in Football's VAR

Although this chapter focuses principally on the issues related to audio, we felt it was appropriate to review the basic characteristics of the image analysis performed by VAR, as released by FIFA. The idea is to compare the information available on video and audio, in order to determine the emphasis and importance that FIFA places on each of these media.

Regarding the video images, FIFA has intensified its efforts to publicize the general characteristics of its use and, in specific cases, some technological details.[5] Accordingly, FIFA provides information regarding the number of signals to be operated and their distribution in the Video Operation Room (VOR), also specifying the number and location of the video signals the VAR system should have (in principal, 33), eight of which should be capable of super-slow recording and four with ultra-slow recording, with access to the cameras allocated for offside calls. In addition, there should be two ultra-slow cameras behind the goals. FIFA specifies that the VAR team should be able to access all TV images that are being used to transmit the match, barring personalized signals or those restricted by audio-visual rights.

The FIFA webpage about innovations[6] provides information regarding refereeing procedures, as well as specifying some of the signals that referees should use, such as pointing to their ear to communicate with VAR or drawing a square to signal an official request for revision (see Chapter 10). It also mentions the need to inform the crowd in the stadium and the television viewers when a decision is being reviewed. As such, FIFA has developed a system for television channels, commentators, and other related info-entertainment users that covers the different steps in the review process. This system includes making reference to the reason for the revision and the result. According to this webpage, data could be made available that could be used in television graphics and on the stadiums' scoreboards. The webpage also provides videos explaining the VOR, describing the distribution and roles of each operational post, along with images and basic diagrams of the panel of screens, without entering into a detailed explanation about the technological demands in terms of layout and design. As can be inferred from this summary, the information provided by the webpage is succinct and it does not offer a detailed description of all the distinct elements that make up the VAR technology.

By contrast, according to the information provided by Francisco José de la Fuente Ruiz (De la Fuente, 2017), the protocol for combining the signals integrated in the VOR should be:

- The integration of the television channel's camera signals that produce the content;
- The transmission of the video signal to the Referee Review Area (RRA) for visualization by the match official;
- Integration with the communication system used by the refereeing team on the pitch;
- Recording of all the camera signals, Program Feed and VOR production (including the audio signals);
- Transmission of the video review signal to the television to inform the viewers.

As for the audio signal, to date FIFA's webpage makes no reference to the expected standards of design, performance, conception, technological requirements, and interactive language, either for the communication between the officials or for the management of the signal by VAR. This may appear somewhat incongruent in the light of the fact that the audio signal is the basis of all communication within the VAR network. Indeed, according to the declarations of Francisco José de la Fuente, there have been problems when the officially recognized intercommunication system has been implemented in some countries, making it more difficult to understand this apparent lack of information and documentation. On the "Football Technology" webpage,[7] FIFA dedicates a single section under the heading "Resources", which in turn is divided into two subsections: "Technical Resources" and "Certified Product Database". The second subsection includes some references to technology providers, among whom and in relation to audio, is Crescent Comms[8] (a company whose webpage did not allow third parties to contact them when requests were made on various occasions in July and August 2018).

It is not only FIFA that fails to provide information regarding the audio and communication requirements for VAR. A number of articles published between March and May 2018 in both the general and specialized Spanish press, as well as in Spanish-speaking American media, failed to make reference to the procedures regarding audio and communications in VAR. We presume that this generalized deficit in information regarding the application and the audio systems involved in VAR originates from FIFA's own silence. Such an approach would explain why FIFA does not give the same weight to the visual and audio components, either due to ignorance or a premeditated motivation, even when it's apparent that VAR could not exist without audio files, or at least not as we know it.

Accordingly, and considering the possible requirements FIFA may establish for audio signal processing and analysis, different types of signal flow must be contemplated. On the one hand, the communication signal between pitch officials and VAR, where we believe the information provided by FIFA regarding the strategic verbal language to be used is ambiguous and inconclusive. On the other hand, the signal captured on the pitch by the broadcasters, particularly given the current technology is capable that it can record audio from virtually every individual player on the field, which could help resolve ambiguous or confusing situations. However, in both cases there are considerable gaps in the information available to the general public. Nevertheless, from the small amount of information that we have been able to gather, we understand that:

- VAR is part of the refereeing team's communication system, allowing the main official to communicate with and access conversations between the referees on the field when discussing an incident susceptible to revision. Clearly, VAR must be able to hear and speak to the refereeing team.

- The VOR's audio circuit should be integrated into the radio frequency circuit of the field officials.
- The Review Operator (RO) and Review Assistant (RA) should also be integrated into the system, but only allowing them to listen to the main official and not to speak with him.
- Only the VAR and AVAR can speak to the main official, although the line of communication is not open all the time but rather, it must be activated by a button so as to avoid distracting the referee with the exchanges between the VAR – AVAR – RO.
- All communications between referees and the VOR should be recorded to safeguard ethical integrity and for training purposes. These recordings should be handed in to FIFA and the pertinent leagues at the end of matches.
- The television channel covering the match will provide FIFA with the audio recordings from the pitch, including both the signal with the commentator (channels 1 and 2) and that without any commentary (channels 3 and 4). This latter signal is particularly important, as the isolated audio recorded from the pitch and the ambient noise from the ground can be very informative when trying to resolve doubts or assess controversial situations. This signal may be in stereo or in multichannel format, such as surround sound 5.1 (surround sound in the horizontal plane) or surround sound 7.4.1 (surround sound in two planes: vertical and horizontal), which can help more precisely locate the sounds and their source. Nevertheless, the quality of the recordings will depend on the standards of sound production implemented, since the retransmission does not always follow the same requirements.

The audio recordings from the pitch could potentially be much more relevant than they are currently considered in the VAR, since while the video signal shows what the eye can see, it cannot pick up on any non-visible events. In the light of our expertise and experience, technological advances in audio capture and recording allow almost all the different individual sources of sound on the pitch to be isolated. Hence, most of the sound coming from the players and the refereeing team can be captured, generating a valid record for both the live VAR process, and for its subsequent analysis and study.

In fact, the contribution of audio and its associated technologies could provide solutions that take VAR to a higher level. In the same way as video footage can be reviewed openly to correct decisions, the audio may not only provide a vehicle to aid refereeing decisions, but it should also be considered as an element of transparency that can be used for consultation and verification when needed, potentially providing objective evidence to aid decision-making. The fact that this signal is not reviewed and its content is not readily available may be due to the concerns of certain bodies in making public the conduct and behaviour of the players, refereeing team, and VAR, which when judged externally could bring into question the validity of some decisions. We only have to remember the controversy sparked in La Liga, between Zaragoza and Barcelona in 1996. During the match a controversial refereeing decision led to a discussion between the main official and one of the assistant referees (Relaño, 2016), the result of which could have directly affected the outcome of the game. The conversation was recorded and broadcast live by the media covering the match, revealing a refereeing process full of doubts to this day; the way the final decision was taken still remains controversial. It should be stressed that in this case, the audio transmission contributed evidence that would not have been accessible from the video footage alone. The repercussions of that soundtrack still fuel heated TV debates regarding the validity of that refereeing decision 22 years later (Sánchez-Flor, 2013; Bárcenas & Martínez, 2016). Yet aside from the TV spectacle generated, the audio recordings did provide transparency as to how the refereeing team act, information that would not have been known otherwise.

As a result of the above, there are clearly benefits that can be had from potentiating the documentary value of the triple signal composed by:

- Communications between the refereeing team and VAR;
- All audio captured from the pitch (players and referees);
- General atmospheric sound (from the crowd).

This audio signal should help analyse the elements of a situation that are not accessible through video alone. Therefore, in light of the importance of audio as an element of communication and its crucial role in conflict resolution, we consider that the audio information available through FIFA and through the media is extremely scarce, and only covers vague generalities rather than contributing to an in-depth understanding of this new and promising tool. Conversely, it's understandable that there are detractors averse to incorporating sound as a source of information in the VAR process, mainly on the grounds of its time-consuming operational management. Nevertheless, as with the visual component, audio can be perfectly well handled within a short timeframe, it being necessary to have in place a technical team with all the necessary knowledge and operational skills to ensure optimal delivery that won't delay the progress of the match.

Aspects of Audio to Be Considered in Football Refereeing Communications

What are the implications of refereeing inter-communications in professional sport and in football in particular?[9] A number of issues should be taken into account when designing the operational elements of referees' inter-communication systems. However, regardless of the issues specifically associated with different settings, it is important to establish standardized procedural requirements so as to generate a functional dynamic capable of resolving particular difficulties that might arise in different pitches and stadia. Among the many different aspects to evaluate, it is important to distinguish between the technical features relative to the communication system technology, the health and safety issues, and those related to transmitting the messages between the refereeing team and VAR. Based on contributions by Sánchez Cid and Pueo (2017), we can list the following aspects:

- It is important to account for the number of people who will intervene and their location. In this regard, it is necessary to consider the location of the VOR when establishing an optimal radio frequency signal, as well as the position of any other elements that might cause interference, such as any additional radio frequency signals in the area, the distance between the pitch and the studio where the VAR is situated, any structural elements that might interfere with the signal, etc.
- The transmission-reception signals that constitute the communication system must be effective in covering a minimum perimeter, which should be normalized for all stadiums and venues.
- The mobility of the refereeing team in the field is fundamental, and it can complicate the transmission and reception of the signal. As each venue might have its own specific characteristics, testing should be carried out beforehand to avoid any surprises and to establish a safe circuit bounded by the touchline, and a zigzag trajectory criss-crossing from side to side.
- The inter-communication system should automatically avoid radio frequency interference, making use of intelligent processing capable of switching and synchronizing the connection frequency for all the participants in the event of any interference with the transmission or reception of the signals.
- The system must be capable of simultaneous reception and transmission, selectively prioritizing sources if necessary. Differentiating between the signal from the refereeing team on the pitch and the VAR team will help streamline the communication between these parties.
- The system should offer a dual-signal transmission-reception mode, permitting a switch between continuous or selective. The communication from the VOR to the team of officials on the pitch should be non-continuous, only transmitting when necessary. However, the pattern of communication from the main official on the pitch to VOR is less clear, given that communication among the team of

officials on the field and their interaction with the players could be relevant.

- For the purpose of consultation, analysis, training, and monitoring, the audio signal should be recorded in its entirety. Ideally, the audio recording system should be able to register each member of the refereeing team individually and together, making it easy to clarify the messages transmitted in moments of confusion.
- It is necessary to consider how to ensure the radio frequency signal is secure in order to avoid accidental and/or premeditated interference.
- The intercommunication system must include "hands-free" functionality. There are many options to automatically activate communications without the need to press a button, as could be achieved by using key stimulators, such as: activation by intensity impulses, frequency range, by reiteration or incidence, through the use of a key word, etc.
- The system should be fitted with an optional noise gate as an automatic filter for any signal that could disrupt or adulterate the communications.
- Similarly, reception of the audio signal must be impeccable, with no interference, creating a clean and clear message.
- The system must be comfortable, ergonomic, and stable, without constraining the referee's mobility or capacity to do his job, and thus, it must be as light as possible.
- The system must be shock-proof, and temperature-, humidity-, and water-resistant.
- The headset must be stable and designed in a way that it won't cause any damage to the user's hearing in case of any unexpected blows or impact.
- Likewise, the headset must also be compatible with sweat and rain, and it should be personalized for purposes of hygiene, complying with health and safety requirements.
- In order to avoid any potential health risk of exposure to radio frequencies (Foster, 2014; World Health Organization [WHO], 2014), the system must adhere scrupulously to international regulations on the Specific Absorption Rate (SAR).
- Given the level of Sound Pressure Level that can be generated at a first or second division ground (Antich & Laguna, 2015) (reaching around 88 dB with peaks of 115 dB, according to different measurements), care must be taken so that the system obtains a clear and intelligible signal on the pitch from the referee's intercommunication system, without surpassing the maximum recommended sound pressures for hearing (Tardón, 2016). According to the WHO, prolonged exposure to certain levels of audio can irreparably damage an individual's hearing (WHO, n.d.). For example, exposure to 100 dB should be kept to a maximum of 15 minutes a day in order to avoid producing severe damage to hearing.

- The system's autonomy should cover at least twice the length of a match, including the time possibly allocated to extra time and penalties, etc.
- A decision should be taken as to whether the players should be made aware that all their comments and queries may be recorded while they are on the pitch, and of the relevant consequences of this fact. The somewhat overused argument against this suggests that the pressure and stress on the field can lead football players to react in an extreme and uncontrolled manner, which may perhaps not be appropriate to record. However, in the same way as physical aggressions are susceptible to sanction when they are seen or detected, verbal aggression should also be sanctioned in a similar manner. Good education and respect must be evaluated in all its aspects and in all circumstances.
- In order to keep the time required for the reviews to a minimum and to avoid long breaks in play, intercommunication between referees should employ an unambiguous code of concise and precise expressions. Bearing in mind the importance of timely communication between referees and the general benefits to be had from minimal interference in play, it makes sense to put in place communication protocols to shorten the stoppage time associated with the use of VAR. In this sense the mean time for review was 80 seconds during Russia 2018, although previous studies situated this figure at 60 seconds (Cortegana, 2018), which was itself an argument used when considering its implementation at the World Cup. Codes based on numbers and letters could be used, assigning them to each particular category and case, constituting a maximum of 16 variables plus the final approval or rejection value.

In the absence of other information from official sources, these points establish interesting features to be considered when compiling a public standards document regarding the use and implementation of VAR. Obviously these suggestions are neither definitive nor complete, and they can be adapted according to particular needs and any further developments in their use.

Analysis

Crucially, the implementation of an effective communication system should not produce inequalities in terms of fairness. That is, access to the facts and the resources that best help resolve conflicts should not be exclusive to those professional categories that generate or manage the largest amounts of money. As such, implementation should be generalized and it should be affordable without compromising its aforementioned characteristics.

Introducing a complete VAR system (video and audio), as suggested by FIFA, could dramatically limit its global implementation and uptake, producing direct or indirect differences in terms of normalization that could be considered unjustifiable in some cases. It is true that the implementation of VAR in national leagues would be under the direct control of their local governing bodies, which may be financially limited. This is undeniable, in the same way as the improvement in fairness and equanimity in all refereeing processes is undisputable. In this sense, we could argue that implementing a basic communication system based only on audio would minimize the costs, helping considerably to establish VAR in the lower leagues and in grassroots football, helping to extend the improvements in objectivity and actively assisting referees who find themselves sometimes in undesirable situations as a consequence of the lack of resources. We advocate this as one of the main priorities for national and international football organizations: to maximize the general interest in normalizing refereeing fairness for all, in pursuit of a cleaner game and enhanced educational values.

Conclusions

Through analysing and studying the treatment in audio as a part of the VAR, it has become clear that the information available regarding this system is generally in line with its name, FIFA placing most emphasis and relevance on the visual component, overlooking the value that can be obtained from the audio signals and their potential benefits for the further development of this system.

The documentation related to audio signals and VAR published in both the general and specialized press in Spain and Latin America, as well as that present on FIFA's own platform, is currently extremely limited and, in fact, almost non-existent. The material we found does not deal with relevant issues in sufficient depth to allow a balanced debate of the procedures and protocols that should be implemented. Likewise, no documentation has been found that describes any model of the codes or language that would help to speed up the communication between referees. In footballing circles, there seems to be an excessive preoccupation about the time required for each VAR intervention but to date, no studies have been carried out on streamlining communications through the use of protocols based on a precise and structured language.

Likewise, it is evident that visual images are attributed greater value as evidence of the events, while little regard is paid to the significance of the audio transmissions from the pitch. It should not be forgotten that audio can provide significant information that is not available from video alone. While images confirm what can be seen, audio can also draw attention to what the eye might miss.

The level of detail conferred by FIFA and other bodies to the visual component of VAR is reflected by the fact that in some cases, they even specify the number of cameras that are required for VAR (see Chapters 3 and

10), as well as stipulating their position and other specific characteristics regarding the production of the images. By contrast, as far as the documentation regarding the audio component is concerned, issues as crucial as coverage, which sources to capture, the distribution of the audio captured, the existence, purity, and clarity of the signal, as well as the structure of the registers are not addressed. As discussed, current technology virtually permits individualized audio signals to be captured from all the individuals that take part in the game, including all members of the refereeing team, making VAR analysis much more insightful and effective.

Alternatively, if we accept that refereeing communication and VAR promote greater objectivity and fairness in decision-making (which undoubtedly directly benefits the sport itself), its implementation on a global scale should be seen as a necessity, at least in the top categories in each country, and over and above its use only in those televised matches that have the strongest repercussion in terms of both economic value and public interest. In the lower professional categories, implementation of a basic system for the refereeing team involving audio intercommunication and recording should be contemplated. An intercommunication refereeing aid based on such a simple set-up could be established and implemented at a relatively modest cost, as effective low-cost alternative solutions are currently available.

It is also worth noting the incongruent behaviour of FIFA in making the video signal from VAR public, while restricting the audio signal from the associated conversations. As such, if FIFA and the other relevant federations are prepared to release images handled by the VOR to television channels and to display them on screens in the stadiums, why do they consider it inappropriate to make public the conversations between the referees and the VAR? If we assume decisions are taken by a professional team following a legal and transparent process, it is not clear where the problem lies. This refusal only generates confusion and represents an apparent obfuscation. Granted, VAR images are considered explanatory, but the conversations associated with them can provide additional details regarding the ability of the assigned team of referees to ratify the referee's decision, underlining their skills and objectivity, and demonstrating how the decisions are reached. This is a very important issue that should not be decided simply by the bulk of the referees, but rather by a higher agency that is not likely to be subjected to external pressures or forces.

In addition and in line with the premises of this chapter, we raise a number of issues that have yet to be tested at the professional level, despite the claim of different international bodies that the impact of VAR on all the parties involved has been assessed (referees, players, coaches, and spectators). For example, in the same way that VAR images are publicly released, would it be considered appropriate for the viewers to be able to listen to the conversations between the match officials, or those between the main referee and the VOR? In addition, would it be of interest to make this sound audible in the stadium itself? How would the crowd react? Similarly, is it a good idea for the public in the stands to be able to listen to a detailed transmission from the players on the field? Would this have a negative effect or, by contrast, could it help clarify issues and make football a cleaner and more transparent spectacle? Would this approach produce a different understanding and enjoyment of live football at the grounds?

These fundamental issues concerning the use of audio in VAR have been summarized in this study, issues we feel worthy of attention in this respect. We have based our assumptions on the understanding that professionalism in football should be founded on a maximum level of transparency and honesty, which we believe is what football represents and can transmit, and what makes it so transcendental.

Notes

1 Francisco José de la Fuente Ruiz is Manager of the International Audiovisual Services Area at Mediapro.
2 For more information, please check the video file (00:04′:00′) inside the DVD.
3 Spanish Public Regional Television channel that is based in the Autonomous Community of Madrid. Affiliated to FORTA (Spanish acronym for Federation of Regional Radio and Television Organism – Federación de Organismos de Radio y Televisión Autonómicos). In the 1990s, it was the Host Broadcaster channel for the grounds of Real Madrid and Atlético de Madrid, as well as other local teams. See also www.forta.es/.
4 For more information, please visit: https://football-technology.fifa.com/es/standards/.
5 For more information, please visit: https://football-technology.fifa.com/es/innovations/.
6 For more information, please visit: https://football-technology.fifa.com/es/innovations/var-at-the-world-cup/.
7 For more information, please visit: https://football-technology.fifa.com/es/resource-hub/technical-resources/.
8 www.crescentcomms.com/.
9 As far as referee communications are concerned, professional football is considered to encapsulate all leagues classified as such, irrespective of the space and seating capacity at the grounds, or other characteristics associated with the sporting action.

References

Antich, A., Laguna, M. (2015). La presión "bombonera". *La bombonera*, 34, p. 9. Retrieved from www.labomboneramalaga.com/ficheros/productos/pdf/34.pdf.

Bárcenas, A., Martínez, D. (2016, September). *El chiringuito de jugones* [Television broadcast]. Madrid: Atresmedia Corporación. Retrieved from www.youtube.com/watch?v=GTeuxURSvQo.

Chiappetta, J. (2018, June 15). Otro golpe a la FIFA en el millonario juicio de los inventores del aerosol. *Clarín*. Retrieved from www.clarin.com/deportes/mundial-2018/golpe-fifa-millonario-juicio-inventores-aerosol_0_rkXdLdWWm.html.

¿Cómo funciona el VAR y en qué ligas está implantado? (2018, March 20). *Europa Press*. Retrieved from www.europapress.es/deportes/futbol-00162/noticia-funciona-var-ligas-implantado-20180320104021.html.

Cortegana, M. (2018, June 24). El VAR dispara el descuento en las segundas partes a 4,4 minutos. *As*. Retrieved from https://as.com/futbol/2018/06/23/mundial/1529791162_146231.html.

De la Fuente. F. J. (2017). Asistente de Vídeo Árbitro. Solución VAR MEDIAPRO. In Sánchez Cid, M., Armenteros, M., Benítez, A. (Eds.), *Jornada sobre Tecnología Audiovisual e Innovación en el Deporte de Alto Rendimiento at the Rey Juan Carlos University, Fuenlabrada Campus. [DVD], 22:00-31:00*. Madrid: Dykinson.

Foster, S. (2014, February 25). WHO knew: The elephant in the room [Blog post]. Retrieved from https://betweenrockandhardplace.wordpress.com/2014/02/25/who-knew-the-elephant-in-the-room-guest-blog-by-susan-foster/.

Fuentes, R. (2018, July 30). El audio del VAR, un audio top secret. *Sport*. Retrieved from www.sport.es/es/noticias/mundial-futbol-rusia-2018/audio-del-var-audio-top-secret-6893425.

García, O. (2018, June 30). Voy a reclamar a la FIFA 15 millones de euros por el VAR. *Marca*. Retrieved from www.marca.com/futbol/mundial/2018/06/30/5b36765b22601da0608b467b.html.

Hurtado, C. (2018, July 9). Fernando Méndez Rivero reclama la autoría intelectual del VAR a la FIFA. *El Heraldo*. Retrieved from www.elheraldo.co/mundial-futbol/fernando-mendez-rivero-reclama-la-autoria-intelectual-del-var-la-fifa-516452.

Kallás, F. (2018, June 19). Brasil envía queja oficial a la FIFA y pide audio y vídeo del VAR. *As*. Retrieved from https://as.com/futbol/2018/06/19/mundial/1529397074_390107.html.

Moñino, J. (2018, June 29). Para la FIFA el arbitraje roza la perfección con el VAR. *El País*. Retrieved from https://elpais.com/deportes/2018/06/29/mundial_futbol/1530285204_509942.html.

Pérez, J. L., Alcalá, J. (2018, July 22). El VAR que viene: en La Liga, la Premier, el Calcio, la MLS, en México, en Australia.... *Marca*. Retrieved from www.marca.com/futbol/2018/07/22/5b4cf5a946163f36648b45a5.html.

Real Federación Española de Fútbol. (2018, March 2). La RFEF y LaLiga impulsan el Proyecto de Implantación de la Asistencia por Video al Arbitraje (VAR) en España. Retrieved from www.rfef.es/noticias/oficial-rfef-y-laliga-impulsan-proyecto-implantacion-asistencia-video-al-arbitraje-var.

Relaño, A. (2016, September 29). «Rafa, no me jodas» (1996). *As*. Retrieved from https://as.com/futbol/2016/09/29/mas_futbol/1475130798_485426.html.

Rosado, B. (2018, July 15). El VAR del Mundial es un invento español y la patente es de Mario Conde. *El Mundo*. Retrieved from www.elmundo.es/cronica/2018/07/15/5b49e987ca4741a9778b459c.html.

Sánchez Cid, M., Pueo, B. (2017). Sonido y Comunicación en el Videoarbitraje. In Sánchez Cid, M., Armenteros, M., Benítez, A. J. (Eds.), *Tecnología Audiovisual e Innovación en el Deporte de Alto Rendimiento* (pp. 6–8). Madrid: Dykinson.

Sánchez-Flor, U. (2013). *Punto pelota*. [Television broadcast]. Madrid: Intereconomía. Retrieved from www.youtube.com/watch?v=lWKA3S1nvkk.

Tardón, L. (2016, April 27). El ruido: una amenaza para la salud. *El Mundo*. Retrieved from www.elmundo.es/salud/2016/04/27/571f7504e2704ed1208b4585.html.

World Health Organization. (2014, October 8). Electromagnetic fields and public health: Mobile phones. Retrieved from www.who.int/news-room/fact-sheets/detail/electromagnetic-fields-and-public-health-mobile-phones.

World Health Organization. (n.d.). Make listening safe. Retrieved from https://apps.who.int/iris/bitstream/handle/10665/177884/WHO_NMH_NVI_15.2_eng.pdf;jsessionid=4655F1AC5D711FCC3E473B6446C8AE7A?sequence=1.

5 Replays in Football

A Technical Resource that Serves as a Narrative and Post-Refereeing System

Alberto Luis García García and Manuel Sánchez Cid

Introduction

Since the birth of television, it has always been closely linked to sports. The immediacy of the live broadcast, along with its ability to reach all parts of the world through images – the main difference from radio – have made this the ideal medium to generate a spectacle around sports. This concept of a spectacle is synonymous with profit motivation and by extension, with obtaining the resources to make this sport a viable professional option.

In the case of football, all these factors have been brought together in a masterly manner, making this sport the master of television shows, both when we consider the number of people on whom it has an impact and the associated financial results. Thanks to the combination of these two factors, synergies have appeared that influence both the way of conceiving this sport, and the way that the rules and regulations structure how football is played. Television is primarily conditioned by broadcasting, since

> the broadcasting of 'live' content is the 'soul' of the business (as a point of contact) and it is a servant to the associated technology due to their effects on its reception, with the content of these media events attracting much attention from the big companies in the audiovisual sector.
>
> (Montemayor and Sobrino, 2016)

Audiences are a key factor when considering the revolution that has taken place in the television media over recent years. Smart TVs have changed the type of consumption, which is in turn altering the traditional programming of the TV channels. Similarly, the synergy between social networks and other types of communication media are creating new structures of communication, whereby the "social audience reflects what the viewers/consumers think about television programming, influencing it and openly manifesting their preferences about content and brands" (Saavedra,

Rodríguez, and Baron, 2015). Thus, today's audience looks for two fundamental things from a television show: quality and interaction.

Sports broadcasts must clearly enhance this dynamic to attract an audience that is increasingly global in its nature, yet more uniform in its preferences and uses. Hence, there is a need to coordinate operating systems and guidelines for television production, which would result in unique formats with a greater capacity for global impact, yet with local content management based on the active participation of viewers. Therefore, incorporating spectator participation into the spectacle must be achieved when producing and broadcasting an event, as if it were a puzzle in which all the elements are connected, with the aim of achieving greater social and media benefits from the event broadcasted.

Football brings these two features together: social interest and media attention. As such, the capture of audiences and the economic profits (more reliable and controllable than in other types of programmes or live events) enable *ad hoc* businesses to be generated that define a new market around that of the broadcasting rights of football teams.

Regarding audiences, football clearly still maintains the leadership as the most watched broadcasts. For example, one local event that generates worldwide interest might be the Champions League final (e.g., Juventus vs. Real Madrid), which, according to the results presented in 2017 by Barlovento (a consultant that measures television audiences in Spain), generated an average audience of 9,620,000 viewers and an audience share of 56.9 per cent on 3 June 2017 for Antena 3 TV, a Spanish private television station that had the rights to broadcast the match. Indeed, it was the most watched programme in the category "linear TV channel plus guests", the latter being "all those viewers who are not usual residents of the household where the television is being watched". Interestingly, if we focus on that month of June 2017 (Barlovento Comunicación, 2017b), it is notable that in ranking of linear TV broadcasts, the five most watched programmes in this period correspond to sporting events or programmes related to them, mostly those related to the world of football (see Table 5.1).

Nevertheless, if we review previous months we will find a similar scenario: that is, the most watched television programmes according to the Barlovento report are football matches and, specifically, the free-to-air Champions League matches broadcasted by Antena 3 TV. However, the same report produced by this audience measurement company indicated that the social audience participation, measured by the number of tweets, was not led by (live) sporting events but rather by reality shows – again live programmes – as well as other programmes related to the world of football.

2018 was a special year, as it was the year in which the FIFA World Cup (FWC) was held, turning football into a global television show during the months of June and July. The audience data provided by the

Table 5.1 June 2017 TV audiences in Spain

Channel	*Event*	*Audience*
Antena 3	Football: Champions League, Juventus v. Real Madrid	8,205,000 – 55.1%
Telecinco	Football: Friendly, Spain v. Colombia	4,652,000 – 27.1%
La 1	Football: World Cup Qualification, Macedonia v. Spain	4,417,000 – 28.9%
Antena 3	Champions Total/A3	4,308,000 – 33.9%
Telecinco	Tennis: Roland Garros, S. Wawrinka v. R. Nadal	4,204,000 – 31.9%

Source: Barlovento report (Barlovento Comunicación, 2017b).

same company shows that audiences rocketed for the TV channels with the rights to broadcast this event. Thus, according to the Barlovento report (2018), the audience data for July 2018 was as shown in Table 5.2.

Russia FWC broadcasts produced the 24 best audience figures for this month of July, making it clear that global events related to football generate a significant increase in audiences. This translates into large movements in terms of revenues related to television broadcasting rights, where the economic returns for these businesses are considerable, explaining the intense trade wars between the larger media groups. For example, Matilla (2016) reports that the La Liga clubs received about 1,573.8 million euros from TV broadcasting rights for the 2016–17 season, although these decreased to a net figure of 1,424 million euros when the corresponding commission to Mediapro (holder of the audio-visual rights) was taken into consideration. Of this sum, 149 million euros went to FC Barcelona, 142 million euros to Real Madrid, and 103 million euros to Atlético de Madrid as the third classified team in La Liga. Eibar was the club with the lowest revenue, receiving nearly 42 million euros. Thus, it is clear that the revenue from television is very significant for football clubs.

Table 5.2 July 2018 audiences for Spanish TV channels

Channel	*Event*	*Audience*
Telecinco	Russia 18: Penalties, Spain v. Russia	14,829,000 – 81.1%
Telecinco	Russia 18: Extra Time, Spain v. Russia	13,835,000 – 76.3%
Telecinco	Russia 18: Match, Spain v. Russia	12,729,000 – 72.4%
Telecinco	Russia 18: Extra Time, Croatia v. England	8,429,000 – 53.1%
Telecinco	Russia 18: Penalties, Colombia v. England	8,355,000 – 48.2%

Source: Barlovento Comunicación (2018).

If we turn to the operators and if we look at the results of Telefónica from January to June 2017, the leading telecommunications operator in Spain (Telefónica, 2017, 32), their financial report states that: "in the conventional area of business, the growth of pay TV stands out (+ 5% interannual up to 1.3M logins), with a net gain of 12K logins in April-June, leveraged on its exclusive offer of football". That is, their good results in the television sector are based on the football broadcasts they offer, making this a highly strategic area. In terms of strategy, Vodafone, the third largest telecommunications operator in Spain, has included football as a fundamental part of its offer to attract new customers, launching a package in a press release on 9 August 2017 that included all the La Liga matches played by Real Madrid and FC Barcelona in 4K quality, thereby putting an end to the exclusivity of its main competitor. According to the National Commission on Markets and Competition (CNMC) (Comisión Nacional de los Mercados y la Competencia, 2018, 99), Vodafone has a "22.7% audience share in the residential segment, a figure slightly higher than that reached in 2016". Indeed, and as a positioning strategy to stand up to the competition from other operators, according to Checa (2018), Telefónica has acquired all the broadcasting rights for La Liga for the 2018–19 season and up to 2022 for 2,940 million euros, and those of "The Champions League" for 1,080 million euros until 2021. Accordingly, the profits associated with football have been made patently clear and, as a result, the main telephone operators have situated football at the centre of the battle to attract customers.

In summary, while all these data represent only a small part of that currently available, they clearly highlight that due to both the large audiences and revenues, and the commercial strategies of the TV operators, football broadcasting is sufficiently strong to survive in the new digital era. As a result, the production of these broadcasts becomes the axis articulating this area of the telecommunications industry, through business strategies based on the consumption of content. Indeed, this is the pretext on which the business plans targeting data consumption are established. Notably, replays are a fundamental part of these broadcasts and they help emphasize the important details when establishing opinions. At the same time, replays helps build a storyline that is necessarily linked to the high demands on production (mainly 4K) and how the events occurring at each sporting event are narrated. And it is here where the viewer's expectations must be fulfilled by the way the transmission of the event is produced.

The existence of a common and universal language, both in terms of the guidelines and of the television language, has turned sporting events into spectacles, providing the largest benefits in a transcultural, diverse, and global economy. Therefore, it is crucial to investigate and explore this meta-universe that transcends cultural and linguistic barriers to generate a communication model that can be easily understood by the whole of humankind. Television is the medium that has turned this issue into a reality. Sports broadcasting, and all the elements and variables that are associated with it, is the vehicle through which this universe is articulated.

Replays: Basic Principles and Initial Considerations

Over the years, the broadcasting of football has become more clearly established, spearheading many developments in the broadcasting profession and in the companies dedicated to technologies surrounding television. Since the beginning of sports broadcasting, the best technical resources and tools have been made available to football, except if we consider the Olympic Games, an event that takes place every four years and that has its own very special and specific characteristics in terms of television broadcasting. The Olympic Games can be considered the greatest television event of all (Benítez, 2013), mainly due to the huge audiences it reaches and the enormous diversity of sports that are practised and broadcast, as well as the technical resources associated with these transmissions. Indeed, the Olympics generally serve as a platform to launch new technologies and tools related to the production and TV retransmission of each sport.

The camera positions determine the spectator's view of the event. Initially, a wide angle of vision of the playing field was shown by the cameras and then, as technology progressed and the capacity to incorporate more cameras increased, the camera angles highlighted more specific details in the match and specific aspects of the player's reactions. The latter represent the main difference in quality between the vision of the spectator at the stadium and that of the television viewer. And it is precisely this that has led to the successful creation of the business model targeting football, which has been able to cross borders through the use of a single language. Thus, the arrival of television to football stadiums changed the way of following a football match. Not only was the angle of view of the spectator at the stadium adapted, but the cameras were situated in such a way that the viewer became an active part of the match due to their privileged view of it. Cameras were situated in a position to analyse potential offsides, there were wide-angle cameras that provided more information than that which even the referees had to judge fouls, penalties, etc., all of which helped socialize the game and shift it from the stadiums and onto the street. In addition, the journalists' discourse with its own jargon provided the narrative to the match and supported what the cameras showed in real time. Yet the role of the journalist has extended beyond that of a simple narrator, and, thanks to television technology, they have become part of the discourse of the match, becoming judges of the quality of the match, of the referee's decisions and of anything that has a direct impact on the show. Technical developments have been incorporated at an unstoppable rate, transforming the sporting spectacle into a social event thanks to the validity of the meta-discourse that has been generated through the implementation of technology.

In this sense, replays represent a point of inflection on which the narrative is articulated in sports events. Benítez (2013, 49) states that "if those images are not shown, some may even feel a little uneasy, which might produce mistrust in the broadcasting team, perhaps believing that they are being driven by some obscure reasons of partisanship to conceal match-fixing". Consequently, replays serve to clarify, show, impress, inform, etc. (see Chapter 3).

A sports event on television is based on the game, as indicated in Benítez (2013, 57), which only becomes a spectacle through the audio-visual narrative that is experienced by millions of people in real time. Thus, any football broadcast must comply with two fundamental premises: it must inform by telling the truth; and it must entertain to generate emotion in the viewers. Both these characteristics can be found in the replays of noteworthy actions during the match, which have become a necessity for the viewer in order to highlight the action and enhance the emotions that constitute the sporting spectacle.

We have been able to ascertain how television has definitively opted for football as a story to be told, serving as a deterrent and a reflection of the need for leisure activities demanded by society. As a result, this media outlet has a need for characters and narrative techniques. The public requires and uses the options available to them to delve deeper into each and every aspect of the story, given that the technology at their disposal allows them to access and interact with such content. Therefore, the spectacle, and the ability to understand and consume each part of the narrated event, must necessarily be taken into account when producing and building the television story.

In the construction of this narrative – a generator of social cohesion around a common interest and propagated by the support of a given team or club – no specific element should be dismissed, but rather they should be enhanced. Magnetic recording offers the possibility of showing replays during live broadcasts, making it possible to re-live an event that occurred during the broadcast in a controlled manner, specifically focusing on the points of interest, and distinguishing this reference from other possible options. As a result, television has searched for the best way to take advantage of this technology and, as the technical possibilities of broadcasting advanced, there was an increased need to contemplate and highlight the notable aspects of the story – and, most importantly, to do so from all possible points of view. The number of television cameras has multiplied, increasing from what might be considered the minimum number of cameras to broadcast any football match to the current number of cameras, positioned all over the field and providing previously unimaginable technical resources, like 360-degree images, as used to broadcast any Champions League match. Capturing details is important, but it is even more important to be able to show these to the viewer – and as such, what better tool than the replay?

During a sporting event, the value of contemplating a specific moment can be considered an indicator of the transcendence it acquires in the broadcast. As such, it is crucial that the production team is perfectly aware of the subtleties of the match, as it is they who choose the key time and the angle of vision necessary to fulfil the needs of the spectator. Therefore, the moment chosen to display the replay influences the effect it has, ruling out other moments in the game that are considered not to be worthy of repetition. Replays offer information about specific aspects of the game and they help to construct the story, which is why they are considered more informative, entertaining, and appealing to the audience.

Replays: Selection of Entry and Exit Points

In light of the above, the first issue that the professional responsible for programming replays should bear in mind is that the moment chosen to show the replay cancels out any action occurring at that time, therefore attributing it independent value as a unit essential to understand and explain an occurrence. Thus, the selection of the entry and the exit point must clearly include the fundamental moments of the action being repeated, and in addition, the replay must be long enough for the viewer to become clearly aware of what is going on. But, once the action to be repeated has been selected, it is necessary to be precise in the length of the replay. As such, the lead-in and lead-out should not be so long as to detract the viewer's attention from the specific moment of interest, in addition to avoiding unnecessary loss of the live transmission that might compromise the continuity of the production.

In short, the initial considerations to select the entry and exit points of a replay should be:

- The entry queue time[1] must comply with two fundamental needs: it must be long enough to see the entry mask (TV network logo or that of the holder of the rights or of the sponsor), and adjusted to the time that the viewer needs to understand the reason for the interruption in the narrative discourse, so as not to break up the action.
- The exit queue time[2] must be adjusted to a reaction time, justifying the need to end the repetition in order to return to the action plus the exit mask from the replay of the live event.
- The queue times must also be coordinated according to the speed at which the repetition is going to be played, and thus, three possibilities exist:

- That the entry speed is slowed down: normally the speed is half of the play, that is 50 per cent of normal playback, although there are also other options like 30 per cent and 70 per cent of the playback or play in real time.
- That the entry speed is in play: used when aiming to show the start of the play, which commences quite some time before the action to be highlighted. This is common in goal-scoring moves or in a good build-up that ends in a foul or in a controversial moment.
- That the entry speed is frozen: in this case the information is so relevant that attention of the viewer is required from the start of the replay, such as in questionable off-sides or penalty incidents.

By analysing the classical types of replays used in television production of football matches, it becomes clear how they are integrated into the narrative of the story of the television show and, at the same time, they provide the background from the world of television itself necessary to consider the possibility of implementing the VAR as an inclusive tool within the dynamics of the match. That is why it is necessary to try to establish a classification of the types of replays that do add relevance to the match. Football is totally linked to television, such that it can be thought of as a global mass spectacle. Accordingly, we have tried to draw up categories of replays from the technical standpoint of their broadcasting. The different types are obviously not mutually exclusive, but they serve to distinguish between them:

- Informative replays of an explanatory nature: especially of fouls committed in match-changing situations and that require the use of close-ups from cameras at the level of the pitch.
- Informative replays related to decisions or evaluations of the game: offsides (one replay). There are two cameras (positioned in each part of the field) that are exclusively used for offside replays. An explanation of the offside incident can be made through the manual intervention of the operator, although modern technology makes it possible to achieve 360-degree rotation of the image or to draw the offside line. This type of replay is explanatory and it is the closest precedent of what the implementation of VAR could represent to the world of football.
- Informative replays of on-field incidents: goals – minimum of three replays (first offside, then the camera positioned next to the goal area where the goal has been scored, camera positioned behind the opposing team's goal, close-up from a camera positioned at the stands). Subsequently, the reaction of the team that has scored and from the team that has let the goal in are shown just after the first replays.
- Replays of a spectacular nature: these are replays used to create emotion, generating intensity in the narration. They are used when

there are scoring opportunities, impressive skills, etc. Images from the cameras positioned at the level of the pitch are usually used, which provide a perspective less likely to be seen by the spectators in the stadium, that is, from a camera facing the action and preferably in a close-up position.

- Replays of a spectacular and informative nature: if there has been an obvious goal-scoring opportunity or possible penalty situation, at least two replays are shown. Almost always one of the cameras used is that positioned behind the goal and the other one is that which more clearly shows the action of interest. Likewise, two replays from the two different cameras are also shown when an important incident occurs in the match, for example, when a player is shown a yellow card and it is necessary for the viewer to receive enough information to be able to establish their own opinion about the decision. This type of replay also includes images from a reverse angle, that is, those from cameras that switch from one angle to another, equivalent to the use of cameras positioned at different angles of vision and that are not usually used in the live broadcast. These are cameras exclusively used to clarify these type of actions or to take close-ups of the game to be used in later summaries.
- Replays of a spectacular and explanatory nature: for example, when the referee has not awarded a foul that could mean a yellow card. This type of replay is explanatory and didactic, as it provides the visual resources that can clarify the cause of the infraction, such as images in a forward or reverse mode, so that the narrator and the commentators can try to explain the reason for the referee's uncertainty the moment that the event occurred. Thus, these are usually replays that help clarify certain refereeing decisions and that are extensively used by journalists to articulate meta-stories within their own narrative of the game. Nevertheless, very rough tackles tend not to be repeated more than once.
- Linked replays: these are used for different actions that are shown in the same set of replays. They are used when the speed of the game or the incident itself does not allow the replays to be shown separately. An example of this might be an obvious goal-scoring opportunity followed by a foul. This type of replay gives rhythm to the narration and it is used to create expectation or to provide all the necessary information regarding the narrative.
- Replays of reactions: these are the replays that provide information on the reaction of the coaches, players, managers, etc., mainly associated with the most important moments occurring during the match or event.
- Replays of reactions to substitutions: when a player is replaced, it is common to show some of the most relevant contributions of the player during the match when they are being replaced. These may not

only be technical details or goal-scoring opportunities (in the case that the player replaced was involved in such moments), but rather, what is often shown are reactions, gestures, or attitudes that help define the player's contribution to the game during his time on the pitch. This type of replay may also show fouls committed by an opponent against him, yet not those committed by the player himself. These replays are used to enhance the image of the player in the eyes of his fans, and to demonstrate aspects of his performance related to commitment, courage, charm, etc., creating a positive narrative from the point of view of the construction of the televised message. This aspect does not have to coincide with the view of the spectator at the stadium, nor that of the television viewer.

Length and Space for Replays

The length of replays is always marked by the duration of the breaks in play and, thus, their pace must be set in accordance with the incidents in the match. Replays should not interrupt the flow of the game, so, once the ball is put back into play, it is necessary to make way for the live broadcast, irrespective of whether the entire replay has been watched. Likewise, and logically, the length of the replay also depends on the importance of the action being shown. Thus, the replay of a foul will be shorter than that of an important move that leads to a goal-scoring opportunity or that ends in an offside decision.

It is fundamental to know how to adapt the time of the replay to the transcendence of the action shown. There are tricks and techniques that help marry these two factors, for example, fouls tend to be shown in slow motion from the start in order to indicate to the viewer the importance of the details being shown. However, in the case of a key move or a move in which some technical details relevant to the match are seen, when defining the length of the replay it is more usual to show it in real time until the detail occurs, slowing down the video replay just at that point. In such cases, it is important that the viewer can see the whole move from the best possible angle and, to achieve this, it is essential to choose the camera that best shows the movements. While in most cases it is more interesting to show the replay using a close-up or even a detailed close-up shot, a wide shot is chosen when the most relevant element is the move itself, showing the greatest width of the field and all the possible reference points relevant to the move.

A further factor to consider is the value of showing several replays of the same move but from different camera angles. This seems to add importance to the event shown. The transcendence of the action on the football field is greater when the viewer is offered more angles and, likewise, the use of multiple angles is useful to highlight the most significant aspects, providing the commentators with more information during the broadcast. In this case, the replays are longer and the space in which they are shown usually coincides with longer breaks of the game, such as when an injured player is being attended to or when there is a longer interruption in the match. Obviously, these moments are those watched most frequently in the subsequent analyses of the matches that take place in specialized television programmes.

As indicated previously, the length of the replays depends on when the events occur within the game, as well as the space or time in which they can be shown. We have already indicated that, logically, the replays of specific details that occur during a game should be shown as close as possible to the moment when the event being replayed actually occurred. However, there are narrative space or time gaps that must be taken into consideration to enhance the transcendence of the story evolving as the match progresses.

In the broadcasting of sports events, the breaks in play are used to display advertising. However, it is usual to broadcast a summary of the first half of the game before the start of the second half of the game, using replays of the highlights or outstanding moments in that first half. A similar procedure is followed at the end of the game. In these summaries, the viewers are interested in reviewing the transcendental moments that occur in the first half, or during the whole match when occurring at the end of the match. In these gaps in the game and in the broadcast, the replays can last longer as they are transmitted when the ball is not in play. The aim here is to create a story that can explain the importance of the most relevant events that took place during that first half or during the entire game. The selection criteria for these replays can be grouped into thematic blocks that are related to key or specific events that shift the momentum in the game and its narrative. In summarizing these, the following types of replays are those usually included in these spaces, covering the informative (first block of replays), spectacular (second block), and personal (third block) aspects, although these blocks does not have to be reproduced in the sequence proposed here during the broadcast:

- Goal-scoring plays: all the goals and from more than one angle. It cannot be forgotten that the ultimate objective of a football game is to score goals and, therefore, their importance must be always stressed.
- Highlighted moves: that is, high-quality moves or those that generate a lot of controversy, although the incorporation of the VAR is changing the way in which the latter are presented, as these controversial moves are shown in detail during the match if the VAR is brought into play. The angles chosen to show these are usually those that give the best perspective or those best suited to analyse these highlighted moments, generally close-up shots to give the audience a sense of proximity to the pitch.

- Outstanding technical details: those details that occur during the game and that serve to create the heroic discourse during the game. The length and the number of these replays are varied in order to emphasize the importance of the details shown. This series of replays probably feature most prominently within this short story, given that they emphasize the beauty of the game and not the controversy. As such, these elements are usually situated in the middle of the narrative.
- The reactions of the players who stood out during the first half or during the entire game: here close-up shots are shown of the face or of gestures of the players that have been outstanding in the game, regardless of whether they contributed to the most relevant moments in the game. These usually involve players from both teams, although if more than one of the outstanding players are from the same team, these replays are shown before those of the other team.

The relationship between the length of the replays, and where they are positioned within the broadcast, reflects their nature as continuity threads in the narrative in order to construct a worthwhile story. The length of the replay exploits the emotions generated by the event described and, thus, the better adjusted the timing of the replays, the stronger the feeling of affinity or rejection generated in the viewer.

The space or spot chosen to give replays visibility is directly related to the information they provide about the event. Nothing must be left to chance, and the broadcast should be like a theatre play in which the dialogues are built gradually depending on the different events that take place during the match. However, the premises at the outset are always the same, that is, the spots in which the replays are shown are established previously to give the different TV producers the possibility of using the same language. The storyline in each game must be constructed taking into account basic and predefined narrative premises, yet at the same time, a consensus must be reached on the global language to be used as this is the best way to create a unique discourse with the capacity to generate a brand image. As such, it is important to consider the fact that the television discourse upon which the football show depends is based on the capacity to generate information and/or controversy through technical details and moves that take place during the matches. Creating emotion is a fundamental premise on which the narrative of football is based, permitting sympathies to arise and a sense of personalization in the discourse, the basic and essential aspects of the digital communication in which our global society is immersed.

Aguado-Guadalupe and García-García (2018) state that:

> the new audience metrics that have emerged in the digital era are essentially focused on the behavioural and emotional analysis of users, providing information and data on their interaction with the content

> that is taken into consideration precisely for the process of developing content

thereby defining aspects required to understand the current generation of the television narrative shown in football broadcasts. The VAR will allow the information about emotion and personalization to be consolidated, as it dilutes the personal discourse and the emotional analysis of each viewer by offering the experts' detailed analysis of the different conflictive situations, experts who make a decision based on the fairest assessment thanks to the help of the available technology. However, this does not have to influence the creativity of the narrative, although it will strengthen the consensual resolutions through value judgements made around emotions.

Therefore, the temporal relationship traditionally applied to replays finds a new battlefield in the VAR, with the role of the protagonists changing and the handling of the time/space criterion modified. As a result, journalism loses ground to the game's arbitrators, such that football wins in equality and loses in emotion.

Replays and Post-refereeing

In the new era that is emerging with the consolidation of the VAR as a tool to arbitrate and assess the game, it is necessary to reconsider the television spectacle in order to keep entertaining and exciting the millions of people in different parts of the world whose emotions vibrate through these live transmissions. And that is the crux of the issue: the discourse that has been established must now be converted into a new discourse to achieve the same results, to generate the same emotions.

With the VAR, the referee acquires a more central role in the television narrative, particularly in the moments during the game when this technology is used. Thus, at the same time the viewer is shown the reaction of the referee to the information that is coming from the analysis of the VAR, the replay of the action from the angle they selected to analyse it and the live action from the pitch while the players wait for the result should also be shown. That is, the TV viewer contemplates all the information and the emotion only arrives when the referee makes the final decision as to whether or not to penalize a player. However, during this time delay the narration should reinforce any facets that are not evident in the images in order to continue building up the necessary emotion.

Decisions using the VAR are usually accepted without problems because of the technology used and the number of experts that are involved in taking the decisions. This does not imply that situations will not arise where the judgement made with the help of the VAR will not match that of the fans and even that of the journalists who are narrating the events. It is at this point that the concept of post-refereeing is introduced as a way to define the entire journalistic and television discourse created after the end of the football match. In this post-refereeing, replays play a fundamental role because they are the basic motor that defines the editorial line adopted in each type of programme.

When a key game in the progress of a league finishes, or the weekend's round of matches comes to an end, the television programmes that have flourished around the world of football begin to emerge. They are usually programmes in which a series of experts give their opinion about what happened in a particular match or in all the matches on a particular day or round of matches. Thus, the narrative of the show, which provides vital financial support for television networks that pay for the expensive rights to exploit the images associated with football over a season, continues to be produced, and it must attract the viewers who are yearning to have the details of each match, details that will establish or endorse their point of view on it. In these programmes, the replays of the most relevant plays are the basis on which the opinions from each match are forged.

The capacity to establish criteria better adjusted to the technical issues that support the incorporation of the VAR favours the informative style over the emotional one. To understand this process, it is necessary to establish the context in which replays are used in the different formats of post-match programmes. It is worth noting here that the broadcasters that hold the broadcasting rights own all the images, including the replays, while the other TV networks have to be content with the images from each match day for which they have bought the rights, or with the summaries of the matches, as established in the sales agreements between the different networks – these stipulating factors like the duration of the summaries or the variety of different shots for each move requested by the buyer.

The type of programmes dedicated to the analysis of football varies depending on the country and the league, yet we could consider the following structure to be a common format in all countries. Although there is no clear and agreed definition of the formats in the academic or commercial sectors, there is a general consensus to define televised sport as a macro-genre. The Euromonitor Observatory has its own database where programmes are coded and classified according to its own criteria, and it has established a classification of macro-genres, genres and micro-genres (Prado and Delgado, 2010). According to the Euromonitor Observatory, the macro-genres are: fiction, information, info-show, show, game contest, sports, children, youth, education, religion, and diverse. Within this classification, the macro-genre sports comprises the following genres: sports news, container, sports broadcast, live sports broadcast, sports reports, and magazines. In a comparative, historical evolutionary retrospective, Jaime Barroso (1988, 36) provided a table with the programmes in the catalogue of the European Broadcasting Union, which groups them into: educational, specific groups, religious, sports, news, popular and current affairs, dramatic, musicals, varieties, other programmes, advertising, and test cards and transitions. With regards to sports programmes, they were further categorized as: news, magazines, events, and others. Therefore, we could attempt to include a more precise description of the type of sports programmes to which we refer, without entering into conflict with the general classification on which certain consensus has been reached. As such, we propose the following:

- Programmes offering an evaluation of the day/weekend/round (expert mode): in these programmes journalists and sports professionals give their value judgements based purely on sporting criteria. This does not imply that controversies regarding the facts examined do not arise, but the debate focuses on more technical aspects, and on an interpretation of the sporting action and replays. The treatment of the images is not usually very complex and they are generally treated with the tools used for their live broadcasting. The journalist's status as an authority is generated on the basis of their knowledge and not enhanced imaging resources. Therefore, the images and replays used in these programmes are those already seen by the TV viewers when the match was broadcast. The relationship with the VAR is usually fluid since the decisions adopted by the assistants serve as a basis to establish an opinion about them. These programmes are usually broadcast live.
- Programmes offering an evaluation of the day/weekend/round (entertainment mode): in these programmes journalists and sports professionals give their value judgements according to sporting and emotional criteria, and they are normally broadcast live. Each of them adopts a stance as a supporter defending their particular team, and this must be borne in mind when considering the interpretation of the images. In this type of programme, the images are subject to a more complex and elaborate treatment in order to support the postures adopted by certain journalists/specialists to the detriment of others. In fact, the images that are used for replays are treated independently to address the controversies as specified in the programme's script. This type of programme is usually quite successful among the most loyal fans of a team and, often, they look to generate some confrontation in the supporters' discourses since they are shows that are based on exaggerating the details of what goes on during the course of the matches. These programmes have an increasingly strong impact on the audience and their relationship with the VAR is not very cordial, as this imposes limitations to the discourse that invites a subjective interpretation of the events that took place. For this reason, they seek to overwhelmingly use replays as the element of analysis and the basis on which to construct a reality from different perspectives.
- Sports programmes of a technical nature: these programmes are usually broadcast on a daily basis and analyse different aspects of both the game and the teams, using replays as the basis on which to create the

storyline of the programme itself. There is usually no debate in these programmes and they focus on explanations of the highlights offered by a presenter (journalist, former player, or a personality from the world of football). This is a sub-format with significant technical demands, where the scripts are so highly elaborated that they do not give any opportunity for later changes. These are recorded programmes that give visibility to and enhance the reputation of the TV stations that broadcast them.

- Sports interview programmes: these are usually recorded programmes that show highlights of players, coaches, and even referees, with the aim of presenting a more personal and human side of those involved in the spectacle of football. In this type of programme, replays are used to enhance and colour what the interviewer is referring to, giving visual support to their narrative and helping the viewer to remember all the issues dealt with in the interview.
- Historical sports programmes: programmes that centre on the events of a bygone era. A large number of images are usually used and in some of these programmes, the images on the whole constitute the script of the programme. Here replays have an explanatory and/or aesthetic role in helping to explain and build the whole story. The images archive is the basis upon which these programmes are made and TV channels owning the rights to such images have a greater capacity to create this type of content.
- Summaries for the internet: these are short summaries posted on specific channels and the images used usually come from the programme itself.

Therefore, replays are a necessary and fundamental resource in order to establish a narrative that helps produce television programmes based on post-refereeing. That is, they serve to shape the progress of the event into a narrative that generates a television spectacle.

The sports press needs, in the new iconic universe of digital journalism, this type of audio-visual resource in order to operate comfortably in front of an audience that has opted for images over text. All elements are necessary but replays give shape to the narrative, particularly with respect to setting out keys elements necessary to generate a show capable of reaching the global community. In this regard, it is now essential to be integrated into social networks and capable of reaching any individual using the new tools available, although this was an issue that has always been linked to the world of football. Replays already form part of the discourse itself, a structuring element that is relevant to the event. A video replayed countless times acquires greater notoriety than one which is watched only a few times. The spirit inherit to the very nature of replays acquires a greater dimension, articulating the emotional and personal aspects of the spectacle in the post-refereeing stage, a global show that is directly linked to the elements that structure communication in a digital society ("virality", visualizations, downloads, etc.). This is particularly relevant if we bear in mind the parameters implicit to the game itself and that there is no need to alter its development. As a result, the VAR is already part of that structure of communicative behaviour associated with the dynamic that brings together sport and spectacles.

Conclusions

The introduction of the VAR has established a new way of conceiving football as a sport and as a television show. The power of this sport requires a narrative that generates emotion, an aspect that is achieved through the game's own rules. However, to make it a global event, it was necessary to use mass media like radio and television. Integrating sports rules into television standards was a huge obstacle that was overcome by television producers, directors, and journalists, and the main instrument in achieving the necessary convergence between sport, television, and journalistic shows has always been the replay. This is a tool that has helped to transform televised football into a global event with the capacity to bring together fans from all cultures and social conditions.

Replays are the elements necessary to articulate a narrative that seeks to maximize the emotion through compliance with the rules of the game. Thus, the length, order, and type of replays has become so subject to protocols that it is difficult to see how different TV channels personalize their use. It is true that the position of the cameras and the broadcasting rules imposed by FIFA or UEFA in the main competitions contribute to this situation. However, an analysis of the way in which replays have been implemented in television allows us to understand the connection between narration and the necessary information that is presupposed when broadcasting any event on television. Benítez (2013) reminds us that "the script used in sports transmission provides clear refuges for the TV viewer [...], the twists in the plot, the expectations of change and the uncertainty about what is going to happen are a basic ingredient of all sports modalities". That is, the script has never been fully defined and, as a result, each sports show has its own essence that becomes evident as it is broadcast.

Through replays it is possible to introduce a narrative into the most relevant details of the game with a precision that allows a meta-discourse to be established that extends beyond the dynamics of the game, transcending into the culture generated around it. Time stops with a replay and expectation is generated in the viewer, which has had a direct impact on the way sport is lived and experienced, since it can no longer be understood without replays. As a result, it can be concluded that the creation of football discourse is based on the introduction of replays as the structural axis that provides the necessary continuity to the television show through the characteristics that are defined in post-refereeing.

Furthermore, this concept does not terminate with the end of the game but, rather, it must be exploited in the time between games by creating formats derived from the capacity to establish a narrative through the replays themselves, even in a context created artificially by journalists. That is, the show ends when the expectations generated during the game are analysed from all possible points of view by specialists in highly focused programmes. Post-refereeing makes use of this information to generate a show and, as the event has been broadcast live, it needs the replays and, increasingly, the VAR to generate a variety of scenarios of opinion and confrontation that turn the sport into pure entertainment, on many occasions forgetting the essence of what the game itself is.

The introduction of the VAR has a clear precedent in the issues addressed above, yet it is changing viewers' relationship with football as a sport, making replays a tool for judgement, not evaluation, while increasing the level of information provided at the cost of a loss of emotion. The VAR brings reliability, veracity, and authenticity to the game and the performances of its actors. Thus, the discourse generated addresses more technical parameters, to the detriment of those interpretations that involved the emotionality of a discourse constructed using value judgements derived from the analysis of replays. It should be remembered that the construction of a given discourse is implicit in the actual selection of the action replayed during the broadcasting of a match. Consequently, the technical resource of replays actively helps to create the specific narrative of the global spectacle that sports broadcasting represents. The VAR, which has generated so much debate, has been implicit in the construction of the story since the beginning of time, although its technical materialization has only just been made possible today. The audience, a vital element of the construction of any kind of story, was demanding this, and it is this aspect that will ensure the success of the VAR in sports broadcasts.

Notes

1 Entry queue time: time between the beginning of the narration and the point at which the images are shown when beginning (usually 1–2 seconds).
2 Exit queue time: time between the end of the narration and the point at which the images end (usually 2–3 seconds).

References

Aguado-Guadalupe, G., & García-García, A. L. (2018). Nuevas métricas de audiencia al servicio del Inboud Marketing. *Doxa Comunicación*, 26, 81–98.

Barlovento Comunicación. (2017a). Análisis de audiencias TV mayo 2017. Retrieved from www.barloventocomunicacion.es/audiencias-mensuales/analisis-mayo-2017/.

Barlovento Comunicación. (2017b). Análisis mensual del comportamiento de la audiencia televisiva. Retrieved from www.barloventocomunicacion.es/images/barlovento-audiencias-junio2017.pdf.

Barlovento Comunicación. (2018, July 31). Análisis de audiencias TV julio 2018. Retrieved from www.barloventocomunicacion.es/audiencias-mensuales/analisis-tv-julio-2018/.

Barroso, J. (1988). *Introducción a la realización televisiva*. Madrid: IORTV.

Benítez, A. J. (2013). *Realización de deportes en televisión*. Madrid: IORTV.

Checa, F. R. (2018, June 29). Telefónica se hace también con los derechos de la Champions por 1.080 millones de euros. *Expansión*. Retrieved from www.expansion.com/directivos/deporte-negocio/2018/06/28/5b350ce5e2704ef7728b45cf.html.

Comisión Nacional de los Mercados y la Competencia. (2018, July 26). Telecomunicaciones y Audiovisual. Informe económico 2017 [Report]. Retrieved from www.cnmc.es/sites/default/files/2244727_4.pdf.

Matilla, A. (2016, June 3). La estimación para la 16-17 es de 1.573 millones de euros. *As*. Retrieved from https://futbol.as.com/futbol/2016/06/03/primera/1464911764_155536.html.

Montemayor, F. J., & Sobrino, M. A. (2016). The media in the events. The power of television. *Revista Faro*, 1(23) (I Semestre 2016). Retrieved from www.revistafaro.cl/index.php/Faro/article/view/462/436.

Prado, E., & Delgado, M. (2010). Tendencias Internacionales de programación. *TELOS*, 84, 52–64.

Saavedra, M., Rodríguez, L., & Baron, G. (2015). Audiencia social en España: Estrategias de éxito en la televisión nacional. *ICONO14*, 13(2), 217.

Telefónica. (2017). Resultados enero-junio 2017 [Report]. Retrieved from www.telefonica.com/documents/162467/138879209/rdos17t2-esp.pdf/532a5599-f251-4b51-9b93-3ee25976325c.

Part III

Video-Assisted Officiating in Other Sports

6 Instant Replays in the Spanish Basketball League

ACB

Ángel M. López

Introduction

The quality of the broadcasting of Spanish basketball league games, known as the Association of Basketball Clubs League (Liga ACB from its Spanish acronym) since the 1983–84 season, received a significant boost when private TV channels became involved in the production and transmission of basketball games in the 1990s. Following the unprecedented success of Spanish basketball in the 1984 Los Angeles Olympic Games, winning the silver medal in the men's competition, basketball became one of the most popular sports in Spain during the late 1980s and throughout the 1990s. This popularity stirred the interest of Canal+ in the ACB, the recently created and first private TV channel in Spain. At that time, the rights to the Liga ACB were jointly held by the state television company TVE and the budding public regional channels, the latter broadcasting some of the games that were of particular interest to their more local target audiences. However, no unified criteria or systematic organisation had been established between these channels and the league, and, evidently, without extending the coverage to the whole competition.

In 1990, Canal+ started broadcasting in Spain in both open-access and pay-per-view modes, the latter initially focusing on football but soon adding the Liga ACB, having acquired the rights to televise this competition. This moment marked a milestone in terms of the production budget, as advertisers and subscribers, as well as the paying viewers, not only provided twice as much money but also demanded a higher-quality product. Despite the quality of the existing public television broadcasts, the arrival of Canal+ with more resources and technical means clearly led to a big improvement in the broadcasting of the Liga ACB, even though its audience was more restricted, as pay television was still not common in many Spanish households at the time. Canal+ produced and showed the best matches, although the majority of basketball games were not covered by TV.

During the 1990s Canal+ shared their rights to the Liga ACB with regional public channels, which, in line with the structure of the Spanish market, existed in the different regional autonomic communities. At the beginning of this century these public regional channels together covered the entire country and obtained the rights to the Liga ACB. The majority of the teams in the league come from cities covered by a regional public TV channel, which would broadcast all the teams' fixtures. While the arrival of Canal+ improved the technical quality of the Liga ACB broadcasts, the retransmission of Spanish basketball offered by the regional TV channels represented an important advance as it offered full coverage of the league, laying the basis for the later implementation of an IR system universally across the entire competition, with no exceptions. Nevertheless, this was not accomplished until 2015, principally because, as indicated above, no regional channel was responsible for the audio-visual retransmission of all the games. Indeed, in some areas of the country with a team in the ACB, there was no specific regional TV channel, as was the case for the Fórum Filatélico team from Valladolid, lying in the region of Castilla y Leon. Moreover, broadcasting was contractually limited to one game per league day and some communities with a regional television channel had more than one team in the league (e.g., Madrid).

In 2015, Movistar+ (the television platform resulting from the takeover of Canal+ by Telefónica, and its subsequent fusion with the buyer's own cable television platform) acquired the rights to the Liga ACB broadcasts in pay-per-view format. TVE retained the rights to a single game per match day and, with this distribution of the audio-visual rights, total coverage of all ACB games was achieved. Accordingly, Movistar+ broadcasts every game in HD through multiple channels, via cable and satellite, except for the open-access free view game that is produced and broadcast by TVE, programmed on its main channel (La 1) or its sports channel (Teledeporte), depending on the relevance of the event.

One of the consequences of this circumstance is that it allowed IR to be used in all the games, since this technology is based on images obtained and transmitted by the television production team and from the footage selected from all the other cameras involved in the live broadcast. Indeed, during the 2015–16 season it was decided that IR would be implemented in all league games for the entire duration of the game. A basic SD system (the standard definition that existed prior to the introduction of high definition) had been used in previous seasons to review very specific actions only in the last two minutes of the game. Moreover, this technology was only available for televised games and, hence, not for every league game.

Instant Replay (IR) System in the Liga ACB

The new system implemented by Live Audiovisual for the Liga ACB involves the signals captured by up to seven cameras deployed in the live television production, the images captured by the broadcasting cameras then being fed to the C21 Replay software. This system does not create a computer simulation, such as that produced by the Hawk-Eye (see Chapter 8) refereeing aid in professional tennis or that of the goal-line technology (GLT) in football, the latter recreating a virtual image from the images recorded by the cameras around the ground to determine whether the ball has gone over the goal line or not (see Chapter 10).

The roles of the director and cameramen are fundamental in the IR, as the signal the referee has access to in order to review the action depends ultimately on their skills. In this case, the director's role is not that related to the creation of the audio-visual account, as the programme signal is not that shown to the referee given that it lacks value when reassessing the play. The value of the director in terms of the IR comes from the prior planning of the coverage and the instructions issued to the cameramen, both before and during the match (see Chapter 3). The director is responsible for deciding the camera layout around the court in advance and the angles that will be obtained by each camera, and this person also needs to be familiar with basketball in order to be able to offer coverage of any rule violations and fouls that occur. This is a crucial aspect of IR as, for example, it is necessary for the director to make sure the cameramen include the players' feet in all the angles. Without these, there would be no possibility of using replays to assess *travelling* violations and, moreover, it would not be possible to review actions in which a player may have stepped over a line, such as that which defines *triple shots*, *out of court*, etc.

In the Liga ACB, a standard television broadcast involves a seven-camera layout, although depending on the relevance of the match to the channel that is broadcasting it, the number of cameras can vary from a minimum of five to a maximum of twelve or more for some of the more decisive games in the competition. As far as production is concerned, camera 4 is the most valuable, as this is the master camera which pans the game from a wide angle showing the movement of all the players and the development of the game.

By contrast, the camera that records the clock above the backboard is of little value for the television coverage, as it only serves as a safety backup in case the graphics system that incorporates the time signal from the referees' table fails. Nevertheless, this is one of the key cameras at the heart of IR-based decisions. This camera is labelled camera 7, or the clock camera, in Figure 6.1, and among the reviews most often requested by referees are those related to whether a play was completed in time, whether the referee table stopped the time accurately, corrections on the time remaining, etc. "Time management is essential in basketball: each possession must be resolved in 24 seconds, the attacking team must cross the half way line within 8 seconds, and each referee's decision stops the watch, some adding 24 seconds" (Benítez, 2013, 427) or 14 seconds, according to the circumstance of the game interpreted by the referees.

Conversely, camera 4, the main broadcasting camera, is less relevant for the IR as it offers a wide angle that lacks the detail necessary for refereeing

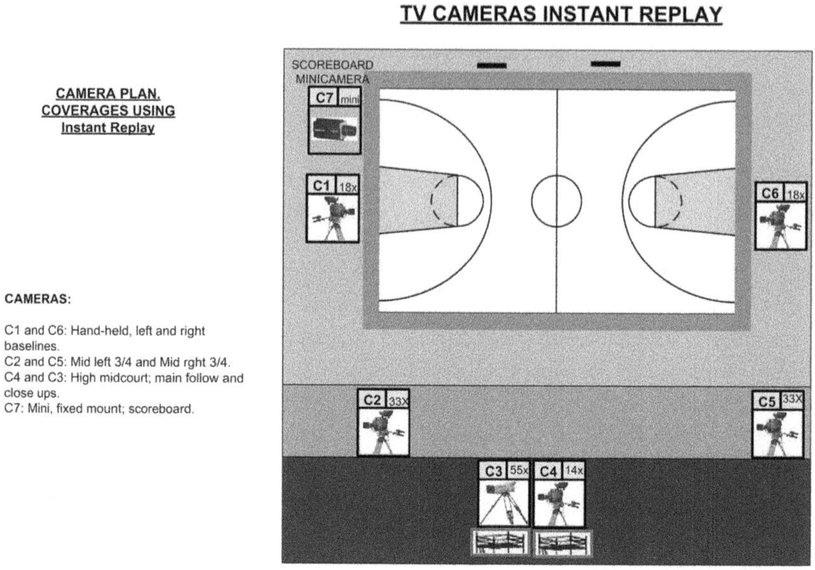

Figure 6.1 Location of the cameras used when the broadcast involves the use of instant replays

evaluation. The team responsible for the transmission must be made aware of the different role each element will play in the broadcast, particularly the main members of the team, the directors, cameramen, and replay operators (ROs), who all need to be aware of the dual use currently assigned to these broadcasting tools. Camera 3 records the game through a short close-up angle. It always tracks the player with the ball, and the whole body of the player must be in frame at all times. For the purpose of the IR, it is crucial to be able to see where the player's feet are and at what precise moment the ball leaves the player's hands. These details are essential for the IR to be able to resolve doubts about, for instance, triple shots or to determine whether a shot was taken on time. In basketball the moment the ball leaves the player's hand marks the end of the play, irrespective of the when the ball enters the basket. If the ball has left the player's hands before the possession time or quarter time has run out, the basket will be valid, and if the player's feet are behind the three-point line but not stepping on it, the basket is worth three points.

Cameras 2 and 5 alternate close-up shots from one side of the court with wider angle shots from the opposite side. The close-up shots follow the same protocol as those recorded by camera 3 and the wider angle shots are mainly characterised by the inclusion of all the players. Basketball fouls are not only committed around the player with the ball and the defenders; they can also happen off the ball in other areas of the court. Thus, proper planning and production of the coverage obtained through these two cameras aid the implementation of the IR in these areas.

Cameras 1 and 6 are located under the basket, offering a good perspective of what happens under the ring. This is a vital area in basketball, although these cameras can also cover movements at other points of interest, perhaps off the ball. As with other wide-angle shots, those captured with these cameras should aim at including all the players possible in order to avoid losing any detail that may be useful in a potential IR review. These shots are also useful to monitor the light that comes up on the board when time is up; this action is complemented in the IR with the superimposed image of the synchronised signals from camera 7 showing the time of the possession and the time left in the quarter.

Signals from all these cameras are integrated through triax, BNC fibre, or cables depending on the technical resources available for each broadcast, and they are transmitted to the outside broadcasting van (OBV) carrying out the television retransmission. All the signals are processed in the OBV and, after going through the camera control units (CCUs), they are sent to the vision mixer, where the director will select the most relevant sequence for each moment in order to assemble the audio-visual account of the action.

Camera signals are fed to the IR system before they are processed by the director. The IR technicians request the signals they require from the seven cameras (1–7) and the retransmission unit's head technician sends these signal via BNC cables connected to the matrix switchboard that distributes all the OBV's signals. The seven signals are then sent through fibre optics from the OBV's transmitting rack to which the BNC cables have been connected, and the signals are then carried by fibre optics (usually over distances between 100 and 200 meters depending on the OBV's location at each sports facility) to the receiving rack located on the courtside scoring table. Once there, the process is reversed until the signal is ultimately transmitted through BNC cables from the OBV to the informatics unit running the C21 Replay software (see Figure 6.2).

The IR post is stationed at the referees' scoring table situated between the two teams' benches, giving the referee easy and rapid access. It comprises two screens, one running the instant replay technician (IRT) software, and the other showing the signal seen by the main official in order to review the action and reach an appropriate decision. As a norm, the screen showing the signal seen by the referee faces the court during the game to allow the official the best and fastest view. Conversely, the screen running the operational software faces away from the court. This arrangement can be modified at the request of the referee or ACB commissioner should there be any interference from the team bench or the audience that impedes the correct operation of the IR protocols.

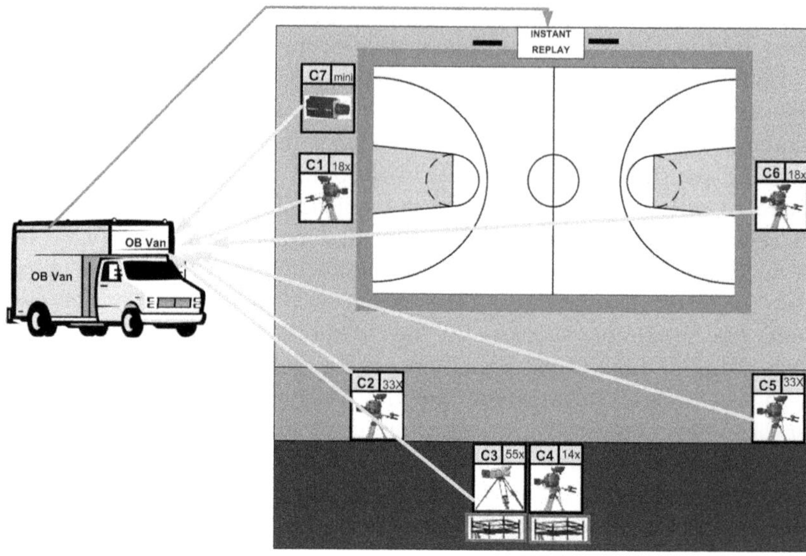

Figure 6.2 Signal flow for the instant replay system in the Liga ACB

Once the camera signals arrive at the C21 Replay software, the program arranges them in a multiscreen mode, with the clock camera superimposed on them. To achieve this, the technician must have already defined the clock camera in the system, although by default it is assigned as the last to be entered. However, to prevent errors, it is possible to attribute any of the incoming signals with the functions of the synchronised clock. The cameras can be visually sorted in the most convenient way, a task carried out by the IRT. This technician commonly starts by placing the master camera (C4) first, followed by the close-up camera (C3), and then the remaining cameras corresponding to their position on the court along the left to right axis. Hence, the main IR screen has a multiscreen configuration with six signals from the main cameras plus the clock camera superimposed in any of the four corners (see Figure 6.3).

The system allows the clock to be moved to any corner of the screen at any time, enabling any play to be conveniently reviewed with the time as a reference, and without blocking out what may be a part of the image that is vital to the referee. It is also possible to omit the image of the clock if the time factor (possession or game time) is not relevant to the foul or call.

Before the match, the referees meet at the scoring table to talk with the commissaries, greet the other officials seated at the table, and check that the IR system is working correctly. At this point, the main referee who will be consulting the system during the match checks with the IRT the

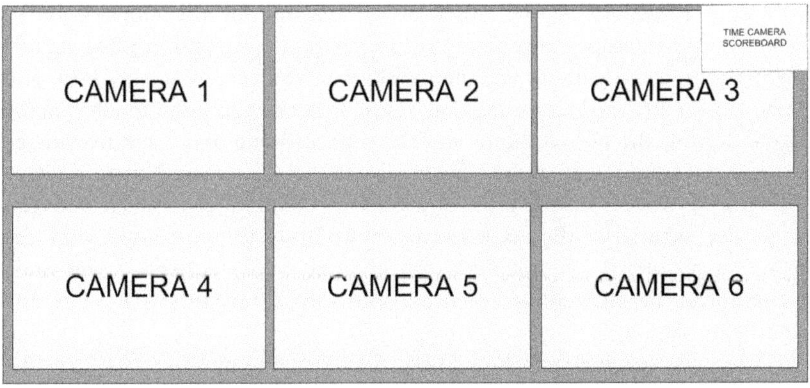

Figure 6.3 Multiscreen view of the instant replay system in the Liga ACB

details of their communication protocol and their organisation as a team. Usually, when a referee requires a review during the game, this is indicated with a sign to the IRT, drawing a (television) box in the air. Fluent communication in this respect is of vital importance, particularly when the game has continued and the review takes place during the first available stoppage. In these cases, the IRT has to mark the review in the time code of the play in order to be able to recover the footage promptly when the play stops and the referee comes back to the table to assess it. However, if the play stops immediately after the action that needs to be reviewed, the IRT only has to rewind a few seconds of recording to show the referee any of the angles needed. Before the game, the main official will confer with the IRT whether to show pre-selected camera angles or whether they will select which camera shot to look at in more detail on each occasion. In either case, evidently, it is the main referee who will ultimately decide which angles to review, at what speed, whether separately or at the same time, etc.

When the referee requests a review, the IRT can offer a multiscreen replay with all the camera angles showing at the same time. The images can be slowed down or fast forwarded as needed, or played backward or forwards frame-by-frame, although high-speed recording is not generally available. This capacity is usually reserved for issues related to timing, to check whether a basket went in within the time limits or to establish the position of feet at the exact moment when a potential three- or two-point basket left the hand. This technique allows all the angles to be viewed simultaneously, freeze-framed or in motion, with or without the image of the clock superimposed, all at the request of the main official.

The IRT may suggest a particular angle for the referee to watch on full screen, disregarding all the other images except for the clock if he so desires. The referees themselves can also request this option. The images from the selected camera are then shown in full screen and in HD, and they can be analysed in motion or freeze-framed. The high quality of the signal permits the use of zooms and the repositioning of a magnified image so that the referee can more clearly discern what occurred from a closer viewpoint. With a simple press of a button, the IRT can switch between angles or return to the multiscreen view in order to compare all the images available. If a simultaneous image of the clock is needed, the clock camera can be moved and resized so it does not interfere with any significant details.

There are no specific time limits for reviews, but the fact that the audience, players, coaches, etc., are awaiting the outcome for the game to proceed is always at the back of the referee's mind. Consequently, it is important to arrive at a sound decision without unnecessary delay, although review times are organically integrated within the normal course of a basketball match due to the fact that the game is not played in real time. By contrast, in other sports where the clock is not stopped when the ball is not in play, it is more complicated to insert IR interruptions as part of the game as the clock keeps running and the referee must then make sure the time lost is added on at the end of the half or at the end of the match. In these sports, added time tends to be controversial, as it is ultimately a subjective refereeing decision that is revealed long after the incident responsible for the delay occurred, at the end of the halves. This problem does not exist in basketball, as the clock is stopped in these instances, after a foul, during a timeout, and at many other points during the game. Therefore, the review time is fully integrated into the normal course of a basketball game. However, according to ACB regulations, a potential foul that has not been called by the referee cannot be subject to review and, consequently, the game must continue.

After analysing any particular action using the technical capabilities available, the main referee, who is exclusively responsible for the use of the IR, and only this referee, will inform the scoring table of the decision. This may be to confirm the original call or to correct it in some way. A three-point field goal could be revised to a two-point field goal and vice versa, a touchline call could change direction, etc. Basketball is well known for cultivating a friendly and relaxed atmosphere between referees, coaches, and players, and during the match it's not unusual to see them talking and discussing some technical refereeing detail or the game in general. Indeed, it is common that the decision taken after the replay is explained briefly to the coaches and team captains, but not to the public.

The Copa del Rey is a knock-out competition organised by the ACB in which the top seven teams classified after the first round of the league participate, plus a team from the host city. The competition takes place at a single venue over four days in February. On this occasion, given that it brings together all the refereeing teams who will be involved in the quarter-finals, semi-finals, and final, together with the IRT,

> one of the officials, who will referee another of the games in this final phase, will act as an assistant video referee in each game, reviewing the incidents with the IR technician and helping the main official, although the final decision will always be the sole responsibility of the main referee, who is responsible for reviewing the controversial plays using the C21 Replay system.
>
> (F. Acedo, personal communication, 2017, October 2)

As detailed in Chapter 10, the actions that can be resolved with the assistance of video recordings are always those that involve objective concepts, absolute facts, issues with a yes or no answer (e.g., was the clock at 0 when the ball left the player's hand or not?). There is a fundamental distinction in the two types of procedures that can be followed when using the IR: actions that are reviewed immediately, where the game is stopped for the referee to consult the IR images; and actions reviewed later, the game continues and at the first possible break in play the referee reviews the previous action using IR. These latter events have no consequence on the immediate development of the game, they do not influence the following play, and they only affect the scoreboard, having no impact on the development of the subsequent play.

Immediate review of actions that might require a correction of the clock, variations in game time, or in the possession time can be requested at any time of the game. The game can also be stopped to determine whether a foul was committed on a player who was shooting a two- or three-point field goal, as the free shots granted will vary from two to three. Identifying the free shot thrower when there is doubt regarding who was the player fouled is another reason for an immediate replay, as is the identification of any players involved in a violent incident. Immediate video replays are also required to determine if a foul or the end of possession came first, and if the player who was fouled was shooting a basket and, therefore, an additional free shot should be granted. In these cases, the game will be stopped until the main referee takes a final decision after watching the different angles of the play offered by the IR.

Two types of actions can be reviewed at any time during the game, although their replay is not immediate. If a basket is scored and there is doubt about the exact position of the shooter because the referee could not clearly see whether the player was stepping on the three-point line or not, in this case the game continues with the score decided by the referees at that time, but the officials make a sign, drawing a rectangle in the air, to indicate a later revision will be needed through the IR. At the next break in play, when the game is stopped, the main official will review the images from the most appropriate IR angle and he may also request a zoom of the shot of the player's feet that includes an image of the triple line. The referee will then indicate his decision to the scoring table using a sign for a two- or three-point field goal depending on whether the feet touched the line or not. For these actions, the feet of the shooter must be analysed in the exact frame that the ball leaves their hands, or when they lift from the floor. That is the factor that determines whether a two-point (if stepping on the line or inside the three-point area) or three-point field goal was scored. The scoring table will amend the scoreboard if the IR finding contrasts with the initial refereeing decision. Another delayed revision involves the 24- or 14-second violation when a basket is scored (depending on which half of the court the ball was in at the start of the play). If there is any doubt in this respect, the corresponding points are scored and on the next break in play the referee will review the appropriate camera angles with the superimposed clock image. If the ball left the hand of the player before the end of the possession, the basket is valid and the score remains the same; if not, the referee will let the table know that the points corresponding to the action need to be subtracted from the score.

There are three actions that may require immediate revision only at the end of each quarter, and as a result of which will cause the game to be stopped until the referee takes a decision about them. One is when deciding if a basket was scored before the end of the quarter or not. The referee will review different angles offered by the cameras to determine whether the light on the scoreboard came on before or after the shot was taken. The same is true to determine whether a foul was committed before or after the end of the quarter. The game can also be stopped at the end of a quarter to find out the exact time remaining after an 8-, 14- or 24-second penalty, or if the ball went out of court. These events are very important in the final moments of a game and they can be decisive in one direction or the other, leaving no chance for the rival team to respond as time runs out. For this reason, the instant IR reviews have been extended to address these three specific instances.

In the last two minutes of the game and during extra time, the IR capabilities are extended to review actions that can decide the fate of the match. These are pivotal actions that require precise assessment. As such, IR can be applied to unsportsmanlike fouls, touchline decisions, identifying who committed a foul, and half-court violations. In all these cases, the game can be stopped for the replay and the clock will not be started again until the main official has reached a decision. A half-court violation or out-of-court decision in these final minutes can determine who wins the game, as the shortage of time remaining may prevent the rival from being able to react. These final interruptions are a common occurrence in basketball, and it is not unusual for the last minutes of a game to last for much longer than the rest of the match due to strategic moves signalled from the teams' benches, including repeated fouls to force free throws and avoid three-point shots, etc. The normal progress of the game is not undermined by this delay, as both the audience and the players are familiar with the lag at the end of a basketball game.

The tempo of basketball makes it an ideal sport to apply IR, with repeated interruptions stopping the clock and the existence of plays that can be assessed objectively. Ruling whether a player has stepped on or over the three-point line or touch line is quite objective in the images. Similarly, it can be determined whether the clock was at zero at any particular moment or if it shows three-tenths of a second, a decision that again requires no interpretation. The experience and expertise of Liga ACB referees enables them to can make the correct calls in most cases, only requiring the use of a replay in exceptional cases. Indeed, it is noteworthy that "last season the IR was requested a mean of 2.36 times per game" (F. Acedo, personal communication, 2017, October 2).

Conclusions and Future Considerations for the IR in Liga ACB Basketball

Every year the Liga ACB league incorporates new actions onto the list of those that can be examined by the referee using the IR, taking advantage of the technological advances in this system to improve the rate of refereeing errors. The ultimate aim of these tools is that the game should be fairer and that refereeing decisions do not influence in the final sporting result.

One evolution in the ACB IR would be to include a challenge, a concept that already exists in other sports like tennis (see Chapter 8). Teams could have three opportunities to request an IR replay, extending the decision as to whether to review an action or not, so that it does not lie only in the hands of the main referee. While the final verdict on the play would still be the referee's sole responsibility, the teams might feel reassured knowing that controversial actions have been assessed as objectively and fairly as possible, having been able to intervene in their initiation. Currently, there are actions which the referees consider very clear and, thus, they see no need to review even though the team affected may repeatedly request for an IR, albeit unsuccessfully. This review modality will help include all the parties involved in overseeing the game and it would help reassure all parties that sporting justice is being carried out at all times. In order for this to be a realistic proposal during a match, a limit of three reviews has to be set for each team, as has been successfully tested in other sports like tennis (see Chapter 8) and American football (NFL) (see Chapter 9). If the verdict after the IR is favourable to the team who requested the review, they will preserve the opportunities they have remaining, while one option will be lost when the IR review concurs with the original refereeing decision. This modality is entirely compatible with the current system and it will complement (and improve) it without substituting it.

Showing replays of IR reviews on the screen is an aspect which also requires improvement in the Liga ACB, and adopting production protocols similar to those used in the NBA could provide greater transparency (see Chapter 7). The television broadcast and the on-court screens could show the images being shown to the referee and the decision could be superimposed on these to help the audiences better understand what is going on in the game. It would be opportune to let the audience know the outcome of the decision, as the only information available to them at present are the signs from the referees to the table and any ensuing changes in the scoreboard, with no further explanation. Providing such explanations would contribute to a better understanding of the progress of the game by both the live spectators and television viewers.

In addition, a video repository of controversial actions and of the refereeing outcomes would be very useful, not only for supporters and journalists, but also for the players themselves, for managers, and as referee training aids.

In technical terms, remote television production of some matches during the 2018–19 season will be tested out. The cameras will be located at the court, but the production will be carried out in a central studio. This was a decision made by the host broadcaster and driven by different financial concerns, one of which was the low cost of sending images by fibre optics relative to the cost of satellite transmission. This raises the possibility of opening a centre exclusively dedicated to IR from which IR can be technically operated for all matches, sending the signal from the cameras to the centre and the signal produced back to the particular court where the referee can see the footage of the action on the screen on the scoring table. This approach offers advantages in terms of economics and proficiency, at the expense of taking some importance from the efficient relationship between the referee and the IRT, who, being in the same location, can coordinate efforts to improve the review process, particularly in terms of timing. More investment could be used to improve such a central facility that would control all games, dedicating funds to train the specialist staff that would run it. This is already the case in the NBA (see Chapter 7) and La Liga, for example.

The next few years will prove crucial for the evolution of the IR, ideally with the advent of cameras solely dedicated to assistant referees (see Chapter 3). This would introduce the possibility of reliably recreating plays in a manner that exclusively addresses the needs of the IR, leaving broadcasting cameras to carry out their original purpose: the creation of the most engaging and realistic audio-visual storyline possible.

References

Acedo, F. (2017, October 2). Personal communication with Francisco Acedo. Technical Coordinator of Instant Replay at Live Audiovisual, Madrid.

Benítez, A. (2013). *Realización de deportes en televisión*. Madrid: IORTV.

7 Instant Replay in the National Basketball Association

Jacob Tingle

Introduction

At its heart basketball has always been an inclusive sport. A game popular among rural Midwesterners as an indoor refuge from the brutal winters, popular amongst inner city dwellers because of the relatively small space and limited equipment it requires, and a game popular among twentieth-century American immigrants because, unlike American football or baseball, it was never deeply ingrained into the cultural psyche of the political and economic elite (Mandelbaum, 2004). As such, it has been a sport quicker to innovate or change in order to fight for its place among American popular past-times. Some of those rule changes included requiring the ball to be advanced past the mid-court line in under ten seconds, restricting some defensive formations, and the introduction of the shot clock in 1954 (Rovell, 2016).

The free-flowing nature of the game, coupled with its popularity among such broad groups as well as the wildly popular college game has led the National Basketball Association (NBA) to maintain its place as one of the most popular sports in the United States, with over 22 million fans attending games in person during the 2017–18 season (ESPN.com, n.d.).

The NBA began in 1946 and at present 30 teams play 82 games (for a total of 1,230 games) during a six-month regular season. The game is played between two teams of five players and squads have one head coach with an average of 5.5 assistant coaches. The game lasts four 12-minute quarters, and each team is allowed to have 15 players on its active roster. Of these 15, at least 8 must suit up for each game.

The NBA's 24-second shot clock, which requires teams to shoot the ball in less than 24 seconds, first appeared in the 1954–55 season. The NBA was not yet a popular televised sport, and the owners knew that to improve the game's watchability – and thus improve its fan base – they somehow had to prevent teams from consuming one or two minutes on every possession (Rovell, 2016; Turner, 2013).

Technological Improvements and NBA Officiating

The process for becoming an NBA referee is detailed, demanding, and difficult. The NBA specifies a six-step process on its web portal, which begins with potential candidates being reviewed from one of eight officiating systems; the top 100 candidates are then scouted and reviewed. A much smaller number move to the more formal training and evaluation phase (NBA Official Instant, n.d.). A select group are hired to work in NBA developmental leagues and an even smaller number are added to the WNBA or NBA staff. Highlighting how difficult the pathway is, in 2018 only five new officials were added to the NBA staff (Reynolds, 2018). The current 65 NBA referees on staff have an average of 14.1 years of experience in the league. Based upon a combination of tenure and number of playoff games worked, NBA referees earn between $150,000 and $550,000 a year (Interbasket, n.d.; Lewis, 2019).

Joe Borgia's comment about replay being a necessary evil is indeed how those officiating in the NBA in 2002 felt about replay's introduction. Referees are like officials in all sports in that they desire is to be right, to be accurate, but they also have a psychological bent towards certainty (Weinberg and Richardson, 1990). One must possess higher than average doses of self-belief and self-confidence to be comfortable knowing that 100 per cent of your decisions will be despised by at least half those watching a game (MacMahon et al., 2015; Weinberg and Richardson, 1990). Even still – especially with the increased competition for top-level NBA jobs – more than anything, referees want to get the call right.

Former NFL official and current broadcaster Mike Pereira summed it up well:

> There is not a sports official around who wants to leave a field, court or rink thinking they didn't get it right. If you give them a tool that allows them to get it right, that's what they want. It's about getting it right, and the expectations are higher about getting it right than back in the old days before we had the technology.
>
> (Saunders, 2013, para. 39)

NBA referees, like officials from all sports, don't want to be wrong, but the reasons might not be clear to those who have never "worn the stripes". As longtime NBA referee Courtney Kirkland said:

> I'm not trying to please this fan, I'm not trying to please this player – I'm trying to please the game itself. That's the way I look at it: What is best for the game? What does the game need? That's who my friend is: The game.
>
> (Forgrave, 2019, para. 14)

Given that mindset, despite initial misgivings, it's not surprising the NBA referees have come to appreciate the impact of technology on their ability to do what's right for the game. One such innovation significantly improved the timing errors in NBA arenas.

The introduction of the Precision Timing System (PTS) significantly improved how NBA games were officiated. Developed by former NBA referee Michael J. Costabile (Barkley, 1995), the system connects the ref's whistle to a transmitter worn in the waistband. As an official blows the whistle, the transmitter sends a signal stopping the clock instantly. Referees can then restart the clock using the same device. It is estimated PTS saves close to three minutes of game time.

Despite its impact, even the PTS isn't perfect or fool-proof: (a) game and shot clocks at NBA arenas are still operated, mostly, by humans; and (b) there is a lag associated with the time it takes for an official to see a play, make a decision, and to actually put air in the whistle.

With its history of being an early adopter for technological innovations to improve training and development of players and make the game more enjoyable for fans, it's no surprise that the NBA adopted replay sooner than many other sport leagues. Even still, the decision for the NBA to begin its use of replay was reactive, rather than proactive. There are previous examples of plays in NBA games where replay could have been used (Abdul-Jabbar, 2011; Ryback, 2016), but finally after the 2002 NBA playoffs, a tipping point was reached and the league could no longer ignore that replay was essential to the future integrity of the game.

After three high-profile missed clock-related calls in the 2002 playoffs (LeBron, 2014), one of which affected the outcome of a game, the league took swift and decisive action (Broussard, 2002). Highlighting the NBA's decision, Stu Jackson, then the senior vice president for basketball operations, said:

> I don't think there was any one event that drove this decision, but certainly during the past season and the playoffs, there were a number of instances where, quite frankly, for any human being it would've been nearly impossible to determine whether the shot got off in time.
> (Broussard, 2002, para. 6)

Furthermore, at each iteration of updating replay's use, there was a catalyst that forced the league into action (Perrin, 2010). Table 7.1 includes all 15 of the replay triggers used by NBA referees. The year the rules were updated to permit each trigger is also included.

To be certain, since the NBA began using instant replay in 2002–03, the game and its narrative arc have changed significantly (Allen, 2010; Ryback, 2016). Yet, despite the incredible accuracy of modern game replay systems, there is simply no way to remove human error from the equation.

Table 7.1 Replay rules updates

2002–03	Referees will: • Automatically review shots taken at the end of each quarter or overtime period. This review will not take place at any other time. • Review whether a player's foot is on the 3-point line or out of bounds. • Review whether a 24-second shot-clock or 8-second backcourt violation has occurred before a shot is taken. • Review foul calls, but only to determine whether a player was fouled before time expired. • Have a maximum of two minutes to review plays on a courtside video monitor.
2007–08 (Limited NBA replay, 2007)	**Player altercations and flagrant fouls** • Referees will call either a Flagrant Foul 1 or 2 on the floor, but can only review Flagrant 2 calls. For replay purposes means any situation where: • Two or more players are engaged in a fight or a hostile physical interaction that is not part of normal basketball play. • A player is ejected from the game for committing a hostile act against another player, for example, when a player intentionally or recklessly harms or attempts to harm another player with a punch, elbow, kick, or blow to the head. • This is an automatic trigger and referees are required to conduct the review regardless of score or time remaining.
2008–09 (Description of the, 2008)	A. 2-point/3-point shots Game officials will be permitted to use instant replay at any point during a game to determine: i whether a successful field goal was correctly scored as a 2-point or 3-point field goal; ii whether, for purposes of awarding the correct number of free throws, an unsuccessful field goal attempt on which the shooter was fouled was a 2-point or 3-point attempt. B. Game clock situations Referees will consult instant replay if: 1 A play concludes (i) with no time remaining on the clock (0:00) at the end of any quarter or overtime period, or (ii) at a point when the game officials believe that actual time may have expired in any quarter or overtime period; and

2 The game officials are reasonably certain that a game clock malfunction has occurred during the play.

2009–10

Out of bounds
Referees can review any out-of-bounds play that occurs in the last two minutes of the fourth quarter and during overtime when they are not reasonably certain as to which player caused the ball to go out of bounds.

- A review can only be initiated on a called out-of-bounds play and only those involving doubt as to which player caused the ball to go out.

Shot clock violations
Referees can review plays in which they were not reasonably certain whether or not a 24-second violation occurred on a made basket or prior to a foul being called.

2010–11

Clear-path-to-the-basket fouls
Referees can use replay if they are not reasonably certain that a called clear-path-to-the-basket foul (clear-path foul) actually met all of the criteria of the rule.

Review of correct free throw shooter
Referees can use replay when they are not reasonably certain which player should attempt free throws on a called foul.

- Referees cannot review to determine which player committed the foul or whether a foul was warranted.

Review of 24-second shot clock reset
Referees can review situations where they were not reasonably certain whether the ball actually touched the rim and are therefore unsure if the shot clock was (or was not) reset properly.

- This type of replay is only conducted in the last two minutes of the fourth period and during all of overtime.

2012–13

Flagrant fouls
Referees will review all flagrant foul calls any time a flagrant foul call is made.

- This is an automatic trigger and referees are required to conduct the review regardless of score and time remaining.
- The officials can also review whether any other players committed unsportsmanlike acts or unnecessary contact immediately prior to and/or immediately following the calling of the flagrant foul.

Restricted area block/charge
Referees can review all block/charge calls when they are not reasonably certain as to whether the defender was inside or outside of the restricted area.

(*Continued*)

Table 7.1 (Cont.)

	Goaltending/basket interference Referees can review situations in which they are not reasonably certain whether a goaltending or basket interference violation was called correctly during the last two minutes of the fourth period and during all of overtime. • Possible goaltending violations that were not called are not reviewable.
2013–14	**Off-ball foul timing** Referees can use video to review situations in which they are not reasonably certain whether a player without the ball was fouled prior to: a his teammate beginning his shooting motion on a successful basket; or b his teammate releasing the ball on a throw-in. **Delay of game** Referees can review to determine whether a delay-of-game penalty is called whenever a player or team commits an act that causes a postponement or interruption in play, whether or not the delay results in an unfair advantage.

Source: NBA Official Instant Replay Guidelines (n.d.).

The Impact of Gambling on Professional Basketball

Gambling is nothing new to the world of basketball in the US. During the mid-1940s to early-1950s there was pervasive game-fixing that led to over 30 college basketball players being charged with point-shaving by the New York City district attorney. The scheme involved gamblers bribing college players to keep scores within an expected range. The scandal led to a year-long suspension for then powerhouse University of Kentucky and other universities to cancel their basketball programmes all together (Gems, Borish, and Pfister, 2008). So appalled was the NBA by the actions of those players that it banned them for life from the league. With one glaring exception, the NBA – and its referees – has managed to stay above the fray that has impacted college basketball and other professional sports.

Enter 2007

In 2007, the worst-case scenario for the league hit the headlines of every major news outlet in the country. Tim Donaghy, a long-serving veteran NBA referee, admitted to and was convicted of conspiracy to commit wire fraud and conspiracy to transmit wagering information over state lines (Beck and Schmidt, 2007; Sheridan, 2007).

During the previous two to four NBA seasons (depending on who is telling the story), Donaghy was paid between $2,000 and $5,000 to impact the point spread of games. He accomplished the illegal activity through a variety of techniques, such as calling more fouls to ensure a team would get to the free-throw line more often. At the time, some reporters said the NBA's integrity took a major hit and the league would be deeply scarred by the news (Adande, 2007; Jones, 2008).

Then NBA commissioner David Stern said that, in three decades working for the league, it was "the most serious situation and worst situation that I have ever experienced" (NBA Commissioner David, 2007). He also indicated that:

> No amount of effort, time or personnel is being spared to assist in this investigation, to bring to justice an individual who has betrayed the most sacred trust in professional sports, and to take the necessary steps to protect against this ever happening again.
>
> (Wojnarowski, 2007, para. 4)

Given the swift and immediate response from the league, it's not surprising that for years the NBA took a position that legalized gambling in the US was ultimately bad for business. What did, in some circles, come as a surprise was NBA Commissioner Adam Silver's response to the watershed US Supreme Court case which ostensibly paved the way for more legalized gambling.

Embrace of Supreme Court Ruling

In 2018 the US Supreme Court issued an important ruling that seems to open the doors for more legal sports wagering. Some important background: the US Congress passed the Professional and Amateur Sports Protection Act (PASPA) in 1992. PASPA prohibited state-sanctioned sports gambling with a few exceptions and also permits any sports entities whose games might be or become the subject of sports gambling to bring an action to enjoin the gambling.

The NBA joined other professional sports leagues and the National Collegiate Athletic Association (NCAA) to prevent the State of New Jersey from legalizing sports gambling schemes in the state. In *Murphy v. National Collegiate Athletic Association* (2018), the US Supreme Court found that PASPA violated certain elements of the 10th Amendment to the US Constitution, thus overturning the law. The ruling then opened the door to legalizing sports betting in the US, which is illegal in all but a very few areas.

The law was originally passed by Congress to ensure the integrity of sporting contests, and the Supreme Court ruling signals a sea-change in the moral relationship between spectators and sports leagues. According to

Gabriel Feldman, the director of the sports law programme at Tulane Law School:

> It [*Murphy v. NCAA*] will likely change how we have viewed sports for the past 100 years. It's called the gamblization of sports. Fans will become much more focused on gambling than following a team. It will make every second of every game of every week interesting to fans as it will give everyone something to root for.
>
> (Liptak and Draper, 2018, para. 10)

Despite the fact that the NBA had joined the other leagues in bringing the suit, as early as 2014 Commissioner Adam Silver wrote: "I believe that sports betting should be brought out of the underground and into the sunlight where it can be appropriately monitored and regulated" (Purdham, 2016, para. 4). While the soil of US professional sports has not yet been radically tilled, looking at the models from across the globe, it does appear that Adam Silver could be on the right track. As for the referees, however, an important question remains.

How does the NBA's embrace of legalized gambling impact the already great pressures league referees face to be *perfect*? Some contend that more money will lead to greater corruption (Eden, 2019). Which begs another important question yet to be answered: will bringing gambling out of the shadows for US sports lead to greater transparency and a reduced likelihood that a future Tim Donaghy will emerge, or will the legal status make NBA referees even more vulnerable to gambling outfits (Ziller, 2019)?

What the Media Say

Nowhere is the debate about replay's value more debated than amongst sport media. Some fall firmly in the "sport is beautiful because of the human element" camp (Cole, 2015; Ryback, 2016; Rymer, 2012), while others claim that "cameras don't lie" (Abdul-Jabbar, 2011) and thus the use of replay is an essential element of the modern game (Clougherty, 2013; De Piccioto, 2017).

Firmly in the "replay is ruining sports" camp is Steve Politi (2018). After Game 1 of the 2018 NBA Finals he wrote: "For the most part, I want replay to go, because it hasn't eliminated the 'human element' from sports at all. If anything, it has simply added another way in which us imperfect homo sapiens can get something else wrong" (para. 4).

Will Leitch (2018) also argues that replay is making sports less enjoyable. He contends that a big factor of sport fandom is arguing, both at the time and after the game, about a referee's decision. Phil Barber from the *Santa Rosa Press Democrat* also calls for a complete elimination of replay. While he extols the notion that replay, in theory, can eliminate all mistakes by game officials, Barber wrote that, "human error has always been a part of officiating" (2017, para. 13) and the imperfect attempt to remove error from one side of the game equation (i.e. the officiating and not the playing) has a negative effect on the reason sports have existed for centuries: the drama of the unknown storyline.

Alters (2016) wrote that the overuse of replay and the constant threat of a next-day rebuke by the NBA offices has turned being a fan of basketball from joyous to laborious. She pled, "For the sake of the sport, let the officials be in charge of their own whistles, and let the players play" (para. 11).

Leitch and Barber also argue that the time it takes to review a play, no matter how short, only exacerbates the pain (agony) for fans and, as such, the use of replay is removing the drama and ruining the narrative arc of a basketball game. Instead of making the experience more enjoyable for sports fans, replay is having the counter effect.

Among his list of replay *cons*, Craig Berman includes the negative impact to the spectators watching the games live. Fans "often have to wait for someone else to confirm what they just saw with their own eyes" (Berman, 2011, n.p.). Dubow (2018) also cited the controversial play from the 2018 NBA Finals, in which LeBron James was called for a blocking foul after the referees went to the monitor based on the restricted area block/charge trigger (see Table 7.1) as the perfect exemplar of Berman's complaint.

Chief among the concerns voiced by the anti-replay media is the nature of what is and isn't reviewable (Barber, 2017; Berman, 2011). According to Leitch (2018):

> One big problem is just how litigious the matter of whether or not to consult replay has become, with each sport having drawn mind-numbingly arcane, labyrinthine distinctions between those things (presumably cut-and -dried questions such as whether someone is out of bounds) that refs can review and those (presumably "judgment calls") they cannot.
>
> (para. 4)

Another important criticism comes from former NBA head coach and current ESPN analyst Jeff Van Gundy. Coach Van Gundy expresses concern that the day-after review of calls, especially those that are contact/flagrant in nature, might "undermine the credibility from the players' standpoint that every player and every team gets the same and equal treatment under the rules" (Ryback, 2016, para. 19).

On the other side of the replay discussion, Hughes (2013) pulls no punches. "Those espousing the value of the 'human element' … in officiating are just clinging to the sorts of necessary evils – relics, really – of yesteryear" (para. 40). Others go so far as to argue that replay use by NBA referees should be expanded. In other words, if the NBA is going to allow referees to use replay, the focus should be helping them get calls right – especially during times in the game when calls are most important. Jim Cavan (2014) summed up this argument:

> If video technology is going to be used at all, it might as well be used toward the best possible outcome … a fair ruling that strikes a balance between accuracy on the one hand, and a timely conclusion on the other.
>
> (para. 20)

Other arguments are used to support use of replay by NBA referees. Among those are increased accountability, quieting the conspiracy theorists (Instant Replay and Sports, 2015), more game stoppages mean more commercial revenue for the league and its broadcasting partners, and a greater likelihood that referees will get calls correct (Hughes, 2013; Instant Replay and Sports, 2015).

Still others take a more nuanced approach (De Piccioto, 2017), attempting to weigh the pros – in essence the technology is there and it should be used – with the cons, that NBA referees are hamstrung by what plays the rules permit and don't permit them to review. As Dan Levy wrote: "Replay can be great, but not when the rules surrounding the process seem to be trumping common sense" (2014, para. 8). Those making this argument contend that the league should allow officials to review *more* plays, if they're going to be allowed to use replay at all.

The media debate about replay's pros and cons often echo that of the fan, but arguments about NBA officials go far beyond how *good* they are and extend to whether they are scofflaws and cheats.

The Game Is Rigged? Bias and Referees

Even outside of the officiating community, there is acknowledgement that NBA referees are great at their jobs.

> So I think I should put in a few words for the guys in the striped shirts. They have an incredibly difficult job; and for the most part, they do their thing very competently. They could succumb to being corrupt; but to the best of my knowledge, there has only been one major league official (Tim Donaghy of the NBA) who has ever been found to have tried to dishonestly affect the outcome of a game. The officials of the various pro sports are overwhelmingly honest, dedicated, knowledgeable and in shape. Their love of the games they officiate is obvious. No one in his right mind would go through all the trouble it takes to qualify for those jobs if they didn't love what they do.
>
> (Abdul-Jabbar, 2011, para. 5)

Even with accolades from an NBA legend, the need for replay is real. A big reason replay exists is that the timing (and other) errors, no matter when they occur, can influence a game's outcome (MacMahon et al., 2015). In addition to the real or perceived threat of future Tim Donaghy situations, there are, however, other reasons for expanded use of replay in NBA games.

A study in the *Quarterly Journal of Economics* (Price and Wolfers, 2010) found that NBA referees called more personal fouls against players when the game officials were primarily opposite-race than when the referee crew and players were primarily the same race. They further indicated that the biases were so significant as to impact the outcome of games. While the results didn't distinguish whether the bias stemmed from the actions of white or black referees, it nonetheless caused a massive public relations problem for the league.

The NBA responded with its own analysis, but independent experts reviewing both studies thought Price and Wolfers' research to be more credible. Yale University Law Professor Ian Ayers told the *New York Times*:

> I would be more surprised if it [implicit association bias] didn't exist. There's a growing consensus that a large proportion of racialized decisions is not driven by any conscious race discrimination, but that it is often just driven by unconscious, or subconscious, attitudes. When you force people to make snap decisions, they often can't keep themselves from subconsciously treating blacks different than whites, men different from women.
>
> (Schwarz, 2007, para. 9)

NBA referees' decisions are accurately described as "snap", but interestingly enough in a follow-up study years later, Price and Wolfers found the bias had disappeared. It's not as if the decision-making process, or time to make them, improved. When asked for an explanation about why the bias seemed to disappear, Price said: "Racial bias is a malleable trait. Large-scale public focus on a specific type of racial bias in a specific group can make it go away" (Ingraham, 2014, para. 8). One wonders: has the expanded use of replay, coupled with an increased awareness of their own bias, had an impact on current referee decisions?

Either way, despite the reports of potential in-group racial bias and the continued fallout from the Donaghy scandal, fans aren't staying away. More than 22 million fans attended games in person (ESPN.com, n.d.) and viewership is up or constant across the four major NBA television partners after the 2107–18 season. So, what has been the long-term impact of the scandal? Broadly speaking, the so-called Donaghy effect hasn't gone away and won't soon dissipate.

Is the Donaghy scandal indeed a permanent stain (Ziller, 2019), a Scarlet G (gambler) or Scarlet F (fixed) on the NBA? The fact that 12 years later the league and its prominent media partners, e.g. ESPN, are still writing in detail (Eden, 2019) about the situation leans towards that answer being a hard yes. As Ziller (2019, para. 7) concluded, "Regardless, it's clear that the unprovable nature of the allegations will keep the Donaghy scandal alive for years and years."

The broad use of replay and perhaps the push towards embracing legalized gambling, it could be argued, are the NBA's battle armour against a future Tim Donaghy. By providing fans more access to how replay decisions are made, the thought is that the massive number of eyeballs on each decision won't permit Donaghy-level corruption from a future game official. In fact, however, one might argue that even with the greater push for transparency, evidenced by the 2014 creation of the NBA Replay Center, there is an ever more strained relationship between NBA referees and fans. With greater access to information comes greater perceptions that the fans know how to do the job of an NBA referee.

A Brave New World: Big Brother and How Technology Has Changed It All

In his 2015 study, Deutscher found no evidence to support biased decision-making amongst NBA referees, including no home court bias, star bias, or US versus non-US player bias. He further asserted that referees have a built-in financial incentive to avoid bias (i.e. future employment), and contended that the constant monitoring of officials could be an explanation.

Furthermore, as reported in the *Boston Globe*, according to the National Basketball Referees Association, NBA referees get 97 per cent of their calls correct. In that same article, the NBA Replay Center indicated that in the last two minutes of games, arguably the most intense and most difficult, league officials get more than 86 per cent of their calls right (Goldenberg and Bowers, 2015). Bottom line: NBA refs are really good at their jobs. Which then begs the question: where does the animus come from?

What is happening in the relationship between NBA referees and fans can be considered part of a larger societal trend to distrust those in authority. Scandals in the automotive industry (e.g. Volkswagen), major college sports in the United States (e.g. Penn State University), international sports (e.g. United States Gymnastics), and match-fixing in German football have led many to assert there is a growing movement to distrust those in authority (Tingle, 2016). While some of that distrust has been earned (Kihl and Richardson, 2009), others contend this lack of trust might be related to other broad societal factors such as the income inequality gap and the growing tribalism we see in global politics (Lewis, 2019).

All this scrutiny and the greater movement to review, seemingly, every decision coupled with increased levels of abuse from players, coaches, and spectators – resulting from those high expectations and distrust of authority – has contributed to a global shortage of sports officials: from basketball officials in lower-division American leagues (Warner, Tingle, and Kellett, 2013) to football referees in the UK (Cleland, O'Gorman, and Webb, 2018). Enter Big Brother:

> If technology offered the promise of perfect monitoring of movement, a sort of closed-circuit surveillance state in which every play was registered and adjudicated exactly as it happened, with no potential for misuse or arbitrary judgment or inequitable application, would we embrace it? Or would we just want to clobber it with a bat?
>
> (Leitch, 2018, para. 6)

The NBA is betting that the increased scrutiny, i.e. transparency, will actually improve the experience for its officials. Indeed, its response to critics of NBA referees was to create the NBA Replay Center in 2014 (Amick, 2014).

One of the key features of the NBA Replay Center is the "Last Two Minute (L2M) Reports". Introduced in March 2015, the league hopes to provide a more detailed and transparent picture of decisions made by NBA referees. From the league webpage, the L2M reports were created, "recognizing that NBA officials are correct roughly 90% of the time, we felt it important and fair to list all the correctly officiated plays as well" (NBA Officials, n.d.). The league also reports that L2M's help with the effort to "build a greater awareness and understanding of the rules and processes that govern our game … and serves as a mechanism of accountability to our fans and the media who fairly seek clarifications after our games". Reports are issued on all calls and "material non-calls" during the last two minutes of the fourth quarter or overtime for all games during which the point differential is three points or less at any point during those time intervals (NBA Officials, n.d.).

According to Joe Borgia, the "main goal of the NBA Replay Center is to get the officials the exact angle they need when they come to replay" (NBA Replay Center, n.d.). The Center is equipped with a 10 gigabit network and more than 90 screens, which allows referees in every arena to view video at 60 frames per second (Lewis, 2019). Replay officials use a touch screen automated program which allows them to zoom and select multiple angles for referees in the arena, all with the simple "tap of a button" (McCue, 2014, n.p.). The network allows referees, at one time, to see split screens with multiple-angle views of a play. These multiple angles provide referees with more comprehensive looks, and more "clear and comprehensive evidence" that allows decisions to be more accurate and more timely.

Another important outcome of the NBA Replay Center is the enhanced fan experience for those watching games on TV (Golden, 2014). The Center has established direct feeds and lines of communication with broadcast partners. The improved and more direct communication with the TV networks allows the NBA to better frame the narrative about (a) what decisions are made, (b) how they are made, and (c) perhaps most importantly, why they are made. According to Frank DiGraci, YES Network game producer:

> This is going to help our storytelling in a number of ways. One, we're going to actually see the angle used for the review. Two, we're going to see the process. Our fans and our viewers are going to get a live shot into the replay center. And third, we can parlay that now to our audience by speaking directly to the NBA Replay Center.
>
> (NBA Replay Center, n.d.)

The statistics from its first year suggest the Replay Center has achieved both aims: to improve accuracy and speed with which replay decisions are made. During the 2014–15 season the NBA Replay Center conducted 2,162 reviews with an average decision time of 42.1 seconds. Stated another way, in about half the time teams are allotted for a timeout, referees can review one of the 15 "triggers" detailed in Table 7.1. The average game had 1.76 reviews and nearly 81 per cent of those confirmed the officials' real-time decision, while just over 19 per cent of the initial referee decisions were overturned (NBA Official, 2015). As good as those numbers were from its initial season, at the midpoint of the 2018–19 season, the average time for reviews was down to 29 seconds. As of January 2019, over a third of the official reviews were to determine whether a shot was a two-pointer or a three-pointer, while the buzzer beater accounted for 20 per cent of replay requests (Forgrave, 2019).

While still far from removing controversy and making the officiating perfect, the NBA Replay Center has made great strides in improving the level of officiating.

Conclusions

Regardless of whether one is for or against the use of instant replay in the NBA, one thing's for certain: it is here to stay. It has become too deeply ingrained in how fans experience the game, players play, and coaches coach. In short, instant replay is now part of the narrative arc of the NBA. As Patrick Saunders, a reporter from the *Denver Post*, wrote: "Indeed, instant replay itself has become theater as fans – whether at home on their couch, or in the stadium, or in a sports bar – wait on the edge of their seats until the replay call is announced" (2013, para. 12).

The controversy and debate about when and how much replay should be allowed will not go away, because no matter how much technology advances, no matter how much it seeps into the game, it will never be 100 per cent accurate (MacMahon et al., 2015). Referees, like the players and coaches, are humans, and the concept of human frailty and imperfection has and will continue to be an important feature of the sports narrative. For all the cries for perfection, unbiased decision-making, and accuracy, players will miss shots, coaches will make bad choices, and referees will blow calls. But, what the use of replay technology in the NBA has taught us is that, though it might fall short, the moon shot of 100 per cent accuracy and full transparency from the league office is, in Joe Borgia's words, "a necessary evil" (Lewis, 2019).

Acknowledgement

I am grateful to Hailey Wilson, an undergraduate student research assistant at Trinity University, San Antonio, TX. Her contribution to this chapter was invaluable.

References

Abdul-Jabbar, K. (2011, August 12). Officials and umpires deserve the benefit of instant replay [News]. Retrieved June 17, 2018 from www.espn.com/espn/commentary/story/_/page/abdul-jabbar-110812/officials-umpires-deserve-benefit-instant-replay.

Adande, J. A. (2007, July 19). Ref investigation only adds to bad perception of the NBA. *ESPN.com*. Retrieved April 10, 2019 from www.espn.com/nba/columns/story?id=2943704.

Allen, S. (2010, October 13). Upon further review: A brief history of instant replay. Retrieved December 14, 2018 from http://mentalfloss.com/article/26075/upon-further-review-brief-history-instant-replay.

Alters, K. (2016, June 16). In praise of imperfect NBA referees. *The Week*. Retrieved April 1, 2019 from https://theweek.com/articles/630192/praise-imperfect-nba-referees.

Amick, S. (2014, October 23). NBA's new replay center is an immediate game-changer. Retrieved December 14, 2018 from www.usatoday.com/story/sports/nba/2014/10/23/replay-center-officials-officiating-instant-video-review/17794825/.

Barber, P. (2017, December 28). Replay reviews are killing our buzz. *Santa Rosa Press Democrat*. Retrieved December 14, 2018 from www.pressdemocrat.com/sports/7809141-181/barber-upon-further-review-please.

Barkley, M. (1995, March 19). On time: Basketball ref invents system to stop clock instantly. *The Greensboro News and Record*. Retrieved from www.greensboro.com/on-time-basketball-ref-invents-system-to-stop-clock-instantly/article_ffed51d2-1790-5805-823b-c072d9c83ce4.html.

Beck, H., & Schmidt, M. S. (2007, August 16). NBA referee pleads guilty to gambling charges. *The New York Times*. Retrieved April 1, 2019 from www.nytimes.com/2007/08/16/sports/basketball/16nba.html.

Berman, C. (2011, July 3). Pros & cons of instant replay in sports. *SportsRec*, Retrieved June 17, 2018 from https://www.sportsrec.com/pros-cons-instant-replay-sports-8681745.html.

Broussard, C. (2002, July 30). N.B.A. will use replay to review buzzer shots. *The New York Times*. Retrieved from www.nytimes.com/2002/07/30/sports/pro-basketball-nba-will-use-replay-to-review-buzzer-shots.html.

Cavan, J. (2014, April 19). Should NBA change current replay review rule to allow referees more leeway? Retrieved December 14, 2018 from https://bleacherreport.com/articles/2035389-should-nba-change-current-replay-review-rule-to-allow-referees-more-leeway.

Cleland, J., O'Gorman, J., & Webb, T. (2018). Respect? An investigation into the experience of referees in association football. *International Review for the Sociology of Sport*, 53(8), 960–974. doi:10.1177/1012690216687979

Clougherty, K. (2013, December). How instant replay benefits professional sports. Presentation presented at the Prezi. Retrieved from https://prezi.com/60wcq1-dfmxg/how-instant-replay-benefits-professional-sports/.

Cole, C. (2015, October 16). How instant replay is ruining professional sports, one coach's challenge at a time [News]. Retrieved June 17, 2018 from http://nationalpost.com/sports/how-instant-replay-is-ruining-professional-sports-one-coachs-challenge-at-a-time.

De Piccioto, E. (2017, October 18). Let's watch it again: Should instant replay be used in sports? *The Perspective*. Retrieved June 13, 2018 from www.theperspective.com/debates/sports/let-watch-instant-replay-used-sports/.

Description of the NBA's new instant replay rules. (2008, October 23). *NBA.com*. Retrieved December 14, 2018 from www.nba.com/2008/news/10/23/102108videoreplayrules/.

Deutscher, C. (2015). No referee bias in the NBA: New evidence with leagues' assessment data. *Journal of Sports Analytics*, 1, 91–96. doi:10.3233/JSA-150012

Dubow, J. (2018, June 1). Replay overturn plays key role in Warriors' Game 1 win. Retrieved December 14, 2018 from http://hosted.stats.com/nba/story.asp?i=20180601025033670152708&ref=rec&tm=&src=NBA.

Eden, S. (2019, February 19). How former ref Tim Donaghy conspired to fix NBA games. *ESPN*.com. Retrieved from www.espn.com/nba/story/_/id/25980368/how-former-ref-tim-donaghy-conspired-fix-nba-games.

ESPN.com. (n.d.). National Basketball Association: Total attendance at regular season games from 2006/07 to 2017/18 (in millions). *In Statista - The Statistics Portal*. Retrieved April 21, 2019 from www.statista.com/statistics/193423/regular-season-attendance-in-the-nba-since-2006/.

Forgrave, R. (2019, January 11). At the NBA Replay Center, the league's most scrutinized employees aim to get it right. *CBSSports.com*. Retrieved April 1, 2019 from www.cbssports.com/nba/news/at-the-nba-replay-center-the-leagues-most-scrutinized-employees-aim-to-get-it-right/.

Gems, G. R., Borish, L. J., & Pfister, G. (2008). *Sports in American history: From colonization to globalization*. Champaign, IL: Human Kinetics.

Golden, J. (2014, December 8). A look inside NBA's instant replay center [News]. Retrieved June 17, 2018 from www.cnbc.com/2014/12/08/basketball-a-look-inside-nbas-instant-replay-center.html.

Goldenberg, R., & Bowers, R. G. (2015, June 5). In close games, how often do NBA referees get it right? *The Boston Globe*. Retrieved March 20, 2019 from www.bostonglobe.com/sports/2015/06/03/close-games-how-often-nba-referees-get-right/KbpT1kG3Sp5ywtUqY4cCbN/story.html.

Hughes, G. (2013, June 13). Pros and cons of NBA's potential expanded use of replay. *Bleacher Report*. Retrieved June 17, 2018 from https://bleacherreport.com/articles/1672374-pros-and-cons-of-nbas-potential-expanded-use-of-replay.

Ingraham, C. (2014, February 25). What the NBA can teach us about eliminating racial bias. *The Washington Post*. Retrieved April 1, 2019 from www.washingtonpost.com/news/wonk/wp/2014/02/25/what-the-nba-can-teach-us-about-eliminating-racial-bias/?utm_term=.f5ba9dcd3cc8.

Instant Replay & Sports [Blog]. (2015, September 25). Zabala's top five list. Retrieved June 17, 2018 from http://sites.psu.edu/zabala/2015/09/25/instant-replay-is-not-killing-sports/.

Interbasket. (n.d.). Here's how much NBA referees are paid per game and their annual salary. Retrieved March 20, 2019 from www.interbasket.net/news/18135/2016/12/how-much-do-nba-referees-make-per-game-year-salary/.

Jones, B. (2008, July 20). Corruption on the court is bad news for NBA. *ESPN.com*. Retrieved April 19, 2019 from www.espn.com/espn/page2/story?page=jones/070720.

Kihl, L. A., & Richardson, T. (2009). "Fixing the mess": A grounded theory of a men's basketball coaching staff's suffering as a result of academic corruption. *Journal of Sport Management*, 23, 278–304. doi:10.1123/jsm.23.3.278

Koppett, L. (2007, December 7). The NBA – 1946: A new league. *NBA.com*. Retrieved from www.nba.com/heritageweek2007/newleague_071207.html.

LeBron, S. (2014, December 9). An exclusive inside look at the NBA Replay Center. *Sports on Earth*. Retrieved December 14, 2018 from www.sportsonearth.com/article/103634604/nba-instant-replay-center-david-stern-joe-borgia-steve-hellmuth.

Leitch, W. (2018, June 5). The case against instant replay – All instant replay. *New York Magazine*. Retrieved April 15, 2019 from http://nymag.com/intelligencer/2018/06/the-case-against-instant-replay-all-instant-replay.html.

Levy, D. (2014, April 21). Is replay review helping sports or helping to ruin sports? [News]. Retrieved May 25, 2018 from http://bleacherreport.com/articles/2036386-is-replay-review-helping-sports-or-helping-to-ruin-sports.

Lewis, M. (Producer). (2019, April 1). Ref, you suck! Against the Rules. [Audio podcast.] Retrieved from https://atrpodcast.com/.

Limited NBA replay OK'd. (2007, September 6). Retrieved December 14, 2018 from www.chicagotribune.com/news/ct-xpm-2007-09-06-0709060080-story.html.

Liptak, A., & Draper, K. (2018, May 14). Supreme Court ruling favors sports betting. *The New York Times*. Retrieved March 30, 2019 from www.nytimes.com/2018/05/14/us/politics/supreme-court-sports-betting-new-jersey.html.

MacMahon, C., Mascarenhas, D., Plessner, H., Pizzera, A., Oudejans, R., & Raab, M. (2015). *Sports officials and officiating: Science and practice*. New York, NY: Routledge.

Mandelbaum, M. (2004). *The meaning of sports: Why Americans watch baseball, football and basketball and what they see when the do*. New York, NY: Public Affairs.

McCue, M. (2014, October 29). How the NBA fixed its instant replay problem? *Fast Company*. Retrieved December 14, 2018 from www.fastcompany.com/3037743/how-the-nba-fixed-its-instant-replay-problem.

Murphy, Governor of New Jersey, et. al. v. National Collegiate Athletic Assn et. al 584 U.S. (2018). Retrieved March 30, 2019 from www.oyez.org/cases/2017/16-476.

NBA commissioner David Stern press conference. (2007, July 24). *NBA.com*. Retrieved April 19, 2019 from www.nba.com/news/sternpc_070724.html.

NBA Official. (2015, April 17). NBA Replay Center reduces review time to 42 seconds. Retrieved April 15, 2019 from https://official.nba.com/nba-replay-center-reduces-review-time-to-42-seconds/.

NBA Official Instant Replay Guidelines. (n.d.). Retrieved December 14, 2018 fromhttps://official.nba.com/replay/triggers/.

NBA Officials. (n.d.). Officiating opportunities. Retrieved April 7, 2019 from https://nbaofficials.com/?AspxAutoDetectCookieSupport=1.

NBA Replay Center. (n.d.). NBA replay center video. Retrieved from www.youtube.com/watch?v=YP3fkfOZHEg.

Perrin, S. (2010, June 9). Instant replay in the NBA. *Clips Nation*. Retrieved December 14, 2018 from www.clipsnation.com/2010/6/9/1509216/instant-replay-in-the-nba.

Politi, S. (2018, June 4). Are instant replay reviews ruining sports? *Ask Politi!* Retrieved December 14, 2018 from www.nj.com/sports/index.ssf/2018/06/are_instant_replay_reviews_ruining_sports_ask_poli.html

Price, J., & Wolfers, J. (2010). Racial discrimination among NBA referees. *The Quarterly Journal of Economics*, 125(4), 1859–1887. doi:10.1162/qjec.2010.125.4.1859

Purdham, D. (2016, February 16). I'm not pro sports gambling. I'm just a realist. *ESPN the Magazine*. Retrieved March 30, 2019 from https://abcnews.go.com/Sports/im-pro-sports-gambling-realist/story?id=36572316.

Reynolds, T. (2018, November 15). NBA promotes 5 refs, including 2 women, to full-time status. *AP News*. Retrieved March 18, 2019 from https://apnews.com/bdf1714aa25e4cf2b8091b105b57382a.

Rovell, D. (2016, July 7). NBA unifying timing setup across league. *ESPN.com*. Retrieved April 15, 2019 from www.espn.com/nba/story/_/id/16852164/nba-unveil-new-shot-clock-timing-system.

Ryback, A. (2016, May 27). How expanded instant replay has hurt the NBA. *The Sports Quotient*. Retrieved December 14, 2018 from http://thesportsquotient.com/nba/2016/5/27/how-expanded-instant-replay-has-hurt-the-nba.

Rymer, Z. D. (2012, April 23). Breaking down the pros and cons of instant replay in MLB [News]. Retrieved June 17, 2018 from https://bleacherreport.com/articles/1156392-breaking-down-the-pros-and-cons-of-instant-replay-in-mlb.

Saunders, P. (2013, November 16). How instant replay has revolutionized sports on TV. *The Denver Post*. Retrieved June 17, 2018 from www.denverpost.com/2013/11/16/how-instant-replay-has-revolutioned-sports-on-tv/.

Schwarz, A. (2007, May 2). Study of N.B.A. sees racial bias in calling fouls. *The New York Times*. Retried April 2, 2019 from www.nytimes.com/2007/05/02/sports/basketball/02refs.html.

Sheridan, C. (2007, August 14). What's next? Q&A on the NBA's gambling scandal. *ESPN*. Retrieved April 19, 2019 from www.espn.com/nba/columns/story?columnist=sheridan_chris&id=2948746.

Tingle, J. K. (2016). An exploration of organizational trust in intercollegiate athletic departments. *Journal of Higher Education Athletics and Innovation*, 1(1), 18–37. doi:10.15763/issn.2376-5267.2016.1.1.18-37

Turner, G. (2013, June 4). The shot clock - The 20 biggest tech advancements in sports history. *Complex*. Retrieved December 14, 2018 from www.complex.com/pop-culture/2013/06/the-20-biggest-tech-advancements-in-sports-history/the-shot-clock.

Warner, S., Tingle, J. K., & Kellett, P. (2013). Officiating attrition: Considering the experiences of referees from a sport development perspective. *Journal of Sport Management*, 27, 316–328. doi:10.1123/jsm.27.4.316

Weinberg, R. S., & Richardson, P. A. (1990). *Psychology of officiating*. Champaign, IL: Leisure Press.

Wojnarowski, A. (2007, July 20). Questionable calls. *Yahoo! Sports*. Retrieved April 19, 2019 from https://web.archive.org/web/20070810085051/http://sports.yahoo.com/nba/news?slug=aw-referee072007&prov=yhoo&type=lgns.

Ziller, T. (2019, February 25). Tim Donaghy is a permanent NBA stain. *SBNation.com*. Retrieved from www.sbnation.com/2019/2/25/18237290/tim-donaghy-scandal-nba-referee-david-stern.

8 Video Technology for Refereeing in Other Sports

Tennis and Rugby

Anto J. Benítez, Esteban Stepanian and Ángel M. López

Introduction

In many sports the athletes are the sole active participants and the judges simply observe each activity in turn, subsequently manifesting their scores. For example, in gymnastics the athlete's performance is not normally interrupted, particularly by the judges. However, in other sports it is necessary for referees to intervene in order for the competition/game to progress, and snap decisions are required to judge foul play, scores, or whose turn it is to compete.

By definition, the ultimate aim when practising any sport is to stretch the limits of the individual's physical, technical, and/or tactical abilities. Evidently, adapting our perception to human performance also has its limits when watching/observing sport, whether as an entertainment, or with regulatory or sanctioning intentions. Refereeing relies on a highly trained eye, knowledge of the rules and the skill to interpret them, and the ability to take quick decisions. In addition, more dynamic sports require an appropriate level of fitness and the capacity to be in the right place at all times. Naturally, although this is not a question addressed in this study, a referee must have a clear vision of fair play, and the resolution and authority to implement it.

Most of the sports played in defined areas that involve moving objects present particular refereeing challenges: the pace of the action and the need to be well positioned close to the points of contact. These are very demanding spatiotemporal constraints. Moreover, in many of these sports, the importance of many incidences is influenced by where they occur, and the speed and length of the actions constantly put the skills of perception to the test. Athletes train constantly to become faster movers and thinkers, to perform and display their skills at the highest level, which undoubtedly involves pushing the capacity to perceive and assess these feats to the limit. However, human perception has clear shortcomings, the result of a process of adaptive evolution, with a smaller and slower margin for improvement than motor evolution – at least until now.

In this context of ever faster movements and limited perception, it makes sense for those in charge of overseeing fair play in sport to be aided by technology, albeit the same technology that highlights the limitations of perception. This technology is that which creates the illusion of movement through successive images, that of video, which allows actions to be replayed on a more comfortable scale and the speeding up or freezing of images.

By looking at two sports as contrasting as tennis and rugby, the use of refereeing aids based on video technology will be analysed. The referees in both these sports have been using video technology as an aid for more than ten years, in an official capacity in the first case and as an uncompromised aid in the latter. Along with occasional assistance in other sports, the historical accumulated experiences from these two sports can be a source of valuable learning for the use of a VAR system in football.

The Nature of Tennis and Rugby, and the Restrictions on Refereeing

Tennis is played in a relatively small court by a small number of players (two or four if doubles is considered). The opposing players have no contact with each other, and they follow turns of service and games or sets. The way this sport is played has changed slightly over the years, even though it has always been proud of its history and tradition. The classic scoring system involves the characteristic 15, 30 and 40 points, and the possible advantage in each game. The umpire decides, with the help of line judges, when the ball is in or out, although they "only call balls which are out" (Nadimpalli and Hasenbein, 2013, pp. 3–4). A new way to keep score is currently being tested, known as "Fast4", and it could be introduced soon. Nevertheless, the advent of this system does not alter the issues addressed in this chapter, but for those interested, a description and some statistical comparisons with the traditional method can be found in Simmonds and O'Donoghue (2018).

Tennis rules are simple, the order of serves in the game are strict and well known, and visually, all elements are accessible to refereeing. Logically, a large number of matches are played simultaneously within a tournament and not all courts are in the same conditions. For simplicity, only the individual competitions of the Master 1000 series and Grand Slam tournaments will be considered, in which the best conditions for refereeing are established. In these tournaments, the chair umpire is assisted by a number of line judges, usually between four and nine depending on the importance of the tournament and the court. All of these are active in terms of their perception of the game, yet fixed in their positions: they do not move during play and, thus, no dynamic skills are required. In order to minimise errors, judges are required to keep a close eye on the lines at all times (Whitney et al., 2008, p. 949); nevertheless, it is an accepted fact that "officials make incorrect 'out' calls more often than they do 'in' calls, since the brain tends to think fast moving objects travel farther than they actually do" (Nadimpalli and Hasenbein, 2013, p. 8, see Whitney et al., 2008).

When considering the limits of human perception, the binary decision that the line judges must make can be defined as an IN or OUT call, based on continuous measures that are often inexact and inadequate by nature. Harry Collins and Robert Evans termed *digitisation* to the task of "turn inexact measurements into discrete decisions. In most sports the referees or umpires are the people who do the digitization" (Collins and Evans, 2008, p. 289).

Speed and size are really crucial to the decisions in tennis. The outer edge of the lines defines the limit of the area to be considered as OUT of court. The ball – depending on the type – measures between 65 and 73 mm in diameter (ITF, 2018, p. 5). The court measures 23.7 metres in length and the service box is approximately 6.4 by 4.15 metres. These measurements are taken from the outside of the lines, as any contact of the ball with the line is considered as "IN". The lines around the service box are between 2.5 and 5 cm wide, while the baselines tend to be wider, in some courts measuring up to 10 cm in width (ITF, 2018, p. 41). For example, if we consider the service in the men's individual game, the main characteristic of this shot is the speed of the ball, which is complicated by its ill-defined trajectory. The player returning the serve is faced with a ball travelling at an average velocity of between 180 and 200 kilometres per hour. In fact, the fastest registered serve was by the Australian Sam Groth, who during the Busan 2012 ATP Challenger in South Korea served the ball at 263 km/h, breaking the previous record of 251 km/h held by Ivo Karlovic when serving in a doubles match between Croatia and Germany during the Davis Cup in Zagreb (March 2011). In turn, the Croat had broken the record of Andy Roddick, who had served at 249 km/h in 2004. Nevertheless, players generally attempt to vary the speed of their serve, as well as its direction, depth, and spin. However, when considering a *normal* serve, the ball can be expected to travel 18 or 19 metres at a speed of 180 km/h – 50 metres per second. Thus, when hit from a height of less than 3 metres to clear the net, the ball will take a third of a second to bounce inside the service box. This represents the time available to the judges to fix their attention on the site where the ball bounces and hence, realistically, there is only about a sixth of a second to decide if the serve is valid or not. Some calculations indicated that this decision is about estimating differences of 3–4 cm in the matter of a millisecond (Mather, 2008, pp. 1649–50) and significantly, "line judges are more reliable than players" at making these decisions (Mather, 2008, p. 1645).

Cinema deceives human perception by creating the illusion of continuity through the projection of 24 static images per second to produce the sensation of reasonably fluid movement – a frequency *only* four times that required by the line judge to see the ball bounce. This illustrates how the temporal resolution of human eyesight does not offer strong guarantees for the (minimal) spatio-temporal accuracy of the judgements required in tennis. According to specialists, "while humans can make perceptual judgements with extreme precision, performance ultimately hits limits set by the neural machinery serving vision" (Mather, 2008, p. 1645). In other words, while the eye may be more or less capable of performing the task required sufficiently well, the brain is more likely to mislead. Thus, the speed of the action is the main factor leading to error. However, on surfaces like clay, the ball leaves an imprint when it bounces and the line judge's eyesight can revert to this to check the validity of the shot.

By contrast, rugby is played on a large pitch, with a scale of detail that is unmanageable for a single static observer. The main competitions are played on pitches that are 100 metres long and 70 metres wide, with a 22-metre in-goal area at each end. There are no specifications regarding the thickness of the lines, and these are usually between 10 and 20 cm. The refereeing team comprises a main official and two linesmen, and according to Law 6 of the game, they can be assisted by the Television Match Official (TMO) and other assistants or reserves if the organiser deems it necessary (World Rugby, 2018, p. 38).

In this space, 15 players per team fight for a ball that they need to have in their possession to be able to score. Contact is permitted in direct actions or in organised formations, such as a scrum or spontaneous ruck or maul. A player from the team in possession that is in front of the ball is offside and cannot play the ball. Any player that kicks the ball places all their teammates in front of them in an offside position, until the kicker or any other member of that team that was behind the kicker when the ball was kicked brings them back onside by running forward from behind the position where the kick was taken and overtaking them, as long as the player(s) in the offside position move/s away from where the ball lands. Aside from these exceptions, when there are no static formations set up, the players can play the ball with unrestricted movement. However, as soon as a formation is organised – as constantly occurs with spontaneous rucks/mauls – or the ball is played with the foot, virtual lines are defined and players not participating in the formation are offside, in a similar manner to those in front of the kicker. Consequently, the referee must take note of different points on the field of play that need to be carefully monitored in conjunction.

The rules of rugby are relatively complex and they need to be interpreted on many occasions during a match. The demands on the referees can be analysed taking into account that they are expected to be close to the ball at all times, a constant movement that requires a good level of fitness. The official's physical exertion is not only required to be close to the ball, but also to ensure a good angle of vision and orientation. This is necessary to determine if the ball is being played in accordance with the rules and, at the same time, if the positioning of the defence is correct. As a result, there is an ongoing debate as to how factors like fatigue (Suárez-Arrones et al., 2013), age (Castagna et al., 2005), experience (MacMahon et al., 2007), the pressure of being observed (Baldwin, 2014), and movement and positioning (Mallo et al., 2012) influence a referee's decisions (some of these factors, which are often connected, have been dealt with together, for example in Guillén and Feltz, 2011, and in relation to the use of refereeing efficacy, or officiating technological aids; Kolbinger and Lames, 2017), affecting either their perceptual illusions or their interpretations (Catteeuw et al., 2009).

The problem of getting into a good position in order to assess sporting actions is shared by referees and video capturing systems alike, and it is particularly challenging when based on pin-hole geometry instead of epipolar geometry. The way in which a ball crosses the screen might not depend only on its speed, size, or trajectory; it may also be subject to phenomena related to the angle of vision, framing, follow-up assignments, and other mediatic decisions associated with generating a coherent television narrative and that are coordinated by the director of the HB (host broadcaster; particularly regarding camera location, audio-visual narratives and refereeing complexities (see Chapter 3)). The duration of each action can be defined as *the time between putting the ball in play and the next stoppage* and it is open-ended, sometimes lasting more than four minutes, during which the ball can travel more than 250 metres in different directions across the pitch. Currently, in top-level rugby it is common for more than ten phases to take place in the same play, which translates into the number of formations that are set up and resolved after the ball is put in play (first phase) before the ball goes out of play or an infraction occurs. The referee aims to keep the game flowing steadily for the maximum time possible, in which no infringement is committed or the attacking team retains an advantage, with players moving rapidly in different directions over the 7,000m^2 area (11,400 if the in-goal areas are factored in). It should be noted that by kicking the ball it can travel 60 or 70 metres up the field, and it can be lifted up to 30 metres or more, remaining in the air for up to six seconds. In six seconds an average rugby player can cover 45 to 50 metres, so the scenic snapshot (see Chapter 3) is in constant flux, perpetually modifying the offside boundaries. Therefore, the referee must be aware of two spatial scales on the field and often simultaneously: the fight for the ball where a hand under a pile of players could be committing a foul; and the position of the teams with relation to the offside lines, which can cover the breadth of the pitch.

Despite all these difficulties, the most complex actions by far for rugby referees are those that take place around the in-goal areas. The line on the field that delimits this area is called the goal line and it is part of the in-goal area. As such, if a player places the ball on the line or in contact with the posts, a try will be awarded, whereas if the ball or a ball carrier touches the *touch-in-goal line* on the side, or the *dead-ball line* at the back, the ball is considered in touch and out of play, and it will be judged to be a dead ball. The attacking team will attempt to put the ball down or to fall on top of it inside the in-goal area and the defenders will try to stop them, put them into touch, or recover the ball outright. These actions can take place in tenths of a second and their fleeting nature undoubtedly complicates the decision of the referee as to whether or not they remain within the rules. Often, the speed of the players' arms needs to be considered in addition to their overall speed. In actions that take place around these boundaries, the refereeing team needs to keep track of both where the ball is and any contact of the rest of the body with the lines. An added difficulty arises when the attacking player is tackled and lunges out with their arm in an attempt to reach the try line with the ball. In such cases it may not be clear if any pressure was placed on the ball when it landed or if it was simply falling forwards at the time. It is also common for a pile of players carrying the ball to fall over the try line, the players' bodies blocking the view of the ball from every angle.

In summary, in a sequence of action, perception, decision-making, and official response, referees are being asked to make immediate rulings and decisions as if they had superhuman perception and decisive powers, which is evidently not feasible. While it's true that some perceptive or physiological skills can be trained and improved, as well as the technique involved in positioning themselves optimally, sportspeople's actions are always going to take place at extreme speed and in the briefest of moments. In this hectic setting, it is clearly an opportunity to review some characteristics of video technology that may serve as an aid to referees.

Audio-Visual Technology and Its Evolution in Relation to Sports Broadcasting

Since its origins, audio-visual media and the possibilities it offers in terms of communication for those that use this medium have been tightly associated with technological advances in the field. It has been a long time since baseball was first broadcast commercially in the USA, leaving audiences dissatisfied by the limitations of a television system that did not allow the game to be properly followed. In those days, it was not uncommon for even the cameramen to be unsure of the ball's trajectory. Even at the beginnings of the publicly available audio-visual narrative offered through television, the initial video technology relied on the trust and compliance of the viewer.

It is believed that the first football match broadcast live was England versus Scotland in April 1938, and thus "the tradition of live transmissions is virtually as old as the medium itself" (Raunsbjerg and Sand, 1998, p. 171). Due to the technological limitations in the early stages of its development, television was similarly ill-equipped to capture and represent fleeting actions. During the 1960s, advances like satellite communications, colour systems, video edition, slow motion, and replays were introduced (see, i.e.

Whannel, 1992, p. 86). Before the possibility of electronically recording images existed, permitting their later reproduction (originally known as video), the only option to replay images was based on photochemical technology that required a developing process and that could not therefore be applied to live sport.

In 1975, technological limitations in television affected the immediate narrative, perhaps due to the poor spatial resolution: "watching the television we hardly ever see the whole pitch, simply because if we did all that would be visible would be tiny dots running around" (Buscombe, 1975, p. 30), or due to poor optical quality and unstable images. The complaints extended from one extreme of the scale to the other, where due to the limitations in lenses, the close-ups used in the production of dramatic programmes could not be employed in televised football (ibid). Technological disruptions have since transformed the experience of watching sport. It is now possible obtain images and sound through television that even the spectators at the ground could not experience, achieved through techniques like changing the size of the image, concentrating time-diffuse events, manipulating time to dramatise action, focusing on isolated actions, or providing more statistical information (Birrell and Loy, 1979, p. 12).

From the moment in the 1990s when digital technology was first introduced, the technology has shown relentless progress. Not only can it overcome human limitations, but it can also set new limits to the viewers' sensory experience. In this sense the audio-visual narrative can now be considered an expression of current technology. In terms of the issues dealt with here, this technology is applied in a traditional narrative setting, that of a linear storyline, as part of a fairly classical production. However, it can also be composed of different realities: whether homologous, iconic, or specific to computer graphics, as part of different layers that can be arranged vertically – as opposed to horizontally or cross-sectionally – along the narrative line. These are images that having originated *naturally* or not, require technological systems in order to translate them into the limited scheme of human perception, and hence, the spectators at the ground do not have access to them without additional aids. Similarly, neither the players nor the referees can access these images directly, as evidently these images are not the direct result of live events but rather, they are the result of a visual construction (VC) designed to be transmitted on screen (see Chapter 3), always crafted re-creations of the events.

Key Aspects and Minor Criticisms of the Technology

Audio-visual technology is fundamental to sports broadcasting and our senses could not achieve perceptions like those that have been attained with the multi-perspective evolution of both sound and vision. Such technology allows the different viewpoints to be manipulated, visual effects can be applied that far exceed the possibilities of human perception, including increased resolution – both spatial and temporal – an exploration of movement, an analysis of the past, the use of freeze frame, compositions with superimposed graphics or enhanced reality, sequences produced with arrays of continuous images (manipulating time and space simultaneously):

> The movement of athletes is faster than the eye, our hearing can't discriminate clearly multiple concurrent stimuli, one angle of vision in not sufficient to gain a true perspective of events or to become aware of the whole scene, the position of the spectator is too distant, and as a result the figures and action can become indistinguishable… Thanks to television, audio, video and other techniques developed around this, audiences can become witness to a large number of events at the same time from a distance that they could never have achieved, and [these media] are capable of generating and store memorable images, which in the past were not accessible as they did not exist in nature or in our culture.
>
> (Benítez, 2013, p. 185)

The key technology in sporting coverage continues to be live broadcasting, cancelling out the spatiotemporal difference (Iozzia and Minerva, 1986, p. 61) between the production and the television audience's reception. The audience's sensation is of discovering the event and that of being a first-hand witness to the events. What is screened on television is an *impromptu* narration: in Umberto Eco's words, one that is produced simultaneously with its own conception (Eco, 1990, p. 232), although the importance that prior planning has in this sense is discussed in depth in Chapter 3.

The fleeting nature of events is a reason that itself justifies the use of technology to aid refereeing decisions. Moreover, in rugby the dimensions of the pitch, the number of players that may be involved in simultaneous actions on different parts of the pitch, and the speed of play are factors that complicate the referee's job further. Supernatural powers are required to review or relive the past. Sometimes a replay can reveal unexpected details or change one's opinion of a player whose skills have been overlooked, allowing them to be recognised, as in the case of Jerry Kramer, who claims on page 218 of his book *Instant Replay* (dealing with American football): "All I could think was: 'Thank God for instant replay'" (see Cummings, 1974, p. 432), as a crucial intervention by him that influenced the final result of a game would have gone unnoticed if it had not been reviewed in detail in the video footage. In principle, for one TV viewer:

> the phenomenon of the replay is related to both our penchant for stopping time, reliving great moments, holding them to an immortal tape present (a kind of motion picture album), and to our need for expert analysis, to see it again to make sure we saw what we saw.
>
> (Cummings, 1974, p. 431)

Replay technology is fundamental in sports broadcasting and it is an essential part of the television show (see Chapter 5). This technology has evolved considerably in recent years, improving the ability to build repeated sequences and, therefore, to construct the television narrative in sports broadcasts. It is commonly based on data storage systems currently achieved with hard drives but soon to be replaced by higher-performance, more reliable and energy-efficient solid state disks, with faster access, longer durations, or any other technology that follows them. These storage systems can record up to four simultaneous signals in conventional resolution and they lie in local or remote networks. The equipment is commonly fitted with two out ports – preview and program – which allows a take to be prepared while another is being broadcast. They offer speeds that can be pre-set and an ample capacity to label all actions and organise them in an infinity of distinct playlists. Plays can be subjected to different actions, such as select, mark, review, prepare, reproduce, or label while they are being recorded. Furthermore, images can be reproduced at different speeds, both accelerated – reproducing the recorded fragment at higher speed – or in slow motion. However, when slowing down the images, some factors may be distorted, such as the trajectory, travelling speed, and the intention of the actions, as will be discussed below.

In the *mise en scène* (MES) (see Chapter 3), events occur at a rapid pace, with players moving extremely fast. On the field, everything is fleeting and ephemeral, yet the "replay reminds us that events, although inevitably sequential, are also invariably multidimensional" (Morris, 1987, p. 156). Each time a slow-motion replay is shown on the screen, it is a reminder that viewers in front of their screens have a privileged advantage (Morris, 1987, p. 151). An instant replay fulfils several narrative roles: it highlights a particular moment in the game; it helps clarify and present the events; it exalts the technical and aesthetic value of a sporting skill; it can potentially show the action from a different angle, or show an event during the game that was not shown live, accentuating the drama of the moment (Iozzia and Minerva, 1986, p. 79). Its primary role is to expand temporal perception, multiplying and prolonging those "fleeting moments" (ibid), in other words identifying the moments of ecstasy and prolonging them (Raunsbjerg and Sand, 1998, p. 170).

Spatial and Temporal Resolution

As a starting point to understand video technology, the concepts of resolution in space and time need to be defined, and we must differentiate between the moments of "capture" (through the camera), "recording" (via the video system), and "emission" or "broadcasting" (through television). In the sequence that runs from the capture of images to their broadcasting, resolution is lost due to the need to compress the images to facilitate their handling. Spatial resolution refers to the number of points that make up an image and the current standard used is 1920x1080 pixels for each image broadcast by high definition (HD) television to people's homes. If we consider a more or less conventional wide shot, this resolution means that the upright figure of a tennis player (1.80 m) closest to the camera who is waiting to return a serve will be approximately 8 cm in height on a 32-inch screen (with the slight flattening effect that is characteristic of TV images), while the image of the ball at the moment it hits the service box will measure approximately 0.4 cm. In this same example, the total length of the court from the perspective of the main camera, which displays the rectangle as a trapezoid, would be 19.3 cm (the height of the court on the screen), while the baseline closest to the camera would be 0.3 cm. Alternatively, if the cameras were covering the game using 4K ultra high definition TV (UHDTV1) technology, the spatial resolution would be 3840x2160 pixels. If these images were converted by the HB and transmitted in HD, the ball and the player will measure the same on a 32-inch television if the aforementioned compression were used, with the advantage that HB systems can amplify the image up to four-fold without any visible defects, retaining a viewing quality equivalent to HD. This means the ball can be enlarged to 1.6 cm and framing effects can be used to analyse and show re-framing movements in more detail across the entire original image.

On the other hand, temporal resolution is defined by the number of images captured each second, which is not to be mistaken with the shutter speed that determines the length of exposure of each image. Television transmission systems work on 25 or 30 images per second depending on whether they work on the PAL and SECAM standard, or the NTSC specification, actually analysing each photogram in two fields using an interlaced technique to smooth the sensation of movement (thus, 25 images are equivalent to 50 frames per second in PAL). However, to be certain of capturing the exact moment that a tennis ball hits the ground, both its size and speed must be taken into account. At a speed of 50 metres per second and a camera covering a metre around a service box in line with the axis of the camera, the ball may or may not be captured at any point in its trajectory (from one edge to the other of the camera frame). Thus, a higher frequency to record the images must be used. To obtain the frequency at which it would be almost certain to capture the ball when it hits the ground, the time it remains on the ground can be calculated as twice the ball's diameter (taking into account its deformation) divided by its speed. Thus, the minimum safe frequency would be the inverse of this value, approximately 357, which indicates the number of images per second needed to be captured in order to register the impact of the ball on the ground. The cameras used to cover these rapid movements with any kind of analytical capacity would therefore have to be capable of recording around 350 frames per second. Nevertheless, it's important to understand that any accurate representation of a continuous movement from discrete images requires some interpolation and, hence, some type of estimation, which is in fact done differently by the PAL and NTSC standards.

Rugby players do not move as fast as tennis balls, yet the movement of their extremities and the ball, as well as the cameras themselves when panning an action must be taken into consideration. The summation of all these velocities increases the demands on the systems that capture and emit the images. Shutter speeds can be increased to avoid blurry contours, but this may often compromise the images captured when they involve contact between the ball and a line or hand. When considering the speed at which the players and the ball are travelling, the fastest actions likely to be assessed when broadcasting a rugby game, the relative movement within the frame must also be taken into account. This factor depends on the distance to the object, the scale of the frame (and of the focal distance), the pan, tilt, and zoom movements, and their speed, as well as the previously mentioned shutter speed. In other words, VC itself could add elements that are detrimental to accurate image analysis and as a result the procedures followed should aim to increase the temporal resolution of the system, as described for tennis. If a capture and recording system is capable of processing 350 images per second (following the example), any contact on the field will be guaranteed to be recorded. It is true that recording 350 fps requires the same minimum shutter speed (1/350th of a second), which considerably limits the amount of light reaching the sensor. According to the Bunsen-Roscoe Law of Reciprocity, to obtain sufficient light, the light lost by using a faster exposure must be compensated for by opening the diaphragm and using a faster lens, assuming the pitch is well lit. As a consequence, the bigger the aperture of the diaphragm, the smaller the depth of field, meaning that a shorter distance is in focus along the camera axis, which can give rise to blurry, diffuse, or ill-defined contours.

Nowadays, it is possible to increase the spatial and temporal resolution of video systems to much higher rates than is necessary, and at an affordable price, with sufficient technological guarantees. This evolution has some downsides, as the data bandwidth becomes proportionally much bigger. Relying on the possibility of amplifying images produces less tension when planning the different angles to use. For example, it might be considered that two cameras would be sufficient to replace the goal-line technology system (GLT) (see Chapter 10), given the risk that players could block the vision of the ball, yet this is not actually the case. However, reproducing high-speed recordings will take proportionally longer in terms of visualisation and emission time, particularly bearing in mind that at 350 images per second, actions are recorded with 14 times more images than the standard PAL used by most of the western European television systems (25 images per second). Thus, a scene of action that lasts 2 seconds on the field would last 28 seconds on the screen if captured at 350 photograms per second and reproduced in PAL timing. This is clearly excessive for a live broadcast and will consequently compromise refereeing decisions.

Other Visual Improvements

Another major breakthrough has been the improvement in image quality due to the advances in camera sensor technologies and the enhanced dynamic ranges (high dynamic range, HDR) of both the camera and the broadcasting system. In a wide-angle shot that shows the pitch with areas in the sun and the shade, it is imperative to decide what exposure to use. In normal conditions, exposure is adjusted by taking the point of interest as a reference, e.g., the place where the ball is situated when in play. For any camera location used in a broadcast, the point of interest should be determined by performing at least three operations: framing the shot, which in turn involves defining what actions are going to be tracked, and thus what will remain in frame and what will be out of frame according to the instructions received (see Chapter 3); adjusting the focus; and adjusting the exposure. The faces of the players closest to the point of interest – for example the rugby ball, in this case – should be correctly exposed. If the ball is located in full sunlight, the differences in exposure with the shady areas may be, depending on the time of the year, latitude and time of day, around five f-stops. This means that if the ball suddenly goes from a sunny area into the shade (after a pass or a run in the case of rugby) and there is insufficient time to correct the exposure, the players closest to the camera will appear as mere silhouettes and may be incorrectly recorded, with no possibility of distinguishing details. This undesired effect is as important on the screen as on the monitors of the camera or video operators, because it could compromise the effective tracking of the game.

The advances in dynamic range are already facilitating, and will evidently improve, both the extraction of details in areas that are not correctly exposed, as well as permitting ex post facto correction for further analysis. Obviously, in well-lit stadiums or in those using artificial lighting to illuminate the areas where the sunlight does not reach (reducing the contrast), any unfavourable effects can be mitigated by shifting the aperture one or two f-stops, a process that can also be achieved using neutral filters in front of the camera's sensor. In any case, the emergence of HDR devices, signals, and monitors allow images with very high levels of contrast to be recorded, broadcast, and watched correctly, hitherto unthinkable in television. That is to say that the details of an image can now be captured and distinguished even when there is a large difference in brightness across the frame.

Another relatively recent development is the use of synchronised camera arrays to obtain sequences for analysis and spectacular replays. These consist of mixed systems for shooting and manipulation, whereby a number of cameras follow the same elements of the action and they are later linked (in the near future almost certainly with minimal delay) to create a single sequence. In these sequences it is possible to follow a continuous *displacement* from one point of view to another, interpolating transitional images and using the combined time of all the shots from the same action as a unit to advance, freeze, or rewind, searching for the best angle and the most appropriate moment for the desired narrative effect. A precursor to this type of system was the so-called *Eye Vision* premiered at the turn of this century in the XXXV Super Bowl broadcast by CBS.

The Narrative, the TV Viewer's Confidence and Justice

The video technology available must also be taken into consideration when constructing the desired narrative. Obviously, any composition based on images of different origins, or subject to any explicit or suggested manipulation, must retain the viewer's confidence in order to be considered valid. In the first explorations of what has been called cinematographic language (this is not perhaps the place to discuss the validity of the term and its content), the zero-degree cinema style paradigm (see Chapter 3) was regarded as that implying that the same action should be narrated in a single shot in order to be completely believable. An editing cut could be considered as an opportunity to manipulate the material inappropriately. Viewers will more readily accept such a cut when the game is live or ongoing, probably in part due to the conviction that the MES and the VC (see Chapter 3) coincide, but doubts may appear when dealing with any analysis or replays. In these latter situations the viewer must be prepared to accept the HB's proposal as valid. When juries have studied the validity of video as evidence in a lawsuit, they warn that some caution should be exercised. In many courts of law, certain factors are taken into account when deciding the probative value of moving images, including the possible distortion that a given angle of projection can produce, as well as the use of certain focal distances, and the decision regarding what is left in and out of the frame, in addition to the consideration that things may be exaggerated by slow-motion replays. Meanwhile, in sports tribunals there is hardly any doubt about their use (Dickson, 2013, p. 20).

A few words should be dedicated here to the different meanings attributed to the term "justice" based on who or where it is contemplated. For example, in ideal terms we can consider the Hellenic view, "justice is the virtue of allocating goods in accordance with desert" (Bordner, 2015, p. 10). For the sports' viewer this could refer to consistency in the decisions, which would be expected to be the same in similar circumstances, yet involving different players on different occasions (Collins and Evans, 2008, p. 299). Other researchers argue that to approach justice from the point of view of video assistant referee systems, *de novo* standard (see Chapter 12) should be introduced, as if who was reviewing the action had not seen it before (Berman, cited in Bordner, 2015, p. 10).

Other systems that, in principle, the fans should take on board unconditionally are those that use some kind of virtual or augmented reality.

These began to be used some decades ago and new versions are emerging as technology evolves. For instance, we can take our mind back to *Telebeam* in the late 1980s and early 1990s (i.e., Alabarces and Duek, 2010). Towards the second half of the 1990s, the *Virtual Replay* system also allowed moments to be reconstructed in a 3D rendering from the real images of different shots taken from the game. The combination of images from real-life action and infographics has gradually become more accepted, offering virtual points of view or those corresponding to positions in which cameras have not been placed, reconstructing the perspective by showing informative or analytic sequences, also adding information that offers many possibilities in a very flexible manner. These are continuations of the proposals initiated at the end of the last century through the Dartfish systems – SimulCam, StroMotion. Another example would be the *Piero* system from Red Bee – developed in conjunction with the BBC – that was used in the Premier League and in the Spanish league, as well as at tournaments in other parts of the world. The combination of big data and graphics engines to create different layers from, or using real images, highlights the various possibilities, and this has great promise for the future.

In relation to refereeing, the most well-known of these combined video and infographics processing devices is clearly the Hawk-Eye (HE) technology. In this system, data is collected from an array of synchronised high-speed cameras. These have been reported to have a capture rate of 2000 fps (Baodong, 2014, p. 400), although the lack of evidence for this sheds some doubt on this claim. Indeed, and as mentioned above, the use of such a recording frequency would probably generate some problems in terms of exposure and depth of field.

In the case of tennis, the system is previously calibrated so that it is integrated into a virtual scenario with the position of the court's surface and each of its lines. According to its authors, the system is based on visual tracking from two or more cameras of those available in the array to triangulate the position of the tennis ball at all times. Some authors have called this type of apparatus a "reconstructed track device" (Collins and Evans, 2011), as the data obtained are entered into a computer system that can interpolate the trajectory between two recorded positions. This system is always used to establish if the ball bounced inside or outside the area of interest at any time. Once the system has calculated the trajectory of the ball and the intersection with the plane of the court, based on the video data captured at the established frequency, the impact of the ball is shown in an animated image provided in 3D that includes the footprint of the ball and the nearest lines, as well as a verdict: IN or OUT (as indicated previously). The same or similar principle applies to sports like cricket or football, in the latter case being used to make decisions about whether or not the ball has fully crossed the goal line.

One of the main criticisms of this system is that the reconstruction of what is likely to have occurred is projected as though it were actually what happened, with no question of doubt (Collins and Evans, 2008). Indeed, there is no allusion made to this fact in the 3D images provided, which are obviously created through interpolations, estimates, and interpretations from the data collected from the cameras, and there is no mention of the acceptable and proven average error. A well-known discrepancy that occurred through the use of this system came in the 2007 Wimbledon final between Rafael Nadal and Roger Federer. In this case, the former challenged a call to check the validity of a shot and the system reversed the decision by one millimetre, a time at which the acceptable error publicly allowed was much higher, specifically 3.6 mm (Collins and Evans, 2008, p. 294; Dickson, 2013, pp. 19–20). This type of occurrence raises questions regarding the tendency of many courts of law and sports tribunals to accept as accurate the evidence coming from more or less automated devices: "the large majority of evidence produced technologically (i.e.: without human 'interference') is uncritically accepted as accurate" (Dickson, 2013, p. 19).

The Use of These Systems

The Stakeholders and the Implementation of a Video Refereeing Aid in Rugby

In an environment where sport and business go hand in hand, any technological system that aids refereeing must pursue different values: effectiveness, that is, helping referees make better and more valid decisions, beyond any doubt, reducing the need for interpretation to a minimum, and without affecting the normal flow or spirit of the game; efficiency, the implementation of an economically and temporally viable system; precision, reliability, and unobtrusiveness, such that its use does not disrupt the show or favour the appearance of unsportsmanlike behaviour or strategies but rather adds an element of intrigue and emotion. Faced with these goals, the opinions that in one way or the other should be taken into account are those of the individuals involved in the practice of the sport (athletes, referees, technicians), the organisers and those responsible for running the competitions (federations, associations, leagues, tournaments, sponsors, owners, broadcasters), those who are the beneficiaries of the show (supporters, fans, attendants, TV viewers), and, naturally, the media.

It is crucial that players accept that the assistance offered by such systems is valid. In this sense, rugby is a not complicated sport, as within its rules, it includes a document named the Playing Charter in which it is stated: "The charter covers the basic principles of rugby as they relate to playing and coaching, and to the creation and application of the laws" (World Rugby, 2018, p. 4), and furthermore:

> Rugby owes much of its appeal to the fact that it is played both to the letter and within the spirit of the laws. The responsibility for ensuring that this happens lies not with one individual – it involves coaches, captains, players and referees.
>
> (World Rugby, 2018, p. 5)

The values encompassed by these guidelines include integrity, passion, solidarity, discipline, and respect. It is clear that through this commitment, the protagonists in rugby were not going to object to the adoption of technology that could help referees.

Through simple observation, it could be argued that the increase in the passion generated by this sport has potentially endangered some of the characteristics of fair play, for example compromising the notion of respect. In the most important rugby encounters during the 1970s and 1980s, when large masses invaded the fields after the big test matches, the players who scored tries hardly celebrated these with their teammates as a matter of deference to their opponents. By contrast, in this century the players huddle together to congratulate themselves, almost like in football, while controversial discussions about the matches in the media have grown continuously. Nevertheless, the players and coaches are still aware of all the difficulties faced by the referees, even in the light of the technology currently available.

It is assumed that a system has not yet been designed that is capable of determining beyond all doubt what happens to the ball when a pile of players collapses with the ball in possession. Video is still incapable of seeing through a player's body and there are as yet no systems approved based on the use of localisation through radio beacons or thermography. Similarly, in such a melee it would be difficult to determine whether the line has been touched or not when the ball was being placed on the ground if it forms in a corner of the field. In the world of rugby, video aids are not yet considered suitably precise in these situations, as video does not always provide "indisputable visual evidence" like it may do in other sports (Oldfather and Fernholz, 2009, p. 52), raising questions about its efficiency.

There are useful examples that raised doubts about the contribution of the TMO when it was introduced, before its use was standardised, coming from some of the most important matches at the 2007 Rugby World Cup (RWC). In one of the semi-finals of these championships, in which Argentina faced South Africa, the Argentinian player Manuel Contepomi dived to ground the ball on the South African goal line in the 47th minute. The Springbok players Montgomery and Steyn challenged him to prevent the try, and there were doubts as to whether the ball had dropped out of the Argentinian's hands before he had grounded it. The referee, New Zealander Steve Walsh, called for assistance from the video. Revising the two takes selected and passed by the television crew from the frontal camera, it clearly looked like a try, but from the side camera, it looked like a clear *knock-on*. The TV viewers watching the broadcast were stunned, as depending on the replay offered it did not look like the same play. It was astonishing.

In another example from the same championship, the English winger Mark Cueto dived to ground the ball in the South African in-goal area, close to the left corner, after an amazing run by Mathew Tait in the 42nd minute of the final in Saint Denis (October 2007), a match that finished without tries. The various video takes did not fully clarify the action, but the Irish referee, Allain Roland, revoked the try after consulting the TMO, Australian Stuart Dickenson. The refereeing team felt that Dannie Rossouw's tackle had forced Cueto's foot to touch the line a moment before he grounded the ball, resulting in the try being disallowed. With the replays that were offered at that time, even today it would be difficult to establish the validity of the referee's decision.

With respect to the accuracy of the TMO, the HE Smart Replay system was used in the 2015 RWC, in order to take advantage of synchronising the replays and the capacity to zoom over them, as well as its possible use to "protect players from themselves" (former player David Rees, cited in Wilson, 2015) in the case of injuries. These features of the system are tried and trusted, but the technology to carry out these tasks has for a long time existed in the outside broadcasting vans (OBV) of the HB: the synchronisation of different takes in replays and zooming. It should also be borne in mind that the main strategy regarding the coverage by the cameras and the planning of this is always established by the HB. This means that as long as the raw material used in these systems or refereeing aids comes from the images produced by HB, the HB team will always be in a better position to make decisions based on them (see Chapter 3) than the TMO.

Interpreting the rules of any sport is always difficult, and applying them without making errors is not easy, even with the help of the TMO. One such example comes from the England vs. Wales game in the 2018 Six Nations Championship. The score was 12–6 to the English, and in the 23rd minute, the Welsh player Gareth Anscombe was faster and more skilful than the Englishman Anthony Watson in chasing a ball into the in-goal zone to ground it. However, after consulting the TMO it was decided that Watson had touched the ball first. In a case that received great media attention, Allain Rolland, who had refereed the final of the last RWC, asked the Welsh technical team to forgive him in the name of World Rugby when it became evident that Clint Newman, the TMO during the match, had made a mistake.

Another aspect of the game that is difficult to interpret, even with the aid of the TMO, is when there are doubts about a forward pass. According to point 2.6 of the protocol of application: "For forward passes, the match officials must adjudicate on the direction of the ball leaving the hands" (World Rugby, 2019, p. 3). It is very difficult to determine with certainty the relative positions of the hands and the ball, as it is a fleeting movement that is often hidden by other elements, for example when the ball carries an inertia that, in absolute terms, means that it ends closer to the opponent's try line when it is received than when the pass is made, especially when considering a long pass at great speed. One example of how difficult it is to judge this kind of pass took place in the Stade de France on 24 November 2018 in the 30th minute of the France vs. Fiji match. In a brilliant set piece from a scrum, and after a fast run by Fiji's number 13, Semi Radrara, he passed the ball to the winger Josua Tuisova (who incidentally played for the French team Toulon that season), who scored a try. While on the pitch, and based on other factors like positioning, orientation, the position of the rest of the players, etc., the impression was that there had been a forward pass; the aerial shot clearly showed that while the ball was passed 2 metres forwards, when it left the player's hands it was clearly going backwards.

The largest number of complaints regarding the TMO centre on the time it takes to apply. Although this is an issue that is constantly being reviewed and debated, the normal time of a match remains the same: 80 minutes, 40 minutes for each half. It has been noted that the time of play is less than this and, on occasions, it has been suggested that the use of the TMO is one of the factors responsible for this. For example, when the evolution of the Super Rugby matches between 2008 and 2013 were studied, it was seen that sometimes the TMO took 3 minutes to reach a conclusion, leading some teams to lose their concentration (Kraak et al., 2016), with the ensuing negative effects both on the sport as a spectacle and with the possibility of an increase in injuries. Nevertheless, the time taken by the players to get up from a collapsed melee and resume play, particularly when close to the in-goal line of one of the teams, may be more or less the same as that taken by an experienced television team in reviewing the best 10 or 12 takes available for a move in order to select the most adequate fragments that will help reach a decision and have it ready for broadcasting: between 25 and 40 seconds.

Nevertheless, this same objection of the excessive length of TMO interventions has also been made at more important competitions. For example, before and during the first part of the 2015 RWC, the media suggested that the longer length of the matches – total time but not the playing time – was due to the use of TMO and that it might cause the public to get bored and look for alternative entertainment. Immediately, Rugby Union (RU) counterattacked through action, on the one hand working with its referees to try to make their consultations faster, and also through a media campaign establishing that only 28 per cent of the time that play was stopped during the opening match of the championship was due to the TMO (Rugby World, 2015). In this same communication, the RU reiterated the principle goals of the TMO:

> The objective of the television match official (TMO) system is to ensure accurate and consistent decisions are made on the field in a timely and efficient manner. The TMO is a tool to help referees and assistant referees with their on-field calls and the referee remains the decision-maker who is in charge of the process.
>
> (Rugby World, 2015)

The problem not only resides in the time that the play is stopped while taking a decision, but also in the discontinuity in the flow of the game. There is also a positive aspect to the interruptions as the players can take a rest and recover to then continue with the match. However, rugby mainly looks to avoid disturbing the balance that exists between continuity and the battle for possession. If this is necessary, it is an objective that can even be achieved by adapting the rules, which in this respect are very clear:

> as one team attempts to maintain continuity of possession, the opposing team strives to contest for possession. This provides the essential balance between continuity of play and continuity of possession. This balance of contestability and continuity applies to both set piece and open play.
>
> (World Rugby, 2018, p. 9)

This is one of the reasons why rugby maintains the possibility of referral to the TMO in test phase, despite the time that has passed since its first implementation, such that it can study how to best use such aids without altering the essence of the game. This is reflected in Law 6.15 Television Match Official, Global Law Trial, together with the refinements to its application over the years:

> A match organizer may appoint a television match official (TMO) who uses technological devices to clarify situations relating to:
>
> - The grounding of the ball in in-goal.
> - Touch-line or touch-in-goal during the act of grounding the ball or the ball being made dead.
> - Where there is doubt about whether a kick at goal has been successful.
> - Where match officials believe an infringement may have occurred in the playing area leading to a try or preventing a try.
> - Foul play, including sanctions".
>
> (World Rugby, 2019, p. 41)

It is important to highlight two issues related to this way of working: in principle the specific norms regarding the TMO do not exclude any type of technological aid, such that it does not have to involve video or only video, and that the match organiser (who hire the HB – see Chapter 3) provides the TMO the tools to carry out the analysis, even with the participation of third parties and, in principle, revoking its independence in decision-making. Proprietary systems or other kinds of technology can be used, depending on those available. Without doubt, two fundamental reasons for this choice are fair play, as already mentioned, which is taken for granted in the world of rugby even by the HB, and the economic demands, even on national leagues with more limited budgets, although in certain places and on some occasions, such aids are of little use as there are few games covered by TV or there are limited resources dedicated to such broadcasts.

The TMO protocol specifies that they can rely on the collaboration of the HB's team: "the TMO will then liaise with the TV director and look at any specific footage in order to support the referee's specific question" (World Rugby, 2019, p. 2). This includes indications for the communication between TMO and the referee, while the director must allow time to offer an angle through which everybody can see the referee making a gesture to indicate his final decision.

It is therefore clear that rugby considers the TMO as an aid, yet for some reason, perhaps due to the time in which it has been applied, it has yet to be fully accepted or integrated into the spectacle (in the MES), neither by the spectators at the event nor by the followers at home. Concerned about the acceptance of all the stakeholders, World Rugby prepared a poll to assess the opinions regarding the application of the TMO at the 2015 WRC, creating a Law Revision Group to act on the results. The questions released at the beginning of 2016 asked the opinion of players, referees, the TMO, journalists, and spectators, identifying to which of these collectives they belonged. The advantages and disadvantages of the system were addressed, as well as the issues that the referees should consider, and an overall opinion was given. Prior to completing the questionnaire, some data was offered with respect to the activity of TMO during the tournament. There had been 132 consultations in total (77 on tries and 55 on foul play), only slightly more than in the regular competitions that season. The match that required most consultations was Argentina vs. Namibia with seven, and there were six matches in which no consultations were required. The average time for each consultation was 70 seconds, although in the qualifying round 13 seconds more were required than in the latter games.

From the results obtained, alternatives are being considered, such as how to incorporate the Decision Review System from cricket, where a field player asks the umpire to review a decision. Each team has three opportunities for review and it is retained if their review is upheld, as occurs in tennis. This procedure helps to eliminate the existing trend in rugby, where all the referee's decisions are questioned.

The Stakeholders and Implementation of the Review Systems in Tennis

Unlike rugby, the HE system was not well received by tennis players, some of whom were used to frequent and long discussions with the chair judges on the validity of a shot. As technology improved, the television coverage was often able to provide relevant evidence of some of those polemic shots, while the judges were limited to their own perception and memory. The HE system was first applied in an experimental form and its results could be seen on television. As a result, the debate regarding its incorporation snowballed.

Despite having tried different systems to aid the referee, like the Cyclops (using infrared rays) or the Auto-Ref, until 2006 the International Tennis Federation (ITF) did not decide to officially adopt any system until the use of the Electronic Line Calling (ELC) system was approved, which employed HE. From that time on, the verdicts of the referrals became indisputable and the players generally had to accept them, albeit with some exceptions, such as the notable moment in the aforementioned Wimbledon final in 2007. Due to the nature of its use, this system is more directed towards decision-making rather than as a decision aid (Collins and Evans, 2008, p. 297), and as such it may be decisive, as in the women's final at Wimbledon 2017. In the midst of the last point, Garbiñe Muguruza considered that Venus Williams' shot had passed the baseline and she stopped playing, asking for a challenge. HE confirmed the ball was out and Muguruza was awarded the point, the game, the set, the match, and the championship.

At the same time, it should be noted that from the onset of its implementation, the HB's replays that might contradict the system's decision were not shown to TV viewers as an alternative (Collins and Evans, 2008, p. 298). The tournaments applied this technological aid according to their own criteria until the ITF normalised its use a couple of years later, and today three unsuccessful challenges are established per player and set (initially there were only two), plus one more if the set enters a tie break. There is no limit to the number of successful challenges (ITF, 2019, p. 30).

One consideration in favour of the use of this type of system is the very fascination that technology creates in people. The mere fact of seeing these sportsmen and women close up on a screen thanks to the powerful telephoto lenses and to contemplate all their movements in slow motion and at high speed produces "a scientific aura. Slow motion replays are treated as part of a hermeneutic process or scientific discovery, which, among other things, allows the viewer to out guess the referee and see what 'really' happened" (Morse, 1983, p. 49). Certainly, although this may not actually be the case, the public is prone to believe that technology has the capacity of eliminating the element of error from refereeing decisions (Nlandu, 2012, cited in Kolbinger and Lames, 2017).

Video technology itself establishes some limits that mean it cannot be considered as infallible in tennis. There are several notable reasons for this, the first of which is related to photographic difficulties, as already discussed: the fact that it is impossible for the whole system to be in focus and on an adequate scale, that of establishing a ratio of frames per second that is not free from the problems related to exposure and the resolution of extreme contrasts. In second place, there are optical effects that are related to the analysis of objectives not within standard ranges: telephoto lenses flatten the objects that are separated along the axis of the camera, and the use of wide-angle lenses exaggerates the perspective, sometimes producing the effect of acceleration when crossing the frame. Besides the intrinsic errors, which will become more complicated to manage as the advances in technology parallel those in television broadcasting by the HB, which will come closer to having the same capacities as the ELC and, consequently, could contradict some of its decisions. There are also other considerations to be taken into account: the interpolation already mentioned, due to the discreet nature of the system, and therefore, the interpretation of reality that this implies, given that a continuous trajectory must be constructed from a finite number of frames.

Alternatively, some detractors of this technology present other arguments. For example, that it could motivate less sportsmanlike behaviours in players, speculating with their challenges to have a rest or to break the rhythm of their opponent. In this respect there are three different possibilities after the petition for a challenge: the player is somehow manipulating the game; the line judges have made a mistake; or there is some perceptive uncertainty (Mather, 2008, p. 1645). It must also be considered that the referees might get used to the system and they might stop taking decisions, or they might simply avoid taking difficult decisions, rather than face the possible shame of the ELC contradicting them. It is noteworthy that some research has already focused on the best strategies to use challenges, the most efficient possibilities for both players and stakeholders (Kovalchik et al., 2017; Nadimpalli and Hasenbein, 2013).

Other objections are related to the corporative dependency. It is a proprietary system that can reduce the official's autonomy in decision-making to some degree. With respect to tennis, it is being applied in a sporting circus (see Benítez, 2013, pp. 41–2) or circuit, which means that there are never more than two or three tournaments being played at the same time, even though it might be necessary to use the system in several courts at the same time. However, it would be much more difficult to reach an agreement to make a system of these characteristics official in all competitions that depend on the WR, as it would mean having to implement the system at hundreds of pitches at the same time. This is not a problem at specific tournaments like the RWC. Other arguments against the use of such systems are related to the cost of purchasing or contracting the service, which is rejected by its supporters, who indicate that it is a cost that is assumed by the sponsors and that it is a very good platform to sell publicity. Likewise, it is argued that these systems to aid the referee imply dedicating a certain amount of the transmission time to the decisions rather than the spectacle of the sport. This, it is claimed, might generate more disinterest in the TV viewers, who might become bored and search for some other type of entertainment.

In favour of these systems, many of those interested in their use believe that greater fairness in the game is always preferable, and in their opinion, the companies concerned will make efforts to generate value for their systems. For example, starting with the same technology on which HE is based (which was bought in 2011 by a multinational in the sector of electronics for consumers and broadcasting), the SMART (Synchronised Multi-Angle Replay Technology) has been launched, which works at high speed (340 fps) and with six cameras controlled by a single operator. This system is said to not only reduce the cost for smaller-scale sport broadcasts, but also to help in refereeing, the diagnosis of injuries, or in coaching. An application has been proposed to involve the public in the decisions of the ELC, SMART Vote, which invites individuals to bet on the result of any challenge (or on any other issue that the fans may be interested in). Contests, prizes, or other marketing initiatives can be established around these decisions as incentives for the public, who, rather than getting upset about the delays the reviews imply, will await those moments anticipating their participation. Of course, the TV viewer can also participate in these activities through social media or SMART TVs (or internet-connected TVs, to avoid being confused by the naming of the application). As such, one aspect of the negative arguments can be turned into a positive aspect of the sporting experience and of the decision-making process.

Logically, the pauses when taking decisions shouldn't just oblige the participants and spectators to simply await the decision, which could engender other negative consequences, like a loss of concentration and patience by the players, as well as the possibility of injuries, and general tedium. From the point of view of the players, coaches, spectators, and TV viewers, an extraordinary effort should be made to somehow integrate this break in the action into the show, by extending the narrative. Likewise, it does not appear to be an adequate strategy to ask the HB to wait until a decision is taken to continue the play without emitting replays, for example. In sports where the broadcast of the referee's call is not so fully defined as it is in tennis, the protocol should include an explanation, an audio-visual one if possible.

Conclusions

Top-class sporting performance reaches the limits in terms of execution and in terms of perception, and referees must make immediate decisions on such actions. Their fleeting nature demands superhuman perceptive qualities and efforts from the team of referees, and while video technology can help resolve difficult situations, it is still not sufficiently conclusive, despite the great advances made. These advances are mainly related to spatial and temporal resolution, but also to the high dynamic range, both in capture and broadcasting.

Analysing the experience gained from the application of the refereeing aids based on video in rugby and tennis offers interesting conclusions in terms of efficiency, effectiveness, viability, or modes of application, among other aspects. Despite the false belief that these technologies are infallible, they still have many limitations. Indeed, there are defects that remain unresolved, some due to intrinsic problems and others related to the way they are used. Some facets are not properly used, such as slow motion, pin-hole geometry, the effect of zooms, depth of field, or dependence on the Law of Reciprocity, or they are attributed an accuracy that they are unable to achieve to provide just and indisputable decisions.

In fact, video participates in two types of refereeing aid: those based on the illusion of discovery – related to live broadcasting – and those based on the illusion of reconstruction or on technological principles that still need to be properly evaluated. Regarding the time required for referral and the wait for a decision, there is still some debate as to whether the system should be automatically applied or upon query, and regarding its integration into the show. In some cases, some imagination has been applied to its use, promoting public participation with the approval of stakeholders.

Acknowledgements

The authors wish to thank Miguel Moscoso, full professor at the Mathematics Department, Carlos III University of Madrid (Spain) for revising the calculations and for interesting suggestions.

References

Alabarces, P., & Duek, C. (2010). Fútbol (argentino) por TV: entre el espectáculo de masas, el monopolio y el estado. *Logos 33*, *17*(2), 16–28.

Baldwin, C. F. (2014). Video killed the sports referee: The role of video replay technology in creating anxiety and stress amongst rugby league referees. *International Journal of Arts & Sciences*, 7(6), 305–312.

Baodong, Y. (2014). Hawkeye technology using tennis match. *Computer Modelling and New Technologies*, *18*(12C), 400–402.

Benítez, A. J. (2013). *Realización de deportes en televisión*. Madrid: Instituto RTVE.

Birrell, S., & Loy, J. W. (1979). Media sport: Hot and cool. *International Review of Sport Sociology*, *14*(1), 5–19.

Bordner, S. S. (2015). Call 'Em as they are: What's Wrong with Blown Calls and What to do about them. *Journal of the Philosophy of Sport*, *42*(1), 101–120.

Buscombe, E. (1975). Cultural and televisual codes in two title sequences. In Buscombe, E. (Ed.), *Football on television*, London: British Film Institute Monographs, pp. 16–34.

Castagna, C., Abt, G., D'Ottavio, S., & Weston, M. (2005). Age-related effects on fitness performance in elite-level soccer referees. *Journal of Strength and Conditioning Research, 19*(4), 785–790.

Catteeuw, P., Helsen, W., Gilis, B., Van Roie, E., & Wagemans, J. (2009). Visual scan patterns and decision-making skills of expert assistant referees in offside situations. *Journal of Sport and Exercise Psychology, 31*(6), 786–797.

Collins, H., & Evans, R. (2008). You cannot be serious! Public understanding of technology with special reference to "Hawk-Eye". *Public Understanding of Science, 17*(3), 283–308.

Collins, H., & Evans, R. (2011). Sport-decision aids and the "CSI-effect": Why cricket uses Hawk-Eye well and tennis uses it badly. *Public Understanding of Science, 21*(8), 904–921.

Cummings, R. (1974). Double play and replay: Living out there in television land. *Journal of Popular Culture, 8*(2), 427–436.

Dickson, C. (2013). Cela n'est pas une passe en avant (That is not a forward pass). doi:10.2139/ssrn.2577097

Eco, U. (1990). *Obra Abierta* (3rd ed., first in 1962). Barcelona, España: Ariel.

Guillén, F., & Feltz, D. L. (2011). A conceptual model of referee efficacy. *Frontiers in Psychology, 2*, 1–5.

Iozzia, G., & Minerva, L. (1986). *Sport e Televisione, un matrimonio di interesse.* Torino. Italia: Edizione RAI.

ITF. (2018). *ITF Approved tennis balls, classified surfaces & recognised courts 2018.* Retrieved from www.itftennis.com/media/278130/278130.pdf

ITF. (2019). Tennis rules. *International Tennis Federation.* Retrieved from www.itftennis.com/media/298557/298557.pdf

Kolbinger, O., & Lames, M. (2017). Scientific approaches to technological officiating aids in game sports. *Current Issues in Sport Science (CISS)*, 2(1), 1–10.

Kovalchik, S. A., Sackmann, J., & Reid, M. (2017). Player, official or machine? Uses of the challenge system in professional tennis. *International Journal of Performance Analysis in Sport, 17*(6), 961–969.

Kraak, W., Venter, R., & Coetzee, F. (2016). Scoring and general match profile of Super Rugby between 2008 and 2013. *International Journal of Performance Analysis in Sport, 16*(2), 786–805.

MacMahon, C., Helsen, W. F., Starkes, J. L., & Weston, M. (2007). Decision-making skills and deliberate practice in elite association football referees. *Journal of Sports Sciences, 25*(1), 65–78.

Mallo, J., Frutos, P. G., Juárez, D., & Navarro, E. (2012). Effect of positioning on the accuracy of decision making of association football top-class referees and assistant referees during competitive matches. *Journal of Sports Sciences, 30*(13), 1437–1445.

Mather, G. (2008). Perceptual uncertainty and line-call challenges in professional tennis. *Proceedings of the Royal Society B: Biological Sciences, 275*(1643), 1645–1651.

Morris, B. S. (1987). Reading replay in «Live» television text. *Journal of Popular Culture, 20*(4), 147–157.

Morse, M. (1983). Sport on television: Replay and display. In Kaplan, Ann, (Ed.), *Regarding television.* Los Ángeles, CA: AFI, pp. 44–66.

Nadimpalli, V. K., & Hasenbein, J. J. (2013). When to challenge a call in tennis: A Markov decision process approach. *Journal of Quantitative Analysis in Sports, 9*(3), 229–238.

Nlandu, T. (2012). The fallacies of the assumptions behind the arguments for goal-line technology in soccer. *Sport, Ethics and Philosophy*, *6*(4), 451–466.

Oldfather, C. M., & Fernholz, M. M. (2009). Comparative procedure on a Sunday afternoon: Instant Replay in the NFL as a process of appellate review. *Indiana Law Review*, *43*, 1–45.

Raunsbjerg, P., & Sand, H. (1998). TV sport and rhetoric. *The Mediated Event. In The XIII Nordic Conference on Mass Communication Research*, pp. 159–174.

Rugby World. (2015). Clarification on the role of the TMO at Rugby World Cup 2015. Retrieved from www.rugbyworldcup.com/news/96575?lang=en

Simmonds, E., & O'Donoghue, P. (2018). Probabilistic models comparing Fast4 and traditional tennis. *International Journal of Computer Science in Sport*, *17*(2), 141–162.

Suárez-Arrones, L., Núñez, J., Munguía-Izquierdo, D., Portillo, J., & Méndez-Villanueva, A. (2013). Impact of several matches in a day on physical performance in Rugby Sevens referees. *International Journal of Sports Physiology and Performance*, *8* (5), 496–501.

Whannel, G. (1992). *Fields in vision: Television sport and cultural transformation*. London: Routledge.

Whitney, D., Wurnitsch, N., Hontiveros, B., & Louie, E. (2008). Perceptual mislocalization of bouncing balls by professional tennis referees. *Current Biology*, *18*(20), R947–R949.

Wilson, B. (2015, September 17). Rugby World Cup: Hawk-Eye ready for action. Retrieved from www.bbc.com/news/business-34242861

World Rugby. (2018). *Laws of the game Rugby union*. Dublin, Ireland: World Rugby.

World Rugby. (2019). Television Match Official (TMO) Global Trial Protocol. Retrieved from https://laws.worldrugby.org/downloads/TMO_Trial_From_January_2019_EN.pdf

9 Instant Replay in the National Football League

Jacob Tingle and Manuel Armenteros

Introduction

American football began in the 19th century with contests between college and university teams and is one of the most popular sports in the United States (US), with over 17 million fans attending games during the 2017 season (Lukas, 2018). The National Football League (2017 NFL Attendance, 2017) was founded in 1920 and at present 32 teams play 16 games (for a total of 256 games) during a 17-week period.

Each team participates with 11 players, has one head coach, and an average of 15 assistant coaches. The game lasts four 15-minute quarters and each team is allowed to have 53 players on its active roster. Of these 53, only 46 players can dress out for the actual game.

When the NFL was founded in 1920, it was modelled after the collegiate game for its first 12 years. In 1932, the NFL broke ranks on a few rules and appointed its own Rules Committee, charged with developing independence from the colleges and increased the number of officials to better ensure compliance and fair play.

To be considered for a position as an NFL official, candidates must have a minimum of ten years' experience officiating football, at least five of which must have been at a varsity collegiate or another professional level. The candidate must be in excellent physical condition (Alder, 2017). For their efforts, officials can earn between $4,000 and $10,000 a game depending upon their experience and the number of playoff games they work (A Tougher NFL, 2014).

Instant Replay was first used on 7 December 1963 during the popular "Army–Navy game" between the US Military Academy and US Naval Academy. Thanks to the innovation of a 29-year-old CBS TV producer, Tony Verna, the world of American football was changed forever (Verna, 2008).

In 1976, Art McNally, then the NFL director of officiating and previously a field judge and referee, wanted to explore how long a video review would delay a game (History of Instant, n.d.). Equipped with a stopwatch and a video camera, he observed the Dallas Cowboys versus

Buffalo Bills game from the press box, and he saw a missed call involving O.J. Simpson that could have been corrected. At that moment, he knew replay would benefit the league, its teams, players, and fans.

At its inception in 1986, video tapes and analogue linear systems were used to move forward and backward in the search for replays. Since then, the NFL has experimented with systems to help referees correct mistakes that have "indisputable visual evidence" (History of Instant, n.d.).

As a result of the technological developments, NFL officials use the same NFL broadcast camera feeds and the system allows immediate labelling and recovery of plays in a non-linear way. Doing so allows officials to make rulings in the allotted 90 seconds.

However, despite having a wide range of cameras and a fast replay system, NFL officiating continues to create controversy, and there are many important voices concerned about the system – some of which propose to eliminate it.

Why does the use of a video replay system that has been supporting officials for more than 30 years continue to create controversy? Is it a technological problem? Is it due to an immature technology? Is it due to the lack of referees' education? Or can it be simply as a consequence of the complexity of the NFL rules?

How are replay decisions between the referee and teams communicated? How many times can replay be used? Is there a limit? In what cases? The understanding of how Instant Replay works and the challenges that the officiating team have to deal with can help football to anticipate some of the challenges that FIFA and the IFAB should expect.

Technological Implementation in the NFL

The NFL is a league that takes advantage of existing technology and an early adopter for technological deployment, both in player development and in the support its officials.

Zieglmeier (2017) studied the different technologies used by football stakeholders. The coaching staff use free action cameras, such as the GoPro, which allow them to analyse the player's angle of view and the position of their players as well as individual advice to perform their play. Other advances in virtual reality, such as Google or Oculus virtual glasses, are used to provide players with a virtual training scenario in 360 degrees.

To monitor the physical and tactical performance, NFL coaches use the "Adidas miCoach Elite Team System", which provides player physical training data such as speed, distance covered, and heart rate.

The NFL has evolved since its inception in ways that help officials "get calls right". As such, replay is a natural development in the evolutionary process. When the NFL began as the first organized professional football league in 1920, only three officials (referee, umpire, and head linesman) were allowed. The field judge was added in 1929 and the back judge in

1947. In response to scrambling quarterbacks, and in particular to tactics of players like Fran Tarkenton, the line judge was added in 1965 to watch the opposite side of the line of scrimmage. The side judge was added in 1978, when the NFL implemented new rules to open up the passing game (Austro, 2017).

When the games were first regularly televised in the late 1940s, NFL officials felt the fans had an advantage because those in attendance could observe the replays on stadium video boards. Everyone but the referee knew what happened on the previous play. In 1975 referees were also equipped with wireless microphones and had the obligation to explain on-field rulings to a much wider audience. "Beyond the entertainment value, the ability for fans to hear the calls added credibility to on-field rulings, increased transparency on rulings for fans and the media, and reinforced the integrity of the game" (The Pursuit of, n.d. para. 28). Without a doubt, TV increased the popularity of the game and enhanced demand from officials to make concessions in order to satisfy fans. Conversely, increased TV presence afforded officials with an important and useful tool: Instant Replay.

First Instant Replay Experiments

Experimentation with Instant Replay began in 1976 (History of Instant, n. d.). McNally, NFL supervisor of officials and key people in the Officiating Department's transformation, tested how long a video review would delay a game. McNally also conducted a survey to know the opinion of teams. They had to grade each foul call on a seven-point scale, from poor (1) to excellent (7).

In 1978, the league tested Instant Replay during seven nationally televised preseason games. During that trial, calls remained inconclusive after lengthy reviews; McNally realized that they needed more cameras than broadcasters used for games. After that first game, Assistant Supervisor of Officials Nick Skorich said: "We still think we need a minimum of 12 cameras to get all the angles on every play. Electronically, I don't know if we are advanced enough yet" (History of Instant, n.d., para. 8).

In 1985 clubs voted to test a review system during eight preseason games. There was no unanimity in the vote. Four clubs voted against Instant Replay, but the 23 votes were more than the 21 votes needed to use Instant Replay in the upcoming year. Also, a new vote would be needed during the following offseason. Cleveland Browns owner Art Modell said after the vote: "Owners didn't want a playoff game decided by a bad call, and so they tried to push it through right there" (History of Instant, n.d.).

On 7 September 1986, Instant Replay was used for the first time in an NFL regular-season game. That first season saw an average of 1.6 reviews per game for a total of 374 plays; only 10 per cent of which reversed the ruling on the field (Flashback 1986, 2017).

The NFL Director of Administration, Joe Rhein, said that "it's possible to review instant replays and get the word to the referee on the field without a significant loss of time" (The Pursuit of, n.d.), but even that lost time was significant enough to make the game longer and slowed the tempo.

At its inception, cameras were less sophisticated than they are now, and relatively few were employed, meaning replay officials often did not get a definitive look at the play. The technology was not very refined and sometimes the field referee misunderstood the message; in one example the referee heard "pass is complete" instead of "pass incomplete" (Goodwin, 1986). Because of these mistakes, the league was forced to replace its walkie-talkies with pagers and radio headsets and it changed the terminology, using clear terms like "confirmed" or "reversed" (History of Instant, n.d.).

In a survey conducted by CBS, nearly 45 per cent of the 175,000 people who voted indicated they thought replay should remain, but with changes. Another 29 per cent said they favoured the rule as it is, while 26 per cent said they opposed the rule (Goodwin, 1986).

A replay official monitored the live game feed from an in-stadium booth, and initiated all reviews, reversing a call only with "indisputable visual evidence". When there was a question about a call, the official replayed the telecast on one of his machines as the head field official was alerted to halt play. The other video cassette recorder (VCR) continued recording the telecast so the official could study any other replay angles shown.

Expressing the unknown nature of the replay booth official, Dennis Lewin, the Head of Production for ABC Sports, said:

> We have no clue about how long he'll wait, or what he's waiting to see. Both the networks and the league emphasized that the official did not dictate which replay angle he wanted to see or what order they were shown in. Nor may he request any particular shot. He merely saw what was broadcasted, replayed what he wanted on his tape machines, made his call, and relayed it to the field.
>
> (Goodwin, 1986, para. 17)

Some executives were also concerned because the dependence on the NFL to provide pictures to settle questionable calls brought with it the potential that the outcome of games could be influenced directly by the producers, who decide where cameras were focused, which ones were attached to replay equipment, which replay angles were selected and chosen for airing, and in what sequence (Goodwin, 1986).

To address that concern, television executives asked the NFL to get its own equipment and highlighted that they had no problem with the theory of improving the game through technology. It was the current practice that bothered them (Goodwin, 1986).

In 1992, six years after its introduction, an opposition group led by George Young got enough votes to eliminate it (King, McDonough and Zimmerman, 1994, in Dudko, 2013). The owners voted 17 to 11 in favour of keeping replay, which fell short of the required number of votes.

According to McCown (2016), at that time there were two main reasons behind the decision to stop Instant Replay: it slowed down the game too much, and it failed to get enough correct calls. Freeman confirmed there was a major problem with incorrect reversals (Freeman, 1992).

Director of Officials Jerry Seeman indicated there were nine erroneous reversals, because there was not the irrefutable evidence necessary to overturn a call made on the field. Seeman also said there were 12 plays that should have been reversed, but weren't. Clearly there was a problem with the system.

Some opponents of replay claimed it caused on-field referees to be timid because they wanted "to avoid the embarrassment of having their decisions overturned before a national television audience" (Freeman, 1992, para. 4). But not all agreed. New Orleans Saints General Manager Jim Finks said:

> I think it's a step backwards for the National Football League. I think it was something we had that was very unique, very effective… I think we're going to regret the day we voted it down and I think we'll have it back in, very frankly.
>
> (Freeman, 1992, para. 5)

Seven years later, Finks' words rang true. In 1999, the NFL decided to try replay again. New advances in technology led them to believe that the new digital system would have a positive impact in the replay operation because there was tape to rewind. Commissioner Paul Tagliabue said:

> Our goal is to take advantage of advanced technology to create the most efficient replay system possible. We believe we have developed that type of system. It uses advanced technology, but it is simple to operate. We did not want to bring back replay with the same system as 10 years ago. There will be a noticeable positive difference in the overall replay operation.
>
> (NFL, 1999, para. 2)

To make a speedy review possible, high-definition signals from all cameras were sent by optical fibre cables to the replay booth. The signals were recorded on two computers and cued up by a technician. A video operator transferred relevant plays to a replay monitor, which the replay assistant examined to determine if a review was warranted. A second person in the booth watched the game action, announcing to the replay assistant whether play has resumed, i.e. if it's too late to commence a review. There were three monitors at field level in open-air stadiums, one of them in a runway if there was inclement weather, and just two monitors in domed stadiums.

In 2013, an agreement between the NFL and Microsoft provided every team with Microsoft Pro tablets. And in August 2017, football referees replaced the traditional monitors with portable Surface Tablets (Darrow, 2017). "Instead of a fixed sideline monitor, we will bring a tablet to the Referee who can review the play in consultation with our officiating headquarters in New York, which has the final decision" (Sake, 2017).

In 2017, there were a total of 39,677 plays, but only 429 of those plays were reviewed. With an average of 155 plays per game in 2017, just 1.1 per cent of them were adjudicated using Instant Replay (These Officials are, n.d.).

The Evolution of the Instant Replay Rules and Principles

Since it began, the NFL's Instant Replay review process focused on expediting Instant Replay reviews and "ensuring consistency".

While it was necessary to use the same television feed that fans see at home, officials also needed access to all camera views in order to isolate the best angle for consulting with replay officials and to allow referees to make the best replay decisions.

Since 1985, "indisputable visual evidence" was the standard and only those calls were reversed. Only "indisputable visual evidence" calls were revised by a replay official who was monitoring the game feed from an in-stadium booth. When replay started, NFL coaches had no way of challenging an official's on-field call, and the decision to review any questionable call occurring inside the last two minutes came from the replay official's booth. The best a coach could hope for was that a seriously blown call might be spotted and overturned by the officials themselves (Long, 2011).

In the 1986 season, reviewable plays included (History of Instant, n.d., para. 18):

- Plays of possession or touching (fumbles, interceptions, receptions, muffs, or ineligible player touching a forward pass);
- Most plays governed by the sidelines, goal lines, end lines, and line of scrimmage (whether a player is out of bounds, forward or backward passes, or breaking the plane of the goal line);
- Easily detectable infractions on replay (too many men on the field).

In 1989, the NFL decided to include every turnover and every touchdown. The protocol established also that the coach could request two reviews, except after the two-minute warning of either half, when the responsibility of calling reviews shifted from coaches to the replay assistant in the booth. If both the coach's challenges were successful, a third challenge was allowed. This method is still utilized 30 years later.

In 1991, after a six-season run, Instant Replay met its demise when 17 owners voted against renewing the system.

In 1996, the new system approved for testing in ten preseason games covered three categories of plays: out of bounds, number of players on the field, and scoring plays.

Also, it continued permitting coaches to challenge on-field rulings. Each coach could challenge three plays per half – at the cost of a timeout per review. The league went away from using replay officials in skyboxes and gave the game official authority to review plays using a small booth equipped with monitors located on the sidelines. And, importantly, referees now had only 90 seconds to make a decision.

Despite the changes, owners voted against implementation for the 1997 regular season. The main concern was that each review cost teams a timeout, even when a challenge was successful.

In 1999, after seven years without Instant Replay, a new system with digital technology was implemented.

The new system with digital technology included four people in the box: a "replay assistant", a "technician", a "video operator", and a "communicator".

The rules were adapted to address some of the main criticisms of previous iterations (History of Instant, n.d., para. 46):

- To minimize delays, the league cut the number of challenges from three to two per half.
- Coaches, unwilling to trade a timeout for any review, would now be charged a timeout only for unsuccessful challenges.
- To allow coaches to focus at the end of each half on calling plays, not on which calls to challenge, a replay assistant initiated all reviews inside the final two minutes of each half.

The system was initiated by challenges from the head coaches in all but the final two minutes of each half. During the final two minutes and in overtime, replay is triggered by the replay official in the booth, with no limits to how many replays he can request. Jerry Seeman, then the NFL Senior Director of Officiating, said: "Every play will be examined in case there is a coach's challenge. The booth will operate under a 'two-minute' mode the entire game, lining up replays of every play in case a challenge is issued (NFL Introduces Instant, 1999) (see Table 9.1).

In 2004, the NFL introduced a more low-tech solution: coaches were given a red flag, similar to the yellow official's flag, to throw onto the field in order to make their challenge known. The Instant Replay rule was slightly changed to allow a third challenge to teams if both of the original two challenges were successful.

In 2016, the NFL voted to expand its current replay system to include more reviewable plays. The reviewable plays under the amendments included penalty enforcement, proper down, spot of the foul, and status of the game clock.

Table 9.1 Rules of Instant Replay in 1999 (NFL, 1999)

Coaches' challenge	Outside the final two minutes of each half, *each team will be permitted two challenges per game* that will initiate referee reviews. Each challenge will cost a team a timeout. If a challenge is upheld, the timeout will be restored, but the challenge will not. No challenges will be recognized from a team that has exhausted its timeouts.
Final two minutes of each half	After the two-minute warning of each half, and throughout any overtime period, any referee review will be initiated by the replay assistant in the replay booth in the press box. He can initiate as many review requests as he thinks necessary, no matter how many timeouts the teams may have. No timeout will be charged for a review initiated from the replay booth.
Referee review	All replay reviews will be conducted by the referee on a field-level monitor after consultation with the other covering officials on the play. A decision will be reversed only when the referee has indisputable visual evidence that the call should be changed. The referee cannot initiate a replay in any instance.
Time limit	Reviews will be a maximum of 90 seconds, timed from when the referee puts on the head phones to the replay booth.
Reviewable plays	The Instant Replay system will cover a variety of plays in three main areas: (1) sideline, goal line, end zone, and end-line plays; (2) passing plays; and (3) other detectable infractions, such as a runner ruled down not by defensive contact, and the number of players on the field.

The league also announced that the replay official and members of the Officiating Department at the league office could consult with the on-field officials during games to provide information on penalty yardage, proper down, and status of the game clock.

In the 2017 season, the league continued to consult with the game's referee and replay official from the Art McNally GameDay Central (AMGC), which was created in 2014. However, all final decisions on all replay reviews come from NFL Senior Vice President of Officiating, Alberto Riveron, or a designated senior member of the Officiating Department with input from the referee (History of Instant, n.d.).

The Coach

Head coaches have not always had the right to review a play. They can communicate with assistant coaches via a wireless headset communication device. A small number of assistant coaches are in the press box, which allows them to have an expansive view of the field. From this vantage point they can see everything happening at once. When alerted by one of the coaches in the press box, head coaches can now activate a buzzer system located in a belt pack to communicate with referees when they want a review.

The head coach also challenges a play by tossing a red flag on the field. At present, he has two per game and, as indicated previously, he can use them in all but the final two minutes of each half and in the overtime period. The challenge requires the use of a team timeout. If the ruling on the field isn't overturned, i.e. the official's call stands, the team loses one of its six official timeouts.

If a team wins the challenge, it retains its timeout and the official's call is overruled. The head coach is allowed one more if the coach wins the first two challenges, according to NFL rules (The NFL Rulebook, n.d.): "A challenge will only be restored if a team is successful on both of its challenges, in which case it shall be awarded a third challenge, but a fourth challenge will not be permitted under any circumstances."

The Debate about Eliminating Instant Replay

Since 1985, Instant Replay has been used to overturn on-field decisions in situations with indisputable visual evidence (History of Instant, n.d.). The replay scope is a controversial issue, with some now arguing that replay should be used even in judgement calls or to reduce the number of missed calls. A missed pass interference call near the conclusion of the 2019 National Football Conference (NFC) championship game between the New Orleans Saints and the Los Angeles Rams only heightened the cry for expanded use of replay. In fact, in March 2019, NFL owners voted 31–1 to expand the use of replay to include pass interference calls (NFL expands replay, 2019).

The debate has been "a hot-button issue since games were first regularly televised in the late 1940s" (History of Instant, n.d., para. 1). With technological improvements and the ability to analyse plays frame by frame, video reviews were supposed to make ruling on plays more objective and less controversial.

The controversy reached a fever pitch after the 2014 Divisional Round playoff game between the Dallas Cowboys and Green Bay Packers. In the fourth quarter, with less than five minutes to go in the game, Cowboys wide receiver Dez Bryant made a sensational catch. After a challenge, the referees determined that "although the receiver is possessing the football, he must maintain possession of that football throughout the entire process of the catch," (Wilson, 2015), and overturned the call. The rules state that if there is even a hint of doubt, i.e. it's not irrefutable, the call on the field stands.

Two Super Bowl-winning coaches, Jon Gruden and Pete Carroll, manifested their disappointment to the Instant Replay system. In the opinion of Coach Gruden, the NFL should "eliminate instant replay and let the officials call the game" and "let the naked eye make the call". Carroll considers "the scrutiny of the officials has become so intense, they don't call the game like they used to. I don't like instant replay" (Boren, 2018).

Those critics notwithstanding, Instant Replay is so pervasive that the entire definition of a catch, a defining element of the modern game, was up for debate (Denton, 2018). It became so controversial that, in March 2018, the NFL changed the catch rule (Patra, 2018, para. 4):

1. Control of the ball.
2. Two feet down or another body part.
3. A football move such as:
    - A third step;
    - Reaching/extending for the line-to-gain;
    - Or the ability to perform such an act.

Getting calls right is important, but so is keeping games exciting and watchable. This balance between keeping the rules simple and the fans, players, and coaches happy seems complicated. Trevor Denton, sport editor of the *Daily Trojan*, explained that the controversy is due to "the video review and its frustrating ability to make games overly complicated and convoluted" (Denton, 2018, para. 8). Even when the NFL tries to simplify a rule, and "remove all the grey areas" as it did with the catch rule, fans, media, and teams can be left with more confusion (Patra, 2018). Fred Gaudelli, executive producer of NBC's *Sunday Night Football*, said, "This is my 29th year doing NFL games. The rules seem to get more complicated and more nuanced every single year" (Seifert, 2018a, para. 16).

Time Control in the Instant Replay

Since the days of Art McNally using his stopwatch to calculate how much time was needed to review a play, time control has been a very important issue when reviewing play situations.

The game itself lasts an hour, split into four 15-minute quarters, but an average NFL game takes more than three hours to play (3:02 in 2008) and in the past decade, that time has gone up nearly six minutes (Seifert, 2018b). But what fills all that screen time is different. The average number of incomplete passes, penalties called, and plays reviewed are all the same or higher, meaning we're spending a larger percentage of those three-plus hours watching referees make decisions, players stand around, or commercials.

In 1986, the first regular season of the Instant Replay, officials reviewed 1.6 plays per game, 374 in total, and only 10 per cent (38) ended with a reversal of the ruling on the field (NFL Introduces Instant, 1999). During that time, reviews could be a maximum of two minutes. The time permitted for reviews was later reduced to 90 seconds, and after the 2005 season, the NFL lowered the time limit for replay reviews from 90 seconds to 60. But the coaches' challenges have the inherent capacity of slowing games down. A big downside to Instant Replay is the effect on game times and the pace of play. "This can serve to cool down a hot team on the verge of victory, or allow an exhausted defender a crucial extra minute or two of rest" (Berman, 2017, para. 3).

Professionalism in NFL

There are some people who think bad calls or missed calls are a result of non-full-time officials. Despite significant rules and video reviews that officiating crews are required to chart each week of the season, all NFL officials meet only once a year, for what's called "The Clinic". There they learn the new rules, review plays, and get their new uniforms. They have video footage material each season that is used by officials to analyse the plays and improve their work.

The fact that NFL game officials are not full-time employees of the league, like those in the National Basketball Association and Major League Baseball, has been recognized as a problem by the media and by NFL executives. Although most of our officials are former athletes or former players, a "lot of them just started working games just to make a little extra money on the side in high school and college," explained NFL Vice President of Officiating Dean Blandino (A Tougher NFL, 2014). Prior to 2017, all officials were part-time and worked in a variety of jobs outside football: from teachers to bankers and insurance underwriters to builders (O'Rourke, 2018). The NFL took action, and in September 2017 it hired 21 full-time game officials from among the current 124-person roster of officials "to promote the common goal of improving every aspect of NFL officiating" (NFL Hires 21, 2017).

Technology Is Not Infallible

The Instant Replay system emerged from (and was a result of) improved broadcast technology (Technology, n.d., para. 2).

In the first experiments, in 1985, the replay official was upstairs using two nine-inch television monitors and two VCRs capable of recording and immediately replaying individual plays. The system worked with National Television System Committee (NTSC) space resolutions of 720 pixels compared with the 1,920 pixels in the current full HD or the 3,840 pixels in broadcasting tests for the new displays in UHDTV.

High-definition and high-speed cameras have certainly altered how fans engage with and watch NFL games, which has led to more demands to get the call right. The resolution of the current cameras are registering speeds of more than 1,000 fps and allow operators to zoom in and analyse frame by frame with more details and determine the accuracy of officials' decisions (see Chapter 8). Some contend technology has changed the game for the betterment of the league, coaches, and players, and even the fans.

The ever expanding and complicated communication systems used by the NFL requires important coordination by the Federal Communications Commission (FCC). In 1975 only referees were equipped with wireless microphones. Currently, wireless communication is used extensively by coaches, players, and officials during games.

The league debuted in 2014 a more efficient wireless official-to-official communications system (O2O). The system speeds up the game by letting the officiating crew talk to each other from a distance instead of face-to-face for every penalty. Previously, the field judge would be required to run across the field to the referee to tell him the specifics of the penalty, but O2O has created greater efficiencies. For example, "O2O saves time when a field judge flags a clear pass interference 45 yards downfield and can tell the referee about the call as soon as the play ends" (The Pursuit of, n.d.).

Despite those improvements, the systems are still not perfect. Frequency coordinators make sure anyone using a wireless microphone, walkie-talkie, or radio is on the correct channel to allow as many people as possible to access the bandwidth they need. "That massive collision is happening in our stadiums every weekend," said Michelle McKenna-Doyle, the league's chief information officer. "That has to get solved—whether we buy, rent or partner with someone who owns frequency" (Technology, n.d., para. 12). Sometimes, weather conditions or stadiums obstructions can make it hard for technology to function flawlessly (see Chapter 4).

Technology cannot always provide an effective solution. According to Brian McCarthy, NFL Vice President:

> We are always looking for ways to responsibly incorporate the latest technology into all facets of the game. For a number of years we have considered various potential first-down measurement technologies but have not found one to date that we were comfortable with to deploy. (Victor, 2017)

The NFL has challenges other sports don't face. A Hawk-Eye spokesman told ESPN in 2015 that the cluster of players around first-down and goal-line plays could make its system useless because 25 per cent of the ball must be clearly visible for it to work (Victor, 2017).

Other challenges inherent to video technology is that an advantage can convert into a disadvantage, as it occurs with slow motion. According to Coach Jon Gruden:

> I think slow-mo replay is the biggest problem with replay. When you're looking at 'is it a catch or isn't it a catch' at that speed it's hard to tell. It really is hard to tell. So I think if you threw that slow-mo out, I think you'd get back to common sense. Let the naked eye determine some of

> these calls. But it always looks like pass interference when you're going that slow; it always seems to look a little bit more dramatic in slow motion. Sometimes it's not realistic, I don't think.
>
> (Boren, 2018, para. 5)

The Importance to Enhance Fans' Viewing Experience

One of the reasons why video replay was implemented was because fans watching games on TV had access to some details of the play that game officials did not. Spectators had the benefit of seeing a play dozens of times from multiple angles and hearing the opinions of the commentators, whose job it is to keep fans interested.

However, with the increased use of replay the fan experience is different for those watching the game on TV at home versus those watching in person. In the stadium, fans are left waiting to know whether the catch in the end zone was a touchdown or an incomplete pass. This time element breaks the act of celebration, as they often have to wait for someone else to confirm what they saw with their own eyes.

In an effort to address the fan experience concerns, on 22 March 2017, NFL Commissioner Roger Goodell wrote an open letter to fans and said the league was going to try and reduce the amount of time spent on replays. This should improve consistency and accuracy of decisions and help speed up the process. The goal behind the change was to "achieve a competitive game with fewer interruptions and distractions from action" (Schwindt, 2017).

To coordinate actions with the officials, the TV broadcaster has a person on the sidelines with a pair of red flags used to communicate to the referee whether to go long or short on the timeout. They also use them to indicate when the network is back from commercial and play is allowed to resume, and when the network is choosing to go to commercial for a stoppage in play such as an injury or Instant Replay challenge.

The role of Instant Replay has had other impacts on its narrative event. Commercial breaks are only allowed to begin during natural stoppages in play. Generally, after scores or possession changes, but also when teams call timeouts. The teams have greatly improved the narrative, not only for viewers at home, but also for fans in stadiums. The introduction of giant video boards in the stadium offers fans close details of players and plays. As such, fans watching the game in the stadium enjoy an experience that more closely replicates that of the fan watching the game on his or her sofa.

Conclusions

The NFL has been one of the pioneering leagues in the use of video technology to help referees. The use of Instant Replay has proven to be an important tool for officials and, despite having been stopped for almost seven years, the new system approved in 1999 has remained and even improved with enhanced digital technology, which is faster than the analogue process. Coupled with greater technological support, such as the creation of the AMGC in 2014, where all the activity of review of plays and communication with the referee team has been centralized, the NFL has a strong, but not perfect, replay system.

As we have seen, not all stakeholders are in favour of NFL officials using Instant Replay. In spite of the criticisms, the NFL replay system has far more supporters of maintaining Instant Replay in all surveys consulted. Further proof of its acceptance is that in recent years, the protocol has included new play situations that were not initially considered (e.g. reviews for pass interference). The NFL has been aware of the dilemma of maintaining the essence of the game without giving up the advantages of technology and has promoted measurements to achieve high professional standards while reducing the impact of the technology on the flow of the game.

Another aspect to highlight using video replay to help referees is the symbiosis between officiating in the NFL and television broadcasting. The officiating profession has benefited by using TV's technology and the visual language developed by sport broadcasting producers. In return, officials have offered viewers and fans more transparency in the refereeing process. TV has been able to integrate Instant Replay as part of the football narrative or story line by making the review process or the timeout time part of the show. They've found additional revenue streams by using waiting times to introduce more commercials. Nowhere is this more evident than the coordination between the TV broadcaster and the referee to manage timeouts.

Instant Replay has been criticized by media and fans for several reasons. One of them is that the rules are still not understood. Most fans genuinely do not know what is – and isn't – a catch. The NFL has developed resources, such as the NFL Rulebook with video examples for the different sections of the rules, and has promoted measures to make the rules easier to understand.

The other widespread criticism has been the lack of professionalism of the referees (i.e. that a vast majority of them are not full-time employees). The NFL has promoted measurements to achieve the professionalization of their officials by hiring full-time referees and preparing an annual clinic to train with the technology and review the latest changes in the rulebook.

References

2017 NFL Attendance Data. (2017). Pro Football Reference. Retrieved from www.pro-football-reference.com/years/2017/attendance.htm.

2018 NFL Rulebook. (n.d.). NFL. Retrieved from https://operations.nfl.com/the-rules/2018-nfl-rulebook/

Alder, J. (2017, August 25). The Path to Becoming an NFL Official. Retrieved from www.thoughtco.com/how-to-become-an-nfl-official-1333790.

A Tougher NFL Job to Get than Becoming a Player. (2014, 18 September). CBS. Retrieved from www.cbsnews.com/news/a-tougher-nfl-job-to-get-than-becoming-a-player/.

Austro, B. (2017, 16 June). NFL has a New Name for the Head Linesman Position: Down Judge. Retrieved from www.footballzebras.com/2017/06/nfl-new-name-head-linesman-position-judge/.

Berman, C. (2017, March 23). Pros & Cons of Instant Replay in Sports. Retrieved from www.sportsrec.com/pros-cons-instant-replay-sports-8681745.html.

Boren, C. (2018, March 28). Jon Gruden Would Eliminate Instant Replay from the NFL, and He Isn't Alone. Retrieved from www.washingtonpost.com/news/early-lead/wp/2018/03/28/jon-gruden-would-eliminate-instant-replay-from-the-nfl-and-he-isnt-alone/?noredirect=on&utm_term=.312ab7263071.

Darrow, B. (2017, August 4). NFL Referees Use Microsoft Surface at Hall of Fame Game. Retrieved from http://fortune.com/2017/08/04/nfl-refs-microsoft-surface/

Denton, T. (2018, May 22). T-Time: Instant Replay Is Ruining Sports. *Daily Trojan*. Retrieved from https://dailytrojan.com/2018/05/22/t-time-instant-replay-is-ruining-sports/.

Dudko, J. (2013, June, 8). The History of Instant Replay in the NFL. Retrieved from https://bleacherreport.com/articles/1666250-the-history-of-instant-replay-in-the-nfl.

Flashback 1986: The NFL Adopts Instant Replay. (2017, March 16). Soundavision. Retrieved from www.soundandvision.com/content/flashback-1986-nfl-adopts-instant-replay#HEKbqRoAtvwm3wop.99.

Freeman, M. (1992, March, 19). NFL Pulls the Plug on Instant Replay. *The Washington Post* [online]. Retrieved from www.washingtonpost.com/archive/politics/1992/03/19/nfl-pulls-the-plug-on-instant-replay/3b4a2d02-45f5-4402-a53a-04927a03e402/?utm_term=.fc659d9d8e9d.

Goodwin, M. (1986, October 11). Instant Replay Rule Troubles Networks. *The New York Times* [online]. Retrieved from www.nytimes.com/1986/10/11/sports/instant-replay-rule-troubles-networks.html.

History of Instant Replay. (n.d.). NFL. Retrieved from https://operations.nfl.com/the-game/history-of-instant-replay/.

King, P., McDonough, W. (Author and Editor) and Zimmerman, P. (1994). *75 Seasons: The Complete Story of the National Football League, 1920-1995*. Turner Publishing, Inc.

Long, T. (2011, November, 3). March 11, 1986: NFL Adopts Instant Replay. Retrieved from www.wired.com/2011/03/0311nfl-owners-ok-instant-replay/.

Lukas, M. (2018, September 17). NFL vs. MLS – Revenue, Salaries, Viewership, Attendance, and Ratings. Retrieved from www.wsn.com/nfl/nfl-vs-mls.

McCown, R. (2016, December, 16). It's Time to End Instant Replay, Again. Retrieved from https://sports.vice.com/en_us/article/bmq788/its-time-to-end-instant-replay-again.

NFL Expands Replay to Include Pass Interference. (2019, March 26). Reuters. Retrieved from: www.reuters.com/article/us-football-nfl-rules-changes/nfl-expands-replay-to-include-pass-interference-idUSKCN1R8044.

NFL Hires 21 Full-Time Game Officials for 2017 Season. (2017, September 6). NFL. Retrieved from www.nfl.com/news/story/0ap3000000839763/article/nfl-hires-21-fulltime-game-officials-for-2017-season.

NFL Introduces Instant Replay Technology. (1999, May 26). NFL. Retrieved from www.nfl.info/nflmedia/news/1999news/InstantReplay52699.htm.

O'Rourke, S. (2018, September 1). Despite All the Technology in the World, NFL Referees Continue to be Terrible. Retrieved from www.the42.ie/nfl-column-referees-browns-raiders-4263019-Oct2018/.

Patra, K. (2018, March 27). NFL Teams Unanimously Approve Simplified Catch Rule. Retrieved from www.nfl.com/news/story/0ap3000000923342/article/nfl-teams-unanimously-approve-simplified-catch-rule.

Sake, P. (2017). Referees with Tablets: NFL Plans Change to Replay Reviews. Retrieved from www.kansascity.com/sports/spt-columns-blogs/for-petes-sake/article140596128.html.

Schwindt, O. (2017, February 22). NFL Plans Adjustments to Commercial Breaks to Improve Game Viewing Experience. Retrieved from https://variety.com/2017/tv/news/nfl-tv-commercials-game-changes-1202014269/.

Seifert, K. (2018a, July 11). Should NFL Worry About Referee Turnover? Here's Why There's Concern. Retrieved from www.espn.com/blog/nflnation/post/_/id/277725/should-nfl-worry-about-referee-turnover-heres-why-theres-concern.

Seifert, K. (2018b, September 18). Hey, Look: The NFL's Average Game Time has Dropped. Retrieved from www.espn.com/blog/nflnation/post/_/id/249052/hey-look-the-nfls-average-game-time-has-dropped.

Technology in the NFL, Balancing Technology with Tradition. (n.d.). NFL Retrieved from https://operations.nfl.com/the-game/technology/.

The Pursuit of Perfection. The History and Evolution of the Official. (n.d.). NFL. Retrieved from https://operations.nfl.com/the-officials/history-of-the-official/.

These Officials are Really Good. (n.d.). NFL. Retrieved from https://operations.nfl.com/the-officials/these-officials-are-really-good/.

Verna, T. (2008). *Instant Replay: The Day that Changed Sports Forever.* California (USA): Creative Book Pub International.

Victor, D. (2017, December 18). Why Doesn't the N.F.L. Use Tracking Technology for First-Down Calls? Retrieved from www.nytimes.com/2017/12/18/sports/nfl-first-down-measurement.html.

Wilson, R. (2015, January 12). Referee Explains Why Dez Bryant's Catch was Ruled Incomplete. Retrieved from www.cbssports.com/nfl/news/referee-explains-why-dez-bryants-catch-was-ruled-incomplete/.

Zieglmeier, M. (2017). *The Potential of Wearable Technology in Amateur Football.* A Qualitative Study. Hamburg: Anchor Academic Publishing.

Part IV

Experiments with Video Assistant Referees

10 The Video Assistant Referee (VAR) Protocol

Carlos Bacigalupe

Introduction

Following the pattern established by other sports, the world of football has reached the conclusion that it is now an opportune moment to apply current technology in order to reach the highest standards of decision-making by referees who take charge of matches in competitions that are watched by large audiences and which have important economic and social repercussions. Right from the outset, there have been opinions in favour and against, which have caused the game's governing bodies to be very cautious about deciding to introduce it. Finally, at its 130th Annual Meeting held in Cardiff on 5 March 2016, the International Football Association Board (IFAB) gave the green light for an experimental phase in which the Video Assistant Referee (VAR) system would be trialled with a view to its possible introduction into the game.

From the very first moment, it was clear that there was a pressing need to keep a tight control over the trials as the only way of effectively evaluating the results obtained. That control would remain in the hands of the IFAB and, consequently, all those national associations, confederations and/or competitions interested in taking part in the experiment would be subject to the only protocol authorised: "one protocol – used by all" (IFAB, 2016a).

A few months later, at an IFAB seminar held in New Jersey, from 19–21 July 2016, David Elleray, Technical Director of the IFAB, reiterated the importance of this issue (Elleray, 2016):

> The on-field experiment today and the whole workshop has revealed to us that this is a complex matter. Quite a lot of people think that video assistant is just putting someone in front of a TV and they can immediately solve all the problems of football, but it is actually quite challenging. What decisions can you review? How quickly can you review them? What is the effect of the review? This is all quite complicated. So, in the end we will produce something which is very effective but it will take some while and a lot of work and education. It's very important that all countries do exactly the same and follow the same protocols, because it will enable us to research and find out what works well, what doesn't work so well.

The IFAB set an initial period of two years (IFAB, 2016b) – the initial phase ended, in principle, in March 2018 – for the trials to be held. Once they had finished, and the results analysed by the Katholieke Universiteit Leuven – one of Europe's oldest universities, based in Belgium – the data and the results obtained would be submitted to the IFAB for a definitive decision to be taken regarding the authorisation of the system. Both the IFAB and the Fédération Internationale de Football Association (FIFA) had expressed worries about the possible negative impact that the application of VAR might have on the flow of the game, with agility in reviewing and taking decisions being key aspects in the final evaluation of the trials, as was observed during the experimental phase, in which players such as Cristiano Ronaldo and Luka Modric expressed their displeasure with the introduction of this technology (Bryon, 2016) (see Chapter 11).

The VAR Protocol

The version of the protocol analysed here is number 8, from April 2017 (IFAB, 2017a), which was used in the VAR trial period and at the 2018 FIFA World Cup in Russia.

The protocol consists of 14 sections: Introduction, Summary of VAR protocol, Reviewable decisions, Scope of the experiment, Review process, Communication between referee and VAR, Experimental timelines, Protocol clarifications, Referee education, Analysing the data, Technology and organisation, Match day organisation, PR and communication, and Glossary.

The protocol has been analysed in its entirety with the aim of predicting possible weaknesses that might make the introduction of the VAR system in all countries difficult.

Guiding Principles

The protocol is very clear regarding the identification of what might be called the "principles" of the whole programme. The first principle – "minimum interference – maximum benefit" (IFAB, 2017a) – evidences the above-mentioned worries regarding the negative impact the application of a VAR system might have on the flow of the game, causing repeated and unwanted interruptions. The intention is to interrupt play as little as possible, only when there is no choice, and where the decision to be taken is potentially a match-changing one. From this starting point comes the following principle: "was the original decision given by the referee clearly wrong?" (IFAB, 2017a, 5), making it clear that the aim is to review only those situations that are decisive for the outcome of the game, not those that are open to interpretation or are debatable, but ones which are clearly mistaken. For that reason, the protocol identifies four categories of situations or incidents as the only ones that may be subject to review: goal/no goal; penalty/no penalty; direct red card (not second yellow card/caution, although it is possible that the review may lead to a yellow card if, for example, it is shown that a player dived or a forward scored a goal with their hand); and mistaken identity when the referee cautions or sends off the wrong player. Obviously, all of these situations have a range of options and conditioning factors, which opens up a wide field for a debate in which, being open to all opinions, a great number of actors take part, including not only players, managers, and referees, but also the media and the fans themselves, as FIFA President Gianni Infantino stated:

> I think that the FIFA and I felt sure that we are leading the way and we are not stopping the progress, we are listening to the fans, we are listening to the players, we are listening to football and we are applying common sense in what we do…
>
> (Infantino, 2016)

Participants and Their Roles

The Referee: the Key Match Official

From the outset, the use of the technological means available now through the VAR system has a clear and definite objective: to help the referee avoid making mistakes that have a significant influence on the outcome of the match, but maintaining at all times their position, namely that of the highest authority. In various sections, it is stated that the referee is the one who must take the final decision, whether it is after watching a repeat of the play on the screen placed beside the pitch (on-field review, or OFR) or after the VAR has indicated a situation – something real and not debatable (when, for example, the repetition shows that a player who scores is clearly in an offside position and the goal must be disallowed) – as contemplated in the protocol. As well as this, the referee always has the power to review the action if they suspect there has been a mistake or that something has happened without their knowledge. Only the referee may initiate a review, as the only thing a VAR can do is recommend that review, and the referee may say that they have seen the action sufficiently clearly and that no review is therefore necessary. In order to guarantee maximum transparency in the whole process, both the images and the conversations between all the participants are recorded. Obviously, the fact that the referee, making use of their power, chooses not to follow the VAR's recommendation to review a given play, means there has been interaction with the other members of the team which may be analysed after the match.

The Members of the VAR Team: Video Assistant Referee (VAR) and Assistant Video Assistant Referee (AVAR)

From the beginning of the programme, great efforts have been made to identify the most appropriate VARs and AVARs and to train them urgently, as they are the key element for the overall correct functioning of the system. It would not be of much use to provide all the technical means and financial resources necessary if the people who are to use them lack the capacity and the experience required to get the results hoped for.

In the selection process it was considered that the fact of being referees of the same category does not necessarily mean that they have the abilities needed to be a VAR, since that job requires paying constant attention over a prolonged period of time and being able to control several monitors.

An important facet in the training, to which many hours were dedicated, was to acquire experience of reviewing the same play from the different perspectives offered by the cameras installed in the stadium, with the aim of speeding up the process of choosing the best repetition – that is, for a given situation, camera X (all the cameras have a name or are designated in some way, depending on the angle they cover) is the one that will give me the perspective I need – thus reducing the time that play is interrupted (see Chapter 5).

VARs and AVARs are considered for all intents and purposes to be match officials, although the Laws of the Game do make a clear distinction between "on-field match officials" and "video match officials".

The identity of the VARs and AVARs, and their respective categories, has a direct influence on their relationship with the principal referee in the sense that if it is a question of referees who have the same refereeing grade and similar level of experience, the principal referee will find it difficult to ignore a recommendation made by other qualified members of the group who have had the possibility of seeing an action replayed, with no pressure from the crowd, and with the necessary time. In practice, the most likely thing is that a principal referee will choose to review a play where a recommendation has been made, which undermines the principle that they are the one who decides whether to review or not.

On the other hand, we must be aware that, regardless of what the protocol says, the reality is that VARs and AVARs are part of the team of officials and are in permanent contact with their fellow team members on the field or with the fourth official in the technical area. It will be inevitable, in fact very convenient, if the whole team – each of them within their function – works together, helping each other even beyond what is laid out in the protocol, to achieve the aim of officiating in the most correct way possible. This discretionary way of working as a team may, however, come into conflict with the definitions of the functions assigned to each of the participants in the protocol which, viewed from a positive angle, could be an excellent tool to continue perfecting the process.

Consequently, a number of questions were initially raised: Is it advisable for VARs to be serving referees of the same category as the referee on the field? Should they be referees of a lower category? Or perhaps former referees with the necessary experience and training? Are the same individuals always going to work together, thus making a team between themselves and/or with the team on the playing field? Would it be more effective if the members of a VAR team carried out only these tasks in order to reach a higher level of specialisation?

The answers to these questions had a direct practical implication, since they affected the number of referees who were to participate in each match, their training in the management of the VAR system, and the logistical aspects regarding travel arrangements to get them to the VOR should it be situated elsewhere.

The answers given have been different:

- In Spain, the VAR system has been used in La Liga's top division (Spanish First Division) since the beginning of the 2018–19 season, and it is foreseen that it will be used in the Second Division starting in the 2019–20 season. Both the VAR and the AVAR are serving referees from the same category as the ones who officiate on the pitch, each one carrying out their respective function. Being designated to act as a VAR or AVAR has meant a notable increase in the number of games referees have to officiate in every season. In Spain, the number of top-division referees is 20, and as there are 10 fixtures each match day, that means all the referees are in action every match day: 10 on the playing field and 10 more as VARs. Usually, the VAR and the AVAR work as a team (each VAR has two AVARs designated to work with them, which improves the interaction between them), but they do not necessarily coincide with the officials on the playing field, as the designations must guarantee geographical neutrality with regard to the teams playing in the game in question. To cover absences in case of illness or problems with licences, a former top-flight referee, who is no longer serving, having passed the age of 45, has also been brought in exclusively for VAR functions.
- In the 2018 FIFA World Cup in Russia, 13 referees were designated exclusively for VAR/AVAR functions, apart from the 36 referees and 63 assistant referees designated to officiate matches on the field.

Players, Managers and Other Team Members

In the same way that it is necessary to provide the refereeing team members with prior training in the use of the VAR system, it is also useful to extend this to the members of teams participating in matches to make them aware of situations that are likely to be reviewed by the VAR, the signals the referee will make when the game is stopped for the review to happen, and the cautions they will be given if they make the "TV screen" sign excessively or enter the Referee Review Area (RRA) or the VOR room.

Spectators and the Media

Given its repercussions, attempts have been made since the project began to give all participants the greatest levels of transparency, be they direct participants (referees, players, etc.) or indirect ones (fans at the stadium, TV spectators, reporters, commentators, etc.). To that end, as well as periodic publications, interviews, training videos, etc., announcements are often made ahead of kick-off by means of the stadium screens, reminding fans of the cases and situations likely to be subject to VAR review and informing them of the names of the members of the refereeing team, both on the field and in the VOR room.

Technical Means

The current protocol does not lay down the obligation to inform the fans in attendance if there has been a breakdown in the communication system or malfunctioning of the VAR system, although it does state that both teams must immediately be notified of such circumstances.

No additional marks are necessary on the pitch itself, and the only recommendation is that the goal nets be exclusively white in colour in order to avoid any distortion in the footage.

Location of the Video Operation Room (VOR)

When the idea of introducing a VAR system to a league or competition is first mooted, one of the first decisions to be taken concerns the location of the Video Operation Room (VOR). According to the protocol, the VOR operations can be located near/in the stadium (in a vehicle or room), or at a centralised replay facility, and requires a VAR and AVAR (IFAB, 2017a, 54). The decision taken regarding this must take into account the facilities available and any technological limitations, such as, for example, the impossibility of centralising VAR operations in a single centre, where an efficient communications network is not available both for the sending and reception of footage and for communication between members of the refereeing team.

The location may affect the kick-off times of games, having them played in different time slots in order to cover all of the games from a single VOR, unless there is a sufficient number of independent rooms to allow several VAR teams to work simultaneously.

The German Bundesliga (in Köln), the Portuguese league (in Lisbon), and the Spanish league (in Las Rozas, Madrid), among others, have chosen to centralise all VAR operations in one location.

An added difficulty lies in competitions at international level, such as, for example, the UEFA Champions League (UCL), in which teams from different countries take part. Some of them have facilities for using a VAR system, as it is used regularly in the national league, while others do not have the necessary infrastructure. In these cases, the most appropriate solution is a mixed system, using the VORs that are set up and equipped where they exist, and mobile units, installed in vehicles parked close to the stadium in other countries.

One curiosity in the protocol is that it requires the lighting inside the VOR to be sufficiently bright as to be able to identify every person who enters the same from 30 minutes before kick-off time in the internal recording of image and sound that is made to guarantee the transparency and the integrity of the process.

Strategic Location of the Cameras

Cameras must be placed in strategic locations to capture the different angles and in such a way that they are not obstructed by any external element (see Figure 10.1). This is to guarantee the integrity of the images. An example of this not being the case was in a Portuguese league game between Aves and Boavista, in which the referee, Fabio Jose Costa Verissimo, was unable to review the VAR footage in order to make a decision regarding a possible offside by the Aves player Vitor Gomes, because the flag that a Boavista supporter was waving in front of the camera meant it was impossible to capture the images ("Una bandera", 2018).

Although this might appear to be just an anecdote, it does show the need to bear in mind any potential interferences in order to ensure the integrity of the system and the credibility of the project (see Chapter 3).

Availability of All the Footage

The protocol states that the VAR system must have access to all the footage recorded, in real time (with a maximum time lag of one second), since the use by a third party, for example, a TV channel, of footage that is different from that made available to the VAR, and which might lead to a different decision, would be a heavy blow to the credibility of the whole system. For transmissions from remote locations, there must be the technical means to guarantee both the reception of the images and an adequate system for verbal communication between the members of the team on the playing field and those who are in the VOR (see Chapter 4).

Referee Review Area (RRA)

This area must have a monitor, protected from the elements, and which is situated away from the dugouts, so the footage cannot be seen from a certain

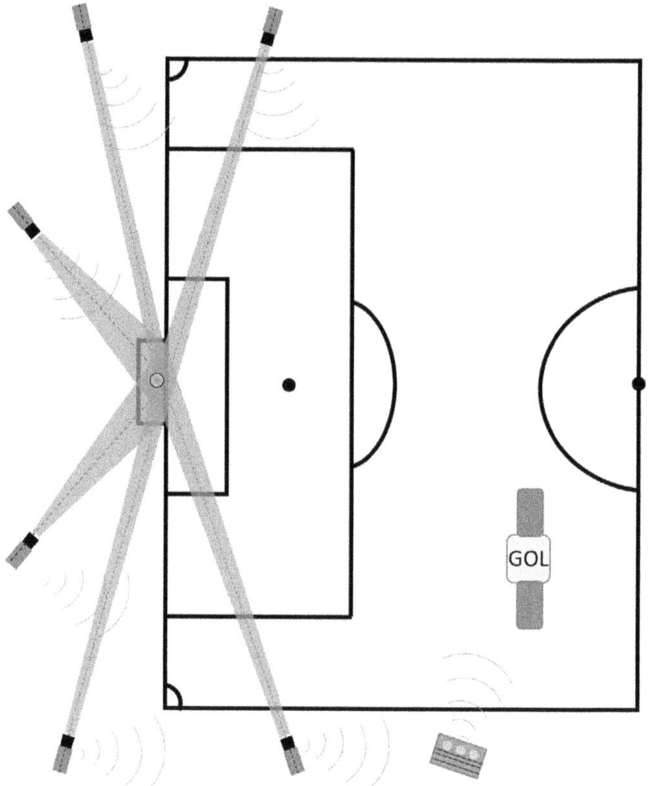

Figure 10.1 Distribution of cameras in Goal-Line Technology

distance, in a clearly marked-out area. To avoid problems of visual adjustment for the referee, who is used to following play from a certain distance, the protocol recommends that, in the review of the footage, the use of the split screen be avoided and that footage from just one camera be shown, with the referee being able to ask for a different angle, or speed, if they think it necessary.

Adequate Communication System

As correct communication between the referee and the VAR is a basic requirement, the system employed must be free of any type of interruption and guarantee the continuity and quality of the connection (see Chapter 4). To confirm that the system is functioning normally, the VARs are instructed to communicate with the referee every 10 or 15 minutes if it has not been necessary to do so for any review before then.

The Challenges of the VAR System

Communication Problems Between the On-Field Referees and the VAR

In order to avoid any possible mistakes in the process of communicating the reviews, some operating guidelines have been laid down, the most important among them being:

- Communication takes place mainly between the principal referee and the VAR, with the AVAR and/or some other team member (ARs, AARs) being able to contribute if the principal referee deems it convenient in order to reach a final decision.
- Short phrases should be used, with negative language being avoided, and the technical terms as fixed in the Laws of the Game being used.
- In order for there to be no interference on the open channel used by the on-field officials, conversations inside the VOR between the VAR/AVAR or the video operator are not heard; to open communication with the referee, both the VAR and the AVAR have a button that they must keep pressed down during the transmission.
- It is recommended that the referee acknowledges receipt of communication from the VAR, repeating the content of the message back, to guarantee reception has been correct.
- The conversation is led by the referee.

The Possible Loss of Credibility of the Refereeing Team

One circumstance that was observed as constant in the matches played during the experimental stage was that, once a decision taken by the referee had been rectified, the players, particularly those from the team that the decision went against, appealed and called for a review every time there was a perceived breach of the rules. This supposed loss of credibility of the members of the refereeing team after a decision has been overturned is critical and should be looked at owing to the negative effect it may have on refereeing.

This situation could turn out to be even more complex if, in the end, it is decided that replays of the footage that leads to a decision taken are to be shown on the screens in the stadiums, something that the protocol advises against.

Version 8 of the protocol suggested showing, at least, a message concerning the review on the screens. Two months before the Russia World Cup, Sebastian Runge, group leader of football innovation at FIFA, explained that, in order to offer a response to the fans in the stadiums, the referee's final decision would be displayed on the screens: "we will have graphics on the giant screens, we will have replays after the decision on the giant screens, and we will also inform the fans about the outcome of a VAR incident and review" (Shread, 2018). During the World Cup, however, replays were not shown, with only a message communicating the final decision being displayed. This was a compromise, a partial solution, most likely influenced by the concern that there might be further controversy caused by showing footage that could be interpreted differently from the referee's decision.

Response Time

This is one of the biggest criticisms levelled at the VAR system: the time that play is interrupted. Football is a sport that must flow as much as possible, with the referee's ability to ensure that it does being one of the most valued features of their performance. This means "playing on" or applying the "advantage rule" wherever it is feasible. Unfortunately, the available information shows that actual playing time in a 90-minute game rarely reaches 57 minutes (54 in the Spanish La Liga) owing to the interruptions that, for one reason or another, happen over the course of a match (see Figure 10.2).

As a result of that, any circumstance that might mean further interruptions – especially when these involve awaiting a refereeing decision – comes in for criticism and urgent solutions are demanded. In its Circular number 10, the IFAB shows its commitment to looking for ways of increasing effective playing time, with proposals such as these (IFAB, 2017b):

- increasing the amount of time the ball is in play (effective playing time);
- linking the stadium clock to the referee's watch;

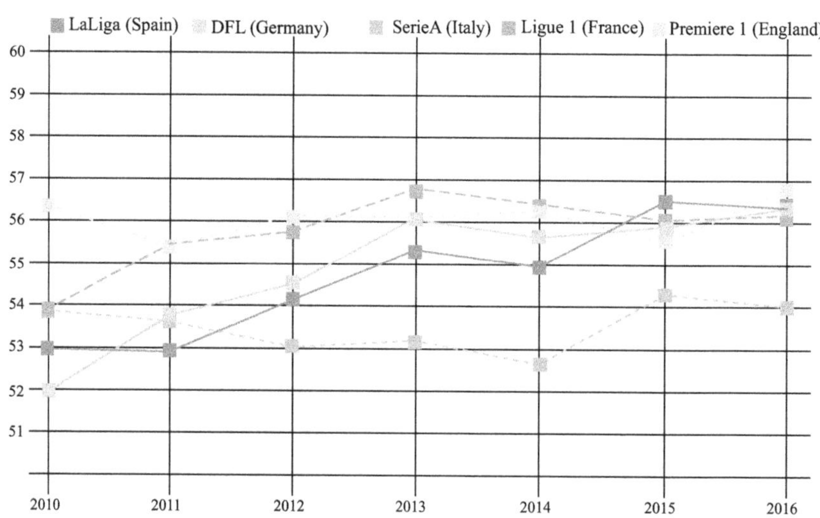

Figure 10.2 Playing time in the five biggest European leagues

- "self-passing" at free kicks/corner kicks, i.e. the player can play the ball a second time (or more);
- allowing the ball to be moving at goal kicks;
- requiring a goal kick to be taken on the side of the goal area where the ball left the field of play.

In order to be effective, the VAR system and those who operate it must have the technical means and the necessary training to be able to provide correct responses in the shortest time possible, without taking shortcuts or giving into additional pressure. As well as this, the protocol states that one of the ARs must control the time taken between when the referee initiates the review until play has restarted to add it on at the end of the game, which means that, despite the criticism, these interruptions do not imply less effective playing time.

It would not be of great use, however, to orchestrate a whole system of reviews, involving a large number of people and dedicating enormous technical and economic resources in order to then give preference to a quick response over a correct response. The proposal to make the check + signal + review + decision happen within a maximum of ten seconds, as Massimo Busacca, FIFA's director of refereeing, proposed during the experimental phase is, to say the least, wishful thinking (Dunbar, 2016) (see Chapters 3 and 11).

Especially noticeable is the delay that occurs when reviewing a play that ends in a goal being scored since, given that the players are fearful of celebrating it as they usually do so as not to look silly in the event of the goal then being disallowed, the VAR breaks the act of celebrating that may happen if the goal is confirmed, as the celebration then loses spontaneity and is somewhat forced.

Potentially Conflictive Situations

The protocol is being used as a document for compiling queries made regarding specific situations so that everyone who takes part in the experiments has the best information available. Obviously, each new version will include new queries, such as "what happens if the referee shows a yellow card to a player who has already been booked but, by mistake, fails to send him off?" This is a situation that has occurred often and is at times sorted out by the assistant referee or, at others, by the refereeing observer, who points out the situation to the referee during the half-time break, for example. With the VAR system, and the protocol as it stands right now, it is not clear if the VARs should alert the referee to this situation, since it is not exactly a case of mistaken identity but could well have a direct influence on the outcome of the match.

Another situation that could give rise to numerous conflicts was the guideline that advised the referee to, in close offside situations in which the AR had already raised his flag to indicate the infringement, delay blowing the whistle and allow the player to finish the move, which may mean that, if the move ended in a goal, and the VAR then confirmed that the AR had been mistaken and the player had not been offside, the goal would then stand (IFAB, 2017a, 49). The theoretical basis for this had a certain logic – the application of the "wait and see" idea – but, in practice, the defenders could argue that, on seeing the AR raise the flag, they had given up trying to stop the attacking player and that, thanks to that, he had been able to score. Evidently, it has been necessary to clear up any doubts here and determine how the referee and the AR should proceed in these situations, with the instruction being that, in case of doubt on the part of the AR, the response must always be to let play go on and the move finish, without flagging anything if there is a chance a goal will be scored and then, once this has happened, flag the possible infringement and leave it up to the VAR to indicate if the attacking player was in an offside position.

> ARs must ALWAYS MAKE A DECISION – there is no option to say 'I don't know so I will look at the video'. However, if a player is within the penalty area and about to score and there is real doubt about offside (position or offence) delaying the flag signal may prevent a major error which can not be corrected if play has been stopped.
>
> (IFAB, 2017a, 46)

Period of Time in Which an Incident Can Be Reviewed

Another of the proposals considered in the VAR system protocol is the possibility of extending the period during which certain plays can be reviewed – specifically those that end in a goal being scored – taking it back to the beginning of the attacking move to check that no infringement has taken place as it has developed. In this way, the review is not limited to the moment in which a goal is scored, or a penalty committed, but contemplates everything that has happened since the previous interruption of play until the end of the move in question.

The current protocol continues on this subject when discussing infringements whose aim is to deny an obvious goal-scoring opportunity, referred to as DOGSO, defining in a detailed way what the "attacking phase of possession" means, and calling on the VAR team and the referee to determine "the point at which the attacking team gained possession of the ball and then the point at which the phase of play that led to the goal/penalty incident started" (IFAB, 2017a, 21–22). As well as this, it goes further in the definition of what should be considered "gain[ing] possession of the ball", differentiating between when the attacking team receives it, whether from a save, recovering the ball in open play, etc.

On many occasions, however, the move leading to a goal may have developed over a long period of time, which would lead to a review time that is too long, especially when the footage to be reviewed has been captured by more than one camera. As well as this, it would need to be specified exactly what type of prior infringement could imply a VAR review being recommended. For example, might a foul throw be enough of a reason to disallow a goal scored 35 seconds later, after the attacking team has completed four passes and the defending side recovered possession of the ball as many as two times as the move unfolded?

Recalling the previous quote from FIFA President Gianni Infantino, "we are applying common sense in what we do" (Infantino, 2016), this extension should be used in a very restrictive way, only affecting those moves that have an immediate and direct influence on a goal being scored. Any attempt to go too far back in the review will possibly have a negative influence, making the review process unnecessarily long and forcing the refereeing team to take decisions about incidents that are long past, and which will confuse players, coaching staff, and fans alike.

From the refereeing point of view, the most complicated scenario might be as follows: during open play, there is a possible penalty for team A, the referee has not seen it clearly and allows play to continue, with team B gaining possession, from which they mount a quick counter-attack and end up scoring. As a promising attacking situation had arisen, the referee has not had the opportunity to stop play to review the possible penalty. After the goal has been scored, the VAR recommends the referee review the original move, which he does, and sees that, effectively, it was a penalty and one that warranted a red card for the same player that then scored at the other end. The correct decision here would be to award team A the penalty and send off the player from team B, having disallowed the goal scored by team B. In a situation such as this, it would seem opportune to recall what is suggested in the VAR protocol itself: "what would football expect?" (IFAB, 2017a, 22).

Initial Experiences

With logical prudence being applied, the initial experiments took place in specially designed training sessions and in friendly matches.

During the friendly fixture between France and Spain at the Stade de France on 28 March 2017, there were two situations in which the German referee, Felix Zwayer, resorted to the VAR system to determine whether a player from the attacking team had strayed into an offside position prior to goals being scored:

In the 48th minute, Antoine Griezmann headed the ball into the net; both the referee and the AR2 initially gave the goal (the flag was not raised, and the whistle was not blown) and the goal was celebrated by the French players and the coaching staff on the bench. However, before he returned to the centre of the pitch, the VAR informed the referee that there had been an offside (two, in fact, one by Kurzawa when he headed across for Griezmann, and the other by Griezmann himself in finishing) during the move leading to the goal, which was then chalked off.

Later, in the 77th minute, the Spanish player Deulofeu scored and the AR1 raised his flag to indicate offside; the ball ended up in the net and then, after consulting with the VAR, the referee gave the goal.

On both occasions, the referee paid attention to what the VAR indicated without recourse to video replay, as marked by the protocol, since they were factual situations, not ones that were open to interpretation.

In an interview he gave the following day, the German referee said he was very satisfied and said the experience with the VAR had been "a positive test" (EP, 2017):

> It was a positive test because thanks to video evidence we could judge all incidents correctly … the co-operation with the video assistant was precise, quick and very professional. When I signalled to the players that I was communicating with the video referee they dealt with the situation in a sovereign and respectful way.

However, and despite the referee's satisfaction, it was also clear that, after the French goal had been disallowed, the Spanish defenders then protested all of the indications made by the AR2, calling for offside every time (when in fact, the AR2 got every call right); the same thing happened with the AR1 after Deulofeu's goal was given: every time there was an offside given against the Spanish forwards, the players protested.

In other matches used as experiments, there have been situations in which, after the VAR has communicated with the referee, he has reviewed the action and, on occasions, adopted measures that are not entirely in line with the protocol. There was an example of this in the Confederations Cup Final, played in Saint Petersburg (Russia) on 2 July 2017, between Chile and Germany (0–1) (see Table 10.1).

Table 10.1 Referees and VARs during the Confederations Cup Final, played in Saint Petersburg (Russia) on 2 July 2017, between Chile and Germany

Referee	MAZIC Milorad (SRB)
Assistant Referee 1	RISTIC Milovan (SRB)
Assistant Referee 2	DJURDJEVIC Dalibor (SRB)
Fourth Referee	SKOMINA Damir (SVN)
Video Assistant Referee 1 (VAR1)	TURPIN Clément (FRA)
Video Assistant Referee 2 (VAR2)	PRAPROTNIK Jure (SVN)
Assistant Video Assistant Referee (AVAR)	DIAS Artur (POR)

Source: FIFA.

On minute 62:21, the German number 11, Timo Werner, advanced down the left wing and challenged the Chilean number 18, Gonzalo Jara, for the ball. When the action reached the touchline, the Chilean defender elbowed the German attacker, hitting him with his left elbow in the jaw. The German fell to the ground, clutching his face. The referee, who had been following the swift counter-attack, was at a distance of some 18 metres, with a direct view of the action, and awarded a throw-in to the German side.

However, before it was possible for the Germans to take the throw-in, and with 38 seconds having passed since the incident, the referee signalled a "review" and went over to the pitch-side monitor. While that was happening, the message appeared on the stadium screen indicating that a review was taking place.

The message was in line with recent modification to the protocol, which lays down that: "It is very strongly recommended that, as a minimum, a message appears on the stadium screen announcing that a review is taking place" (IFAB, 2017a, 18).

After the review, the referee returned to the spot where the incident had occurred, identified the Chilean player, and showed him a yellow card (since the moment the incident had occurred, 2:25 minutes had passed). Clearly in disagreement, Gonzalo Jara applauded the referee, who then warned him verbally. Instead of the throw-in originally given, play was resumed with a direct free kick for the German team. In total, play was held up for 3:10.

The television audience, for their part, were able to view the replays of the action from a front-on camera only 18 seconds after the incident took place, almost 20 seconds before the referee signalled that he intended to review the action.

Supposedly, in accordance with what the protocol stipulates, the television audience were also able to view the footage, given that it is specifically pointed out that all the broadcasts must be made available to the VAR team. One of the recent modifications to the protocol does in fact indicate that: "The integrity of the VAR system would be undermined if the broadcaster could show footage not available to the VAR/referee which contradicts the VAR/referee decision" (IFAB, 2017a, 6).

From the complete analysis of this incident, it can be appreciated that there was a discrepancy between the VAR team's subjective criterion and that of the referee himself since, according to the protocol, the VAR can only make a recommendation to the referee in cases where a direct red card is involved (not a booking offence), for which reason it is to be supposed that, by their judgement, the elbow to the face was a red-card offence for the Chilean player as it was serious foul play; once the play had been reviewed, however, the referee booked the defender, for which it is understood that he considered the action to be a reckless infringement while challenging for the ball. In this sense, the refereeing decision was taken in accordance with what point 5.1.13 of the protocol states: "taking the appropriate disciplinary action, including cautions (YCs), for any offence not originally seen but which is clearly identified by the review" (IFAB, 2017a, 19).

Given the importance of the match, and the large TV audience, the media made much of this particular incident and there were many negative comments regarding the VAR, some claiming the interruption had been too long for a simple yellow-card offence, others arguing that if it had really warranted a red card, then resolving it with just a booking was sending out the wrong message and, as a result, the usefulness of the VAR system was left in question.

The Biggest Test – Using the VAR System in the 2018 FIFA World Cup

For this competition, FIFA designated 36 referees and 63 assistant referees. As well as this, another 13 referees were designated exclusively for VAR duties (see Table 10.2).

Table 10.3 sums up the main situations reviewed by the VAR during the tournament.

Recent experiences in the Spanish La Liga 2018–19 (First Division)

The fact that in any given game there may be several VAR reviews does not necessarily mean that the refereeing team are especially distracted or are getting it wrong, as it is possible that there have been a good number of events of the type that are included in the protocol as reviewable (goals,

Table 10.2 VARs appointed for the 2018 FIFA World Cup

AL JASSIM Abdulrahman	AFC	QAT
SAMPAIO Wilton	CONMEBOL	BRA
VARGAS Gery	CONMEBOL	BOL
VIGLIANO Mauro	CONMEBOL	ARG
DANKERT Bastian	UEFA	GER
DIAS SOARES Artur	UEFA	POR
GIL Pawel	UEFA	POL
IRRATI Massimiliano	UEFA	ITA
LOPES MARTINS Tiago Bruno	UEFA	POR
MAKKELIE Danny	UEFA	NED
ORSATO Daniele	UEFA	ITA
VALERI Paolo	UEFA	ITA
ZWAYER Felix	UEFA	GER

Source: FIFA.

Table 10.3 Summary of the main situations reviewed by the VAR during the tournament 2018 FIFA World Cup

Situation reviewed by the VAR	N°
Penalty that had not initially been awarded	7
Goal given, when the AR had flagged for offside	2
Dive that had led to a penalty being awarded, then cancelled	2
Mistaken identity of a player who was booked	1
Violent conduct against an opponent out of sight of the referee	1
Penalty awarded then chalked off owing to an offside offence prior to the foul	1
Initial decisions (penalties and/or goals) confirmed	5

Source: FIFA.

penalties, no-goal situations, etc.). Consequently, the number of reviews should not be a parameter for judging the refereeing performance, but rather the level of accuracy of the initial decisions. There are, nevertheless, statistics that are beginning to appear concerning the referees who are getting a greater percentage of difficult decisions right and then having them confirmed by the VAR, and those whose decisions are amended by the system most often.

By way of example, Figure 10.3 shows the ten decisions that were rectified by the VAR on the 29 matches of the first three match days of La Liga 2018–19.

Match	Player	Referee	Decision	VAR	Final decision
Rayo - Sevilla	André Silva	Mateu Laoz	No goal	->	Goal
Rayo - Sevilla	Embarba	Mateu Laoz	Fault (outside penalty area)	->	Penalty
Girona - Madrid	Bernardo	Martínez Munuera	Mistaken identity	->	YC to Muniesa
Espanyol - Valencia	Granero	Del Cerro Grande	Ghost goal		Goal awarded
Valladolid - Barça	Keko	De Burgos Bengoechea	Goal	->	No goal
Madrid - Leganés	Benzema	Jaime Latre	No goal	->	Gol
Alavés - Espanyol	Laguardia	Iglesias Villanueva	No penalty	->	Penalty
Alavés - Espanyol	Sergio García	Iglesias Villanueva	Goal	->	No goal (offside)
Celta - Atlético	Cabral	Mateu Lahoz	Goal	->	No goal (offside)
Barça - Huesca	Alba / Suárez	Melero López	No goal	->	Goal

Figure 10.3 Ten decisions that were rectified by the VAR of the first three match days of La Liga 2018–19 (29 matches)

The Approach of Refereeing Teams to the Use of the VAR

Does the VAR Really Help the Referee's Performance?

The reply to this question seems obvious, given that referees have at their disposal costly technical means that allow them to eliminate any serious mistakes they may make (see Chapter 2). However, from the point of view of the referee, the message they receive every time the VAR communicates with them is: "you've made a serious mistake." And it is thus because, from the outset, the VAR is aimed at avoiding clear mistakes at key moments in a match; situations such as bookings (yellow cards) should not be reviewed by the referee nor should the VAR suggest their being reviewed, presumably – it is debatable – to prevent requests for all the bookings in a match to be subjected to review (it should be sufficient for the referee to decide which ones they would like to review and those which they have seen clearly and therefore need no additional information for).

Would it be advisable to gather all the information before taking a decision?

One aspect that has given rise to a good number of conflicting opinions is the convenience of the referee having access to the VAR system in order to get additional information that allows them to judge an action more accurately before taking a decision, rather than doing it *a posteriori* in order to correct a possible mistake.

Those who defend the first option – that of completing the information before taking the decision – claim that, if said information is available, why not use it from the outset and thus avoid a possible refereeing mistake? Against that, those who are of the opinion that the VAR should only be turned to in the case of a clear mistake being spotted and needing to be corrected, argue this with the aim of preventing referees from taking the easy route of reviewing every play to be sure of their decision and thus interrupting play constantly. The concept of "FORGET about the VAR BEFORE giving a decision, REMEMBER the VAR AFTER giving a decision" (IFAB, 2017a, 11) leans towards this second option, indicating that the referee should take a decision with the information available and then, if and when necessary, being able to change it if they see that they have clearly made a mistake.

Once a decision has been taken, can the referee request a review to be sure that it is the correct one, even though the VAR has not recommended a review?

In principle, when a referee takes a decision it is because, with the elements of judgement they have at their disposal, live and on the playing field, it is deemed to be the correct one. The work of the VAR is only to review (or recommend the referee review in the TV screen beside the pitch) a decision when, in view of the footage available in the VOR, it can be seen that the decision was clearly erroneous or that it is highly likely that it was.

Referees, obviously, are human and capable of making mistakes, which implies that a doubt may arise regarding how correct a decision is the moment after it has been taken. In this case, it seems suitable to make it possible for the referee to tell the VAR that they would like to review a particular piece of action.

For the time being, the only possibility in the protocol that allows for a VAR review before a decision is taken is in situations in which there is reasonable doubt regarding whether an offence is worthy of a booking or a sending off, and concerning the identity of the player to be cautioned, although it is only to be used sparingly.

On rare occasions, when it is unclear whether a penalised cautionable (YC) offence is a sending-off (RC), or who should receive the sanction, the referee may consult the VAR (IFAB, 2017a, 5).

Does the VAR System Require Any Type of Special Psychological Capacity on the Part of the Referee on the Field and/or the VAR/AVARs?

While play is stopped for a review, a great deal is demanded of the referee on the pitch in terms of non-verbal language, since it is a moment of high tension, one in which the players, coaching staff, fans at the stadium, and millions of TV viewers are all looking at the referee, who must show the self-control necessary to give an impression of being calm and maintaining authority, despite being the one at the centre of the decision. If it is an action which, according to the protocol, must be settled by the VAR (a factual decision), the referee must simply wait for the decision to be communicated and then apply it. However, when the VAR only recommends an on-field review, the pressure is very much on the referee, who must take a decision knowing that millions of people are hanging on their every gesture, comment, or indication.

On the other hand, not only must the review process and the taking of a final decision be carried out within the shortest time possible, but it must also, necessarily, be right, since no mistakes are allowed after use is made of the technological tool at the referee's disposal. Not every action is black and white, though, as may be the case, for example, of whether a handball is deliberate or not, if there has been sufficient contact to warrant the awarding of a penalty, or those situations in which both players are tugging each other's shirts during a tussle for the ball.

For their part, and despite not being on the field of play, the VAR is also subject to a level of tension that is difficult to measure, as they are the watchdog – and therefore ultimately responsible – who is there to make sure nothing that happens on the pitch goes unseen. Unlike the referee, who is given a certain amount of leeway when it comes to "human" errors for having accepted that they might have read a situation wrongly, there is no such tolerance for a mistake made by the VAR, because it is supposed that they have had made available to them all the footage from every possible angle and have had time to review it. If we add to this the fact that both the communications and the actions of those present in the VOR are recorded, there is a level of tension that cannot be externalised in a closed room when the decision must be taken hastily. In fact, for those who are not familiar with it, what seems like a comfortable job (watching footage on the screens and letting the referee know if something has happened) turns out to be an activity that is highly stressful and emotionally and physically draining.

As well as this, the referee in the VOR must have the special ability to not fall into the trap of trying to justify their partner's initial decision on the playing field. In an interview he gave to SportTV on 26 July 2018, the Brazilian referee Sandro Ricci recognised the mistake he himself had made while acting as VAR for the England–Tunisia game during the group stage of the 2018 FIFA World Cup in Russia. In his words, during an attack by the English team, and inside the penalty area, a Tunisian defender pulled Harry Kane back and prevented him from shooting. The referee on the pitch, the Colombian Wilmar Roldan, did not signal the infringement and, when Ricci reviewed the footage of the incident, he unconsciously looked for an image that would back the decision of the principal referee, noticed a slight contact by Kane on the defender, and informed Roldan that his decision had been right. Later, however, when the play was looked at again, FIFA determined that it was a clear case of a penalty not being awarded to England that should have been, and included this particular VAR decision within the three clear errors committed during the group stage:

> I'm very comfortable because criticizing my own job is easier. There was a hold from the Tunisian defender on Kane. The first feeling I had it was a penalty, then I looked for a better angle, but sometimes it is better not to look again. I should have immediately called the referee (Wilmar Roldan). Instead of checking what I had seen, I looked for an image that would allow me to support the referee's decision. That was when I found an arm of Kane on the opponent and that was what gave me the chance to support the referee, but it was a mistake, it was a clear penalty. It was one of the three errors of the group phase recognized by FIFA.
>
> ("Ricci admits", 2018)

The Importance of the "Silent Check"

Initially, when the first protocol on how VARs were to proceed was being outlined, the possibility was brought up that team managers could be allowed to request a review of certain situations on a determined number of occasions during a game.

Doubts were raised, however, owing to the view that this could be used in a tactical way and, therefore, the principal objective, namely that of preventing serious mistakes, would be undermined.

On the other hand, it was also thought unadvisable for the players themselves to be allowed to call on the referee for a review on the pitch as it was anticipated that there would be a huge number of calls and later complaints about the hypothetical differences in criteria if the requests of one team were attended to while those of the other went unheard.

To put an end to the debate, the decision taken was to assume that nobody should be allowed to ask for anything to be reviewed given that absolutely everything was being reviewed anyway, with the understanding that, right from the kick-off, the VAR team are controlling absolutely everything that happens on the field, notifying the referee immediately of any situation observed that was within the aspects that are subject to VAR review. In this way, the so-called "silent check" guaranteed that nothing worthy of sanction would go unnoticed by the refereeing team.

It is called "silent check" because, once the VAR has looked over a play, if they find nothing worth informing the referee of, they do not engage in any communication in order to avoid distractions; they only communicate with the referee if, effectively, there is a reason to do so. Nevertheless, in practice, if the conditions of the game allow for it and it might be beneficial, the VAR can confirm, briefly, to their colleagues on the field that the latest decision was indeed correct.

Effects on Players and Team Managers

The fact that there is a system that ensures constant checking has had a coercive effect on players when it comes to committing infringements, especially those that happen "off the ball". An example would be obstructing a player who is not playing the ball from advancing into a position where he might be able to receive a pass or a cross from the one who is, or violent conduct far out of sight of the members of the refereeing team.

At the same time, there has been a notable reduction in the number of infringements involving grabbing hold of or jostling opponents inside the penalty area when there is a large number of players from both teams (in situations such as corners or free kicks close to the area), which traditionally have made it difficult for the referee to judge the different actions that occurred simultaneously.

In the same way, the use of the VAR system has been positive in helping to rid the game of diving, especially when done by artful players who, when properly trained, are able to go down in fast moves in the opponents' penalty area and fool the referee into awarding them a penalty or, pretending to have been tackled hard or fouled by an opponent, are able to convince the referee that the infringement is worse than it really was and thus provoke a caution for the player concerned.

Regarding the relations between the referee and the players, the use of the VAR has had a number of effects that may be considered positive. One of them is that it helps to reduce the tension in a match. All participants are, now, subjected to less tension since, up to a certain point, but in a way that is more realistic than refereeing without the VAR system, they are fairly certain that there are not going to be serious, game-changing mistakes. In the past, any player or team official would vehemently protest a decision because it was their understanding that "if I do not get this decision looked at again, then there will be no way of getting it overturned" and, thus, continued protesting to unacceptable limits. For the referee, it is a "safety net"; knowing that they will not go down in history as the referee who took an incorrect decision that wronged one team or the other frees the referee from the tension that goes with refereeing important matches. For the players, there is the relief of knowing that what really happens is going to be arbitrated and cautioned more fairly, with those players who attempt to cheat the referee by means of diving or other unsporting trickery being shown up for what they are.

As well as this, there has been a notable reduction in the amount and intensity of protesting, since, from the moment in which the referee signals that the action is being reviewed by the VAR, the players remain calm and accept the decision. The certainty that comes with knowing that the action has been reviewed by technical means that go beyond what the human eye can see leads to players having more confidence in the decision being right, whether or not it goes in their favour, and they are more inclined to accept it.

Effects on the Fans and the Media

For the fans and the media, the possibility of seeing the footage that backs the referee's decision once the review process is over would put a stop to the subjective opinions, the doubts, and the suspicions, and the decision, given the technical possibilities available, would be accepted. However, in order to avoid incidents in the case of decisions that could still harbour doubts, the protocol recommends not showing the footage on giant screens in the stadium.

Future Developments

Would It Be Advisable to Expand the Situations Subject to Review?

In general, after the experiences of the VAR system being used in several of the world's major leagues and, particularly, in the 2018 FIFA World Cup in Russia, we can say that the level of acceptance on the part of players, coaching staff, and fans has been generally satisfactory, though there is still room for improvement. It is not, therefore, wishful thinking to imagine the number of situations currently subjected to review being increased, specifically those which may be observed objectively by the VARs, without the presence of the referee being necessary for the review to take place. The following infringements, for example, might be considered for review:

- Penalty kicks:
    - Encroachment in the penalty area by an attacking or defending player. In the current protocol, the only aspect subject to review is if the encroaching player in question has a direct influence on the goal being scored or not.
    - If the goalkeeper steps forward off the goal line before the penalty taker kicks the ball. This is an offence, as the defending goalkeeper must remain on the goal line until the ball has been kicked. This would also apply during the kicks from the penalty mark to decide the winning team.
- Goal kicks:
    - When an attacking player is in, or encroaches in, the penalty area before the ball has left it, and attempts to play it. This is an offence: opponents must be outside the penalty area until the ball is in play.
    - When a defender or any other player touches the ball before it leaves the penalty area. This is an offence: if the ball does not leave the penalty area or is touched by a player before it leaves the penalty area, the kick is retaken.[1]
- Unsporting behaviour:
    - On the part of players and/or team officials in the dugout and technical area.
    - Mass confrontations in which it is difficult to identify the participants.
- Ball in or out of play:
    - Confirm whether the ball crossed the goal line or the touch line, not only in goal or no-goal situations, but in the attacking move also.
- Mistake by the referee in not sending off a player already cautioned when showing them a second yellow card.

Can the VAR System Affect Other Technologies and/or Practices?

One of the technological advances that has been partially introduced, after lengthy trial periods, certification and training, and which has meant considerable expense, is Goal-Line Technology (GLT). The system requires a certain number of cameras to be installed, trained on each goal, with the sole and exclusive aim of detecting when the ball crosses the goal line completely.

Obviously, the introduction of the VAR system, with as many as 30 cameras being employed, means that it should be relatively simple to determine whether the ball has crossed the goal line or not – by means of cameras placed in line with the goal line and either side of the goal – therefore replacing the GLT and its sole function, although with certain limitations (see Chapters 3 and 5).

Another practice that may be affected by the introduction of a VAR system is the participation of the Additional Assistant Referees (AARs), who are currently employed in some competitions, positioned just outside the boundaries of the playing field, close to the goals, and whose task is to determine whether or not the ball has crossed the line completely in goal or no-goal situations – and who would be a replacement for GLT – and to keep a close eye on incidents that occur inside the penalty areas. It seems obvious that both functions can be adequately carried out by the cameras used in the VAR system, which means the AARs' days may be numbered.

Conclusions

While awaiting the progress that will, without any doubt, come with successive versions of the protocol, and after the results obtained during the experimental phase and its introduction into the principal European leagues, the outlook for the VAR system appears very hopeful. Football in general, and referees in particular, should be able to enjoy the advantages that modern technologies offer, and, for that reason, the system must be introduced in as complete a way as is possible. It must, though, be open to later updates, based on experiences acquired over time, but must never lose sight of the fact that the main aim of the VAR system is that it is there to help the referee, not to be an additional responsibility, nor to be a source of criticism on the part of other participants in the game.

Leaving behind this fundamental aim, and reaching a point where games are re-refereed live, would mean the system has failed and lead to the rejection of this technological asset. The ideal thing, at least at the current time, would be to keep the system as simple as possible: referees having access to all footage on a pitch-side monitor only when they deem it necessary to clear up any doubts they may have. The VAR would carry out the silent checking function and would only be able to inform the referee of cases of violent conduct that have taken place out of their sight, or in cases of errors being clearly committed. All other doubts should be cleared up by the referee.

Lastly, it should be stressed that, once a definitive protocol has been laid down, it must be followed consistently, in an unbiased way, one that is fair for all, in order to avoid any mistrust and interpretations that are more or less interested, without letting down the expectations raised that all situations marked as reviewable will, in fact, be examined and the relevant decision will be taken, making the VAR system what it evidently can be:

an extraordinary tool in the hands of referees that will help avoid mistakes being made in football.

Note

1 This offence will be modified in 2019/20 Laws of the Game, effective from 1 June 2019.

References

Bryon, A. (2016, December 15). Modric slams video technology: "I don't like it, it's not football". *AS*. Retrieved from: http://en.as.com/en/2016/12/15/football/1481807287_205369.html

Dunbar, G. (2016, December 22). FIFA moves toward goal of video review at 2018 World Cup. *AP*. Retrieved from www.apnews.com/c34cde8217d44a86b378380bef1b27cb

Elleray, D. (2016, July 22). IFAB workshop features first "live" trials with video replays. *FIFA TV*. Retrieved from www.youtube.com/watch?v=h08-QO05crU

EP. (2017, March 29). Zwayer: "Fue un test positivo el VAR del Francia-España". *AS*. Retrieved from https://as.com/futbol/2017/03/29/seleccion/1490791340_711119.html

IFAB. (2016a). Video Assistant Referees (VARs) used live in competitions and leagues. Retrieved from www.theifab.com/projects/vars/principles-practicalities-protocol

IFAB. (2016b). Video Assistant Referees (VARs) used live in competitions and leagues. *Background & Scope*. Retrieved from www.theifab.com/projects/vars/background-scope

IFAB. (2017a). Protocol version 8. April 2017. Video Assistant Referees (VARs). Implementation handbook for competitions conducting live experiments with video assistance for clear errors in match-changing situations. Retrieved from www.knvb.nl/downloads/bestand/9844/var-handbook-v8_final

IFAB. (2017b). Documents. Circular 10. The IFAB strategy and 'play fair!' initiative. Retrieved from www.theifab.com/search/fair%20play

Infantino, G. (2016, March 5). *FIFA TV*. IFAB agrees to introduce experiments. Retrieved from www.youtube.com/watch?v=_Zvn8OQPkcY

Ricci Admits his VAR Error at the 2018 World Cup. (2018, July 27). *Refereeingworld* Retrieved from http://refereeingworld.blogspot.com/2018/07/ricci-admits-his-var-error-at-world-cup.html

Shread, J. (2018, April 18). VAR decisions to be explained on screens in stadiums at World Cup. *Sky Sports*. Retrieved from www.skysports.com/football/news/11095/11336980/var-decisions-to-be-explained-on-screens-in-stadiums-at-world-cup

Una bandera boicotea el VAR en Portugal … y perjudica a su propio equipo (2018, February, 7). *MARCA*. Retrieved from www.marca.com/futbol/liga-portuguesa/2018/02/07/5a7b3e77468aeb45608b4590.html

11 VAR Experiments in International Matches

Manuel Armenteros

Introduction

At its 130th meeting on 5 March 2016, the IFAB gave the green light to experiments with the Video Assistant Referee system. The use of video to assist referees was not something new.

"'IFAB and FIFA are now leading the debate and not stopping the debate,' said new FIFA President Gianni Infantino following IFAB's annual meeting in Cardiff on Saturday" (Patterson, 2016). In fact, the debate about introducing video technology to help referees had begun some years before.

Some years before the IFAB approved VAR experiments, some voices in the MLS launched the idea of using instant replay in soccer as is done in the NFL or the NBA. Don Garber, the MLS commissioner, defended that bad calls could be reversed and some bad practices on the part of players, such as diving, could be corrected, although they were aware of some risks this might entail, especially interrupting the flow of the game (Levy, 2010).

In 2013, the MLS conducted some experiments with referees wearing a head camera (later known as a RefCam) (see Chapter 14). The idea was based on some rugby experiences and the desire to give viewers better footage from close range and to check if the video footage was also useful for the referee (see Chapter 7).

In 2014, ex-president of FIFA Joseph Blatter, who had previously shown his concern about replacing the referee with machines, announced the feasibility of each coach being awarded two video challenges per match (Gibson, 2014). "We've got to talk to US Soccer, we've got to talk to FIFA, we've got to make sure the technology works, but you should know that MLS is a supporter of the idea", Garber said (Murray, 2015).

Jeff Agoos, the head of competition for the MLS, explained that they had been quietly running trials for the past two seasons and the data proved that video replay could work (Murray, 2015). At that time, the trials were focused on using video replay for three types of calls: red cards, penalty kicks, and goals. They accumulated a lot of information about the time consumed with video replay, including the reviewing process and the communication between the refereeing team. They laid down the main principles that later on would be developed in the IFAB and FIFA document about the protocols in VAR experiments.

But they were not alone in this innovation process. On 27 February 2015, Patrick Nelson, as chair of Advisory Panels, explained to the members of the 129th IFAB Annual General Meeting (AGM) that KNVB had conducted an ("offline") experiment, using referee video assistance based on video footage provided by broadcasters' cameras. He reported on the advantages for referees to receive feedback from a video assistant via a headset. Nelson also mentioned another challenge system for team managers and coaches to request a review of referees' decisions, which had previously been presented at the 64th FIFA Congress in São Paulo in 2014. Video replay for match officials was placed on the IFAB agenda.

MLS Commissioner Don Garber, a public supporter of video replay, offered some months later his first hint that the technology might actually be used in MLS competition. Peter Walton, Professional Referee Organization chief, also added:

> It is not there to referee the game and it is not there to re-referee the game. I think it will enhance the reputation of referees because people will actually see how many decisions that referees get right the first time.
> (Murray, 2015)

130th IFAB Meeting: VAR Experiments Are Approved

On 5 March 2016, the IFAB agreed at its 130th meeting to conduct experiments on the use of video as an aid to match officials in certain situations, just four years after approving Goal-Line Technology.

The IFAB accepted that the technology should be tested in two phases and with a protocol that specified in which situations it should be used. The first phase, the "offline" (with no monitor on the field), and "live" (with an on-field monitor), after the IFAB agreed with the organizers of the competition.

In the offline option, the referee could only receive information from the VAR team through the audio communication system. On the other hand, in the live option, the referee had a monitor on the field of play that he could use to review play situations.

In both modalities, the VAR team were in a room called the VOR (Video Operator Room), where all video camera feeds used by the TV broadcaster were received. From this location, the video assistant referees team watched the plays and chose the best camera shot that had captured the incident, so that it could be consulted by the referee in the field monitor, if necessary.

After that experiment, the IFAB conducted several workshops that allowed for information to be gathered from club directors, team managers, players, referees, and fans, and created a protocol to be followed homogeneously during the experiments.

In a meeting with Johannes Holzmüller, Head of FIFA's Football Technology Innovation Department, on 17 May 2016, professor Manuel Armenteros made a presentation in which some relevant points regarding the use of video in refereeing were considered, among them the importance for referees to receive all video camera feeds independently and for new strategic cameras to be placed in specific positions to help them.

131st ABM–IFAB Meeting

The 131st Annual Business Meeting took place on 3 November 2016, and one of the points of discussion and/or decision was the VAR system.

At that time, the IFAB secretary, Lucas Brud, hinted in an interview conducted on the British ITV network ("Video referees", 2016) that VAR could be implemented in the 2018 FIFA World Cup Russia, which would depend on the referees having the necessary training and the results of VAR experiments being satisfactory.

The expectation was not to achieve 100 per cent accuracy in decisions for every single incident, but to avoid clearly incorrect decisions that are predefined "game-changing" situations – goals, penalty decisions, direct red card incidents, and mistaken identity (FIFA, 2016a).

Sample and Methodology

The sample analysed were all the international matches organized by FIFA, and all the play situations where the VAR intervened to uphold or change the previous decision. The so-called check reviews have not been included, because this represents all the play situations where the VAR team has marked a potential clear error or a serious incident/offence, but not communicated to or been reviewed by the referee. This data was collected internally by FIFA during the tournaments, and tagged and classified by the IFAB and the University of Leuven, Belgium, to prepare the FIFA official report.

The search for information has been carried out taking as a timeframe the date when the IFAB gave its approval of the VAR experiments (5 March 2016) to 3 March 2018. Searches have also been extended to previous years to verify some information about older experiments carried out by a few federations that tested the use of video replay in their leagues.

All the comments made by the main football actors (referees and referees' committee, sports journalists, team managers, players, and club chairmen) on TV and radio sports programmes, sports magazines (online and printed), and referees' blogs, have been followed.

To compare the information offered by the official sources, searches have been done in English and Spanish, in Google Chrome and online media using the following keywords: "Massimo Busacca", "David Elleray", "Lukas Brud", "Pierluigi Collina", "Gianni Infantino", "VAR" and "video arbitraje".

The information collected is presented in a chronological order to observe the evolution of the discourse of the main actors and to check how the main problems encountered in the implementation period have been addressed.

2016 Friendly Matches

During 2016, the VAR system was used in three friendly matches between September and December, just before the FIFA Club World Cup.

In an Italy vs France friendly match, on 1 September 2016, the "off-line" VAR was used with very favourable results.

"Chiellini for instance said 'Rosso!' After reviewing the scene, my VAR informed me nine or 10 seconds later that it was sufficient to show the yellow card. The players immediately accepted my decision" (FIFA, 2016b).

On 11 November 2016, in the international friendly between Italy and Germany, two VARs, Manuel de Sousa and Danny Makkelie, were included on the FIFA referees list. Time management was an important issue and FIFA's Head of Refereeing explained to the Italian radio and TV broadcaster RAI that, "In five seconds they assessed that Volland's knee was in an offside position" (Calemme, 2016). At that time, in the opinion of FIFA Head of Refereeing Massimo Busacca, the review of the incident "should not take more than five, six seconds" or "two seconds more if we need one angle more" (Dunbar, 2016).

The president of FIFA, Gianni Infantino, affirmed at the end of the friendly that the VAR was not ready to be tested in the "live" option and that "the 'video-assistant referee' system still needs to be tested further before it can be implemented into live games" (Scott, 2016).

In the last international friendly, an U-21 fixture between Italy and Denmark on 19 November 2016, there were no incidents.

The opinion of the FIFA president was in marked contrast to the view of the technical director of the IFAB, who, a few weeks later, expressed that he was very satisfied with the achievements in training and technological development, and hinted at the use of the live option (with a monitor by the pitch) in the 2016 FIFA Club World Cup in Japan. It is likely that the practices carried out in the seminars motivated so forceful a statement.

"The progress FIFA has made on this project, especially around educating referees and ensuring all technical requirements are met, has been impressive," said David Elleray, Technical Director of the IFAB (IFAB, 2016a).

Although, in mid-November 2016, Gianni Infantino stated publicly that VAR was not ready for experiments with a monitor, the "online" option with a monitor were put into practice in December, in the 2016 FIFA Club World Cup in Japan.

FIFA Club World Cup in Japan and Controversies

Four games had been played and the VAR had not made an appearance. It was in the fifth match, on 14 December 2016, Atlético Nacional vs Kashima Antlers, when the most controversial situations in the tournament happened. Referee Viktor Kassai awarded the first penalty in an official FIFA match with the help of the VAR, but it took about two minutes to make the decision. The play aroused controversy, in addition to the extended use of time to show the final decision, because the Japanese player Daigo Nisho was offside when he was knocked down by the Colombian Orlando Berrío. According to FIFA, it was not offside because the "player was deemed to not be offside, having been unable to challenge the opponent for the ball and the VAR rules did not foresee using the VAR for offside situations" (FIFA, 2016c).

During three consecutive matches, in the semi-finals and final, there were a series of situations that led the FIFA authorities to have to face the media and justify the refereeing decisions.

The semi-final between Club América and Real Madrid threw up a new controversy, perhaps the most important because of the public statements later made by players such as Cristiano Ronaldo and Luka Modric.

The referee, Enrique Cáceres, ruled out Cristiano Ronaldo's goal while he was waiting to confirm his possible offside position. The pressure of the opposing team to restart the match confused Ronaldo, who did not understand what was happening. Finally, the referee gave the goal as valid, with the consequent controversy in sports newspapers: "Real Madrid players unenthusiastic about video refereeing" (Félix, 2016) or "Mondiale per Club, Var delle polemiche. Modric duro: 'Non è calcio'" (La Gazetta dello Sport, 2016), and "Mondial des Clubs: L'arbitrage vidéo 'Crée de la Confusion' said Zinedine Zidane" (L'Équipe, 2016) .

Important players such as Modric and Ronaldo expressed their great unhappiness with the implementation of the repetition technology: Modric, "I do not like it, it's not football, I hope the rule does not continue"; Cristiano Ronaldo, "It creates a lot of confusion!" (Bryon, 2016).

The coaches involved in the tournament also spoke about the technology, especially Zinedine Zidane, who was concerned about the timing of taking decisions.

> We must adapt to what FIFA want to do. In Ronaldo's goal, there was bit of confusion. Things must be done clearer. But we must adapt. Improving things is good for football. But it must be clearer for everyone and it was not in this move.
>
> (West, 2016)

The media reacted with headlines that reflected the confusion generated by the new refereeing system: "Video ref 'confusion' snags trials at Club World Cup in Japan" (Warsaw, 2016), "VAR system clouds the Club

World Cup" (Relaño, 2016), and "New controversy with VAR with Real Madrid as protagonist" ("Nueva polémica", 2016). FIFA and its top officials were forced to explain and clarify to the media the incidents and situations that had occurred.

Despite what happened, FIFA's message was that the experience had been very positive: "The results are extremely positive" [Infantino] (Armstrong, 2016). "When Ronaldo scored the second goal it was a clear situation and there was no offside" [Van Basten]. "So, in the end it was good, and the right decision was made. Only the communication between the video assistant referee and the referee was not optimum" (Armstrong, 2016). "The main point of technology is that no one loses because of a refereeing mistake" [Busacca] (García, 2016).

The controversy and declarations of the top football leaders in FIFA led other important actors in the world of football to voice their opinions, but in this case, not against VAR: "I am in favour of it … I am in favour of everything that helps referees" [Luis Enrique] (Marsden, 2016), and "VAR will not make football lose its essence"[1] ("Del Bosque", 2016).

More prudent was David Elleray, IFAB's Technical Director, who, when interpreting the evaluation of the results, emphasized that this was a trial phase and they would have to wait until the end of that period to decide whether or not to implement the VAR:

> But we have to remember that this is the first experience some of these referees have had with video assistance … Checking replays, communication between the officials and then the review itself can take time when it's a completely new situation for them.
>
> (Warsaw, 2016)

There was another incident in the final between Real Madrid and Kashima Antlers. Referee Janny Sikazwe[2] did not show a second yellow card to Sergio Ramos in the 54th minute, although the official initially reached for the pocket where the card was. The replay showed a clear foul by Ramos, but the referee missed it, and was about to caution Casemiro, which caused great bewilderment among the Japanese players ("El árbitro", 2016) and the Kashima Antlers' team manager, Kashima Masatada, who said, "The referee lacked courage, which was regrettable" (Matthew, 2016).

131st AGM–IFAB Meeting and Second Stage of the Experiments

The 131st AGM–IFAB meeting had, once again, the VAR as one of the main points on its agenda. Its members received updates about the previous workshops and from the data collected from more than 20 matches where the VAR protocol had been tested. In addition, members received information on the main learning areas that would be incorporated into the "live" experiments (IFAB, 2017).

It was also noticed that Australia, Belgium, Brazil, the Czech Republic, France, Germany, Italy, Mexico, the Netherlands, Portugal, Qatar, and the USA had all signed or would sign contracts to be part of the experiment at that time.

They had noticed great progress with regard to the 2016 FIFA Club World Cup Japan, but they were still aware that they could do better (FIFA, 2017a). The use of the VAR, according to Massimo Busacca, should be used "to eliminate clear mistakes – the mistakes that people, years later, still remember", and, as David Elleray affirmed, "It is not going to solve those grey-area decisions which will always be debated, and it's never been intended that it should be that" (Burton, 2017).

FIFA U-20 World Cup, Korea Republic 2017

In the match between Italy and Uruguay, in the first youth tournament where the VAR was used, one of the first plays that was not detected by the commentators and the players themselves, was noticed by the VAR as a penalty once the game was stopped, almost a minute and a half after the foul had been detected. The press did not understand the decision to award a penalty, almost two minutes after the play had occurred.

The Uruguayan player Joaquín Ardaiz went down in the penalty area and, after the replay on TV, journalists understood the decision taken by referee Walter López.

In the match between Argentina and England, an Argentinian player became the first player to be sent off with the help of VAR. After a struggle between Lautaro and Tomori, Lautaro elbowed his opponent and fell to the ground. While he was being attended to on the ground, and when it looked like the game was about to restart, the referee was informed of the aggression, and after consulting the field monitor, decided to penalize it with a red card.

In the match between South Korea and Guinea, the video technology also served to chalk off a goal for offside. The commentators felt that the virtual offside line had been missing on the TV broadcast.

In the match between Venezuela and Vanuatu, the VAR cancelled out an offside that ended in a goal because the Venezuelan player had received the ball in a headed clearance from a Vanuatu defender.

According to data collected from FIFA, the use of VAR in the U-20 tournament, still in the experimental phase, was the "most prominent" test to date: of the 52 games played, 15 decisions were reviewed by the VARs, and 12 decisions were changed based on the VAR input (seven game-changing decisions corrected) (FIFA, 2017a).

Workshop, 30 May 2017

Between 30 May and 1 June 2017, the fourth workshop was held in Zurich. There, the associations and leagues provided the results of the previous experiments. They also participated in drawing up a new version of the protocol and implementation procedures for the future.

At that time, the leagues of Australia, Brazil, France, Germany, Italy, Holland, Portugal, and the United States confirmed their participation in the experiments.

Likewise, other federations such as Belgium, the Czech Republic, Poland, Qatar, Saudi Arabia, South Korea, and Turkey showed interest during the course of 2016 and 2017 in participating in future experiments.

FIFA Confederations Cup, Russia 2017

In the opening press conference on 15 June 2017 before the Confederations Cup started, VAR was one of the main topics on the agenda. Marco van Basten, FIFA's Chief Technical Development Officer, Massimo Busacca, FIFA's Head of Refereeing, and Johannes Holzmüller, Head of FIFA's Technology Innovation Department, were all present.

During the competition, the VAR system was used to change or uphold some refereeing decisions.

First, in the 2–2 draw between Portugal and Mexico, by disallowing a valid goal by the Portugal player, Pepe, who was not in an offside position.

In the match between Cameroon and Chile, after correctly disallowing a goal by Eduardo Vargas, and then chalking off, this time wrongly, another goal by the Chileans, the decision was reviewed and it was decided to give the goal as good.

The match between Australia and Germany, which the Germans won 3–2, is also remembered because of the VAR team's intervention. A shot from Rogic hit his teammate Juric's hand and was parried by the goalkeeper, Bernd Leno. Juric then pounced on the rebound and scored the Australians' second goal of the game. The play was reviewed by means of video replay and, although the ball did make contact with the Australian forward's hand, the goal was given.

The VAR appeared again in the match between Mexico and Russia (Group A), and with the 2–1 triumph in the bag for the Mexicans, the VAR correctly ruled out a goal by Mexican defender Hector Moreno because he was in an offside position.

Finally, in the match between Germany and Cameroon, in the last duel of the group stage, referee Wilmar Roldan sent off Emre Can for a rough challenge on Cameroonian player Sebastien Siani. After a VAR review, the decision had to be reversed, and Ernest Mabouka was booked as the real offender.

Press Conference after Group Stage, 26 June 2017

After the group stage of the FIFA Confederations World Cup Russia 2017, the feeling of satisfaction in FIFA was evident in their statements to the media.

Gianni Infantino said:

> I am extremely happy with VAR so far. We have seen how video assistance has helped referees to make the correct decisions. This is what VAR is all about. The VAR tests during this Confederations Cup are also helping us to improve the processes and fine-tune communication. What fans have been waiting for over so many years is finally happening. This is a milestone tournament. Video Assistant Refereeing is the future of modern football.
>
> (FIFA, 2017b)

Massimo Busacca, the Head of Refereeing, added that, "The tests have been very good so far, the most important thing for us, by far, is that no clear mistakes were made."

According to Busacca, they had to practise in the future to manage different angles under pressure and reduce the response time:

> We need to keep the waiting time as short as possible. However, choosing the right camera angle as quickly as possible, under pressure, is something that is still not second nature. After the conclusion of the tournament, we will work for the future.
>
> (FIFA, 2017c)

In the opinion of FIFA's Head of Refereeing, VAR can reduce many failures, but not eliminate them: "We are convinced [that], if used correct[ly], it can reduce many mistakes. But not eliminate [them]. We are here only to reduce" (Haydon, 2017).

Video replay in the group stage had helped the referee change six decisions and confirmed another 29 decisions involving "major incidents". He also called on member associations to work with VAR every day.

He also expressed unhappiness about the fact that they had only had five days to prepare for the competition when the member federations had doubts about using the technology.

During the group stage, VAR was used ten times to review the referee's decision, and on six reviews the decision was changed.

Semi-finals

In the semi-final match between Portugal and Chile, on 28 June 2017, Iranian referee Aliareza Faghani, although he was in a good position, saw how the Chilean Francisco Silva was brought down by defender Jose

Fonte and did not blow for a penalty or give him a yellow card for having tripped Francisco Silva, stopping an obvious goal.

The media reacted with headlines like: "Portugal vs Chile brings more VAR controversy as clear penalty is NOT given by television officials" (Whaling, 2017), or "Confederations Cup 2017: 'VAR confusion, Russian hospitality & Ronaldo complaints'" (Begley, 2017).

Even FIFA's Head of Refereeing, Massimo Busacca, recognized that officials were still on a learning curve with trials of VAR at the Confederation Cup ("Fifa, Busacca", 2017).

According to the Head of Refereeing, the VAR team reported that for them it was a questionable foul and the match continued. The VAR can make a difference, but when it is not clear, the option taken on the field remains a priority. This is what the protocol of the IFAB indicates, as it also indicates that VAR is used when there are clear and obvious cases. Where it is not clear, the referee's decision on the pitch should be maintained ("Fifa, Busacca", 2017).

Finals

It was quite different for the Mexicans, who did not get a VAR revision when Héctor Moreno was brought down in extra time, and with the score against them.

In the finals, in the duel for the third place between Portugal and Mexico on 2 July 2017, Portugal benefitted from a VAR review after Rafael Márquez had brought down Adrien Silva in the Mexico area; the result was a penalty for Portugal.

Press Conference at the Closing of the FIFA Confederations Cup Russia 2017

President of FIFA Gianni Infantino, Russian Deputy Prime Minister Vitaly Mutko, Chairman of the FIFA Referees Committee Pierluigi Collina, and the executive director of the Local Organizing Committee of the tournament, Alexey Sorokin, spoke to the media and answered questions from journalists.

The VAR was the most important point in the press conference. The FIFA President highlighted the positive results of VAR in his speech:

> We also wrote history here in Russia with VAR and from my side that has been a great success as well. If a problematic tournament looks like this one, well, I want to have many problematic tournaments going forward.
>
> The VAR is a very positive tool for the sport as it helps referees avoid committing mistakes. That is what has happened during this competition, and I share the view of the FIFA President: we are very happy with the results. We are aware we can improve, but it would be very surprising after so few matches if it was perfect.
>
> (Infantino, 2017)

Copa de Libertadores and Copa Sudamericana

In both championships, the Copa de Libertadores and the Copa Sudamericana, the live trial option was used – although it was in New Jersey, from 19–21 July 2017, when the live trial option was used for the first time by referees during a workshop organized by IFAB in cooperation with Major League Soccer (MLS) (FIFA, 2016d).

Copa de Libertadores

The president of the National Commission of Referees of Peru (CONAR), Víctor Hugo, had anticipated in July 2017 that VAR would be used in the semi-finals of the Copa Sudamericana and the Copa de Libertadores ("El polémico", 2017).

The president of the Conmebol, Alejandro Domínguez, announced on 28 September that VAR would also be used in the finals of the Copa Sudamericana on 6 December (first leg) and 13 December (second leg) 2017.

"We believe that it is one more instance of Conmebol making a commitment in favour of justice within the field of play and to promote Fair Play. It is another tool for referees to do their work much better,"[3] said the chairman of Conmebol, Alejandro Domínguez ("Final de", 2017).

The VAR's first outing in the Copa de Libertadores was on 24 October 2017 in the semi-final (first leg) between River Plate and Lanús, with referee Sandro Ricci (Brazil) as VAR. The president of the Conmebol Refereeing Commission, Wilson Seneme, described the first experience with the use of the VAR in the Copa Libertadores, which took place in the first semi-final between River Plate and Lanús as "successful": "It was a great success, there was excellent officiating and the VAR did not need to be used,"[4] commented Seneme ("Conmebol considera", 2017).

The semi-final (second leg) between Lanús and River Plate on 31 October 2017 would be the first intervention of the VAR in a Conmebol competition. The VAR was used for a refereeing mistake that favoured Lanús and its use was excluded in a handball by Silva in the area that would have favoured River in the first part, and a foul on the player Scooco.

"The feeling was that the VAR was not applied equally on one occasion and the other, but that its use favored Lanús,"[5] argued journalist Vernónica Brunati at the end of the game (Brunati, 2017). "Colombian referee Wilmar Roldán received harsh criticism for calling a penalty in favour of Lanús and denying another one to River Plate"[6] ("Fue o no", 2017).

The *Olé* newspaper wondered why referee Wilmar Roldán did not request the VAR review in a supposed penalty by the player Iván Marcone inside the area ("No iba", 2017). Wilson Seneme recognized a mistake in the protocol: "The play should have been reviewed on the field by Roldan, that was the mistake. However, it was unclear, and for the referee it was not a penalty and VAR assistants did not consider this play should be reviewed[7]" ("No iba", 2017). The officiating generated a great discussion in the media and the use of VAR came in for criticism.

In contrast, the VAR was used successfully in two decisive actions that "probably would have qualified the River for the Cup final" ("Cómo hubiera", 2017).

The time came for the final of the Copa de Libertadores 2017 between Grêmio and CA Lanús on 22 November 2017, and the VAR did not appear either, especially in the push from Pasquini on Jael. Chilean referee Julio Bascuán did not decide to review the play ("Hasta Stevie", 2017), which Grêmio's manager, Renauto Graucho, would criticize, as would the journalists, especially after the outlay of $850,000 (Mauri, 2017).

The return game, on 29 November, between CA Lanus and Grêmio, had no actions where the VAR was used either to uphold or change a decision.

Copa Sudamericana

The use of the VAR system in the South American Cup had gone unnoticed.

The replay of the plays on the big screen had been in the background until then, despite the bewildered in-stadium fans' experience that "pointed to the need to inform fans about what was happening" (Armenteros, 2017):

> The IFAB has to consider very seriously what to do with those waiting times and replays on the screens of the field. Soccer is a spectacle, and almost a minute of waiting for the referee's resolution is a radical change in the amateur's emotional experience. Sharing the video replay with the fans can improve the image of transparency in the game and will make them accomplices of a process to seek sport justice.[8]
>
> (Armenteros, 2017)

On 19 December 2017, the *Sport Bild* newspaper (Boßmann, 2017) leaked that FIFA and the IFAB would be considering this measure as an instrument of communication with fans and spectators.

FIFA Club World Cup 2017

In the semi-finals of the FIFA Club World Cup 2017, VAR was once again at the centre of controversy concerning its use. In the game between Al Jazira and Real Madrid, on 13 December 2017, there was a highly controversial incident.

Brazilian referee Sandro Ricci initially ruled out a goal by Real Madrid's Casemiro, adjudging that he had committed a foul on an Al

Jazira defender during the build-up to it. Ricci was informed by the VAR, however, that it was not clear that the Brazilian player had committed a foul, and so he rectified his initial decision. But while he was making the sign to indicate a goal, the VAR team continued reviewing the play and decided that Benzema had been in an offside position, and so, once again, the referee changed his decision for a second time, this time owing to an offside. Madrid manager Zidane, as well as Madrid players Casemiro and Modric, complained that there had been no interference in play by Benzema and that Casemiro himself had been in a legal position. The goal was eventually disallowed after several minutes of uncertainty. It had taken almost three minutes to resolve this situation.

The Spanish commentator and ex-referee Andújar Oliver said, "there must be two regulations, one for Conmebol and one for UEFA. The Brazilian referee disallows the goal for an offside that exists but has no influence. This action in La Liga would have been considered correct"[9] (Oliver, 2017) – a conclusion shared by international Spanish referee Eduardo Iturralde González (Iturralde González, 2017).

After what happened, Gianni Infantino was convinced that VAR was the best solution to minimize refereeing mistakes.

The main headlines on that day:

- "FIFA has no doubts: the VAR will continue in the Club World Cup" (Díaz, 2017)
- "Players criticise VAR at Club World Cup, but opinion split as FIFA say it was 'seamless'" ("Players criticise", 2017)
- "The VAR test time will end in a few months. The results so far have been very encouraging, very positive"[10] ("Infantino considera", 2017).

132nd IFAB Annual Business Meeting, 22 January 2018

In early January 2018, a day before announcing the date of the Annual Business Meeting in Zurich, where VAR would be the main topic (IFAB, 2018a), the press published statements of the FIFA president, who wished to implement VAR in Russia, a declaration that was reinforced by the technical director of IFAB, David Elleray, who claimed that after almost a thousand games with video refereeing, the tests had produced better results than expected (Gabilondo, 2018).

IFAB Annual General Meeting

In the opening of this meeting, Gianni Infantino "stressed [that] the requests for video assistance for referees came from all over the world and that everyone should realise the importance of the decision to be taken about Video Assistant Referees (VAR)" (IFAB, 2018b).

Among other points, experiments about VAR filled the agenda of this annual meeting. A presentation about experiments carried out with VAR was made to explain and summarize all the activities related to this innovation.

IFAB Secretary Lukas Brud explained that the workload was much greater than originally anticipated; the main challenges for participants were effective VAR/referee education, project management and planning, technology and communication, existing technical infrastructure, and financial resources. He also stressed the crucial importance of informing fans in attendance so they understand what is happening when the game is delayed for a check or review.

David Elleray focused on the fact that the "successful" reviews are for situations which are not "bad" mistakes but situations where the referee/assistant could not see the incident clearly or where the decision was extremely difficult.

Regarding the reaction from some players, David Elleray explained that, overall, the reaction to VARs had been that players generally accepted the final decision when replays were used, but preferred on-field reviews if a major decision (e.g., a penalty) is changed. Some wanted more decisions included; some wanted fewer, such as only factual ones.

David Elleray also added that the processes/protocols were not yet fully understood, but that once people were familiar with how the system operates, the general reaction would be positive. He mentioned that "with more practice, the understanding of what can and cannot be reviewed increases, as does the identification of 'clear and obvious errors' and the speed of the review process" (IFAB, 2018b).

Johannes Holzmüller, for his part, highlighted the importance of separating the referee communication system from the VAR system and identified the best solutions for the virtual offside line. He also mentioned the automatic VAR information system developed to keep spectators involved and aware of what is happening on the pitch, "to improve transparency and accuracy".

Conclusions

The VAR system has been one of the most important innovations in the world of football. An idea that some stakeholders clamoured for initially, and that Joseph Blatter suggested partially in 2014, eventually became a project that was led by Gianni Infantino two years later. We can affirm that the VAR has been one of the most important topics in FIFA meetings ever since the trials were given the go-ahead, on 5 March 2016. Since then, FIFA has spared no effort in applying all available resources to provide the necessary technology and teach referees how to use it.

Trials in FIFA competitions were a place to detect possible weaknesses of the system. The first stage, "offline", showed that a period was necessary for the referees to familiarize themselves with the protocol and the difficulty of managing the response time in some plays. It was also crucial to observe the reaction of the players and learn how to manage the communication between players and the VAR team during the game.

It also served as a training ground, an important step to observe when the move to the "live" option could take place. The move from the "offline" to the "live" option was made in December 2016, at the FIFA Club World Cup 2016, and it is important to highlight Infantino's statement one month before adopting this modality, in which he had said that VAR was not ready for this step. And, although he said that, the use of the on-pitch monitor was activated, which means that the staff involved in the trials pushed the next stage forward over the president's head. And it was precisely in this competition where the enormous deficiencies in communication and time management between the refereeing team was observed.

Important players such as Cristiano Ronaldo and Luka Modric expressed their discontent about VAR to the whole world. The 2016 FIFA Club World Cup, held in Japan, was a key moment for spotting and identifying "the worst-case scenarios", mentioned in the points approved at the 130th IFAB meeting on experimentation in the use of the video assistant referee (FIFA, 2016a).

The 2017 FIFA U-20 World Cup, held in Korea Republic five months later, was a good scenario in which to improve communication and practise with the system in the live option in international competitions. The championship finished with the VAR being used on 15 occasions, but with no controversy.

Another important aspect of this innovation was the role of the workshops in the communication and organization of the experiments. Just one year before the FIFA Confederations Cup Russia 2017, there had already been five workshops and at least three information meetings with competition organisers and participants.

The 2017 FIFA Confederations Cup, held in Russia, still highlighted important aspects that needed to be corrected, and Head of Refereeing Massimo Busacca acknowledged in an interview that they were still on a learning curve.

And the Club World Cup, at the end of 2017, helped FIFA to reinforce the VAR's usefulness to reduce refereeing mistakes, despite the fact that Real Madrid players criticized once again the way the VAR system was used. The results did not match with FIFA's positive statements.

In the Copa de Libertadores and the Copa de Sudamérica, 2017, in the Conmebol tournaments, there was a clear dissatisfaction in the media concerning the use of the VAR.

The IFAB ABM on 22 January 2018 ratified FIFA's idea of applying the VAR in the World Cup, which was backed in public by other important members, such as IFAB Technical Chief David Elleray. Time management was also discussed as an important issue, and IFAB Secretary

Lukas Brud reminded those in attendance of the importance of informing fans of VAR decisions.

Video technology has caused a division between supporters and detractors of the system, the latter having found one error after another. A major source of disagreement and media reaction to VAR experiments has been in South America. Future investigations should be conducted to observe if the introduction of the VAR into different leagues and confederations is successfully accepted or not, and why.

Notes

1 Original: "El videoarbitraje no hará que el fútbol pierda su esencia".
2 In November 2018, top Zambian referee Janny Sikazwe was suspended by the Confederation of African Football (CAF) for alleged biased officiating during the Champions League semi-final match between Tunisian side Esperance and Primeiro Agusto of Angola.
3 Creemos que es una apuesta más de la Conmebol en favor de la justicia dentro del campo de juego y de promover el *Fair Play*.
4 Original: Fue todo un éxito. Hubo un excelente arbitraje y el VAR no necesitó ser usado [idem].
5 Original: La sensación fue que el VAR no se utilizó de manera equitativa en una y otra ocasión, sino que su utilización favoreció a Lanús.
6 Original: "El árbitro colombiano Wilmar Roldán recibió duras críticas por pitar un penal a favor de Lanús y negarle uno a River Plate".
7 Original: "La jugada debió revisarse en el campo de juego por Roldán, ahí estuvo el error. Sin embargo, fue una jugada de interpretación que para el árbitro no fue penal y para los asistentes del VAR tampoco mereció ser revisada".
8 Original: La IFAB tiene que considerar muy seriamente qué hacer con esos tiempos de espera y la repetición en las pantallas del campo. El fútbol es un espectáculo, y casi un minuto de espera a la resolución del árbitro es un cambio radical en la experiencia emocional del aficionado.
9 Original: "debe haber dos reglamentos, el de la Conmebol y el de UEFA. El colegiado brasileño anula el tanto por un fuera de juego que existe pero que no tiene influencia. Esta jugada en LaLiga se hubiera dado por válida".
10 La FIFA no tiene dudas: el VAR seguirá en el Mundial de Clubes.

References

Armenteros, M. (2017, August 7). El Videoarbitraje: un cambio en la experiencia emocional del aficionado. Retrieved from https://es.linkedin.com/pulse/el-videoarbitraje-un-cambio-en-la-experiencia-del-manuel-armenteros

Armstrong, J. (2016, December 17). FIFA president defends video replays at Club World Cup. Retrieved from http://bigstory.ap.org/article/b86cf4c041384a91a43f36bc449adc06/fifa-president-defends-video-replays-club-world-cup

Begley, E. (2017, July 1). Confederations Cup 2017: VAR confusion, Russian hospitality & Ronaldo complaints. Retrieved from www.bbc.com/sport/football/40447230

Boßmann, V. (2017, December 19). Mehr Transparenz für die Fans. Für WM 2018: Fifa testet Videobeweis-SMS. Retrieved from https://sportbild.bild.de/fussball/inter

national/fussball-international/wm-2018-fifa-regelboard-ifab-test-videobeweis-sms-54245258.sport.html

Brunati, V. (2017, October 31). Una semifinal marcada por los errores arbitrales y el mal uso del VAR. Retrieved from https://colombia.as.com/colombia/2017/11/01/opinion/1509506539_967594.html

Bryon, A. (2016, December 15). Modric slams video technology: "I don't like it, it's not football". *AS*. Retrieved from http://en.as.com/en/2016/12/15/football/1481807287_205369.html

Burton, S. (2017, July 3). Jara's elbow shows video referees need practice to end controversies. Retrieved from www.theguardian.com/football/2017/jul/03/video-refereeing-confederations-cup-final-fifa

Calemme, M. (2016, November 16). Las pruebas dan resultados satisfactorios y la idea es seguir mejorando: "Todavía no tenemos la tecnología para los fuera de juego, pero llegará". Retrieved from https://as.com/futbol/2016/11/16/internacional/1479250803_704270.html

¿Cómo hubiera sido la Copa Libertadores sin el VAR? (2017, November 30). *Gol*. Retrieved from www.goal.com/es/noticias/como-hubiera-sido-la-copa-libertadores-sin-el-var/fane8n8jmrj41rgx1m2fjtidp

Conmebol considera "un éxito" el estreno del VAR en Libertadores. (2017, October, 25). *Antena 2*. Retrieved from www.antena2.com.co/futbol/conmebol-considera-un-exito-el-estreno-d#ixzz52TNrFBe4

Del Bosque: "El vídeo no hará que el fútbol pierda su esencia" (2016, December 16). *EFE*. Retrieved from www.mundodeportivo.com/futbol/20161216/412669104800/del-bosque-el-video-no-hara-que-el-futbol-pierda-su-esencia.html

Díaz, J.F. (2017, December 14). La FIFA no tiene dudas: el VAR seguirá en el Mundial de Clubes. Retrieved from www.marca.com/futbol/mundial-de-clubes/2017/12/14/5a32a83046163f786f8b45cf.html

Dunbar, G. (22/09/2016). FIFA moves toward goal of video review at 2018 World Cup. *AP*. Retrieved from www.apnews.com/c34cde8217d44a86b378380bef1b27cb

El árbitro le perdonó la segunda amarilla a Ramos con empate. (2016, December 18). *AS*. Retrieved from http://futbol.as.com/futbol/2016/12/18/internacional/1482064251_726920.html

El polémico VAR se podría implementar en la Copa Libertadores. (2017, July 3). *RPP*. Retrieved from https://rpp.pe/futbol/futbol-mundial/el-polemico-var-se-podria-implementar-en-la-copa-libertadores-noticia-1061886

Félix, J. (2016, December 15). Real Madrid players unenthusiastic about video refereeing. *Marca*. Retrieved from www.marca.com/en/football/real-madrid/2016/12/15/5852e672ca474157498b462f.html

FIFA (2016a, March 5). IFAB agrees to introduce experiments with video assistant referees. Retrieved from www.fifa.com/about-fifa/who-we-are/news/ifab-agrees-to-introduce-experiments-with-video-assistant-referees-2768643

FIFA (2016b, September 2). FIFA holds successful first official "semi-live" tests with Video Assistant Referees (VARs). Retrieved from www.fifa.com/about-fifa/who-we-are/news/fifa-holds-successful-first-official-semi-live-tests-with-video-assist-2830365

FIFA (2016c, December 14). Video review used for penalty decision in FIFA Club World Cup Japan 2016 semi-final. Retrieved from www.fifa.com/clubworld

cup/news/video-review-used-for-penalty-decision-in-fifa-club-world-cup-japan-20-2860247

FIFA (2016d, July 20). First 'live' video replays trials underway at IFAB workshop. Retrieved from: www.fifa.com/about-fifa/who-we-are/news/first-live-video-replays-trials-underway-at-ifab-workshop-2812564

FIFA (2017a, May 15). VAR Update. Refereeing Department. May–June 2017. Retrieved from https://resources.fifa.com/mm/document/tournament/com petition/02/89/46/00/varupdate_neutral.pdf

FIFA (2017b, May 19). "A milestone tournament": FIFA President on VAR at Confederations Cup. Retrieved from www.fifa.com/confederationscup/news/y=2017/m=7/news=russia-2017-var-praised-at-closing-press-conference-2899588.html

FIFA (2017c, May 26). Group stage deemed a success by FIFA and LOC. Retrieved from www.fifa.com/confederationscup/news/y=2017/m=6/news=group-stage-deemed-a-success-by-fifa-and-loc-2898663.html

Fifa, Busacca: "La Var non è infallibile" (2017, June 30). *Gazzetta dello Sport.* Retrieved from www.gazzetta.it/Calcio/Estero/30-06-2017/fifa-busacca-la-var-non-infallibile-2001340993768.shtml

Final de la Sudamericana tendrá videoarbitraje. (2017, September 28). *Colombiano.* Retrieved from www.elcolombiano.com/deportes/futbol/final-de-la-copa-suda mericana-contara-con-videoarbitraje-ML7397529

¿Fue o no fue penal? La polémica con el VAR en la Copa Libertadores (2017, October 31). *Espectador.* Retrieved from www.elespectador.com/deportes/futbol-internacional/fue-o-no-fue-penal-la-polemica-con-el-var-en-la-copa-liberta dores-articulo-720891

Gabilondo, A. (2018, January 10). La FIFA tiene previsto utilizar el VAR en el Mundial de Russia. Retrieved from https://as.com/futbol/2018/01/10/interna cional/1515590503_571419.html

García, A. (2016, December 16). FIFA ref's chief defends Club World Cup VAR trial after Ronaldo goal. Retrieved from www.espn.com/soccer/fifa-club-world-cup/story/3020814/fifa-refs-chief-defends-club-world-cup-var-trial-after-ronaldo-goal

Gibson, O. (2014, June 13). Brazil penalty gives Sepp Blatter's video referees plan fresh impetus. Retrieved from www.theguardian.com/football/2014/jun/13/brazil-penalty-croatia-video-referees-world-cup-sepp-blatter

"Hasta Stevie Wonder vería ese penalti y ni sería necesario el VAR para cobrarlo" (2017, November 22). *Marca.* Retrieved from www.marca.com/futbol/copa-libertadores/2017/11/23/5a167c3ee2704e0b4f8b462b.html

Haydon, S. (2017, June 26). FIFA Accepts Video Reviews Must Improve After Russia. *The Associated Press.* Retrieved from www.apnews.com/b87b6daf1d3e411e82f083a16df85a23

IFAB (2016a, December 7). Video replay technology to be tested live at FIFA Club World Cup. Retrieved from http://theifab.com/news/video-replay-technol ogy-to-be-tested-live-at-fifa-club-world-cup

IFAB (2017, March 3). 131st AGM agrees fairer game strategy. Retrieved from http://theifab.com/news/131st-agm-agrees-fairer-game-strategy

IFAB (2018a, January 11). Agenda of the 132nd Annual Business Meeting of the IFAB confirmed. Retrieved from www.theifab.com/news/agenda-of-the-132nd-annual-business-meeting-of-the-ifab-confirmed

IFAB (2018b, March 3). Annual General Meeting, 3 March 2018, Zurich, Minutes. Retrieved from www.theifab.com/document/meeting-downloads

Infantino considera "muy positiva" la incursión del VAR en el fútbol. (2017, December 28). *Caracol Radio*. Retrieved from http://caracol.com.co/radio/2017/12/28/deportes/1514481475_152272.html

Infantino, G. (2017, July 1). Russia 2017, VAR praised at closing press conference. Retrieved from www.fifa.com/confederationscup/news/y=2017/m=7/news=russia-2017-var-praised-at-closing-press-conference-2899588.html

Iturralde González, E. (2017, December 13). El VAR anuló el gol de Casemiro por fuera de juego de Benzema. *AS*. Retrieved from https://as.com/futbol/2017/12/13/internacional/1513186935_449873.html

L'Équipe (2016, December 15). Mondial des Clubs: L'arbitrage vidéo «crée de la confusion» selon Zinedine Zidane. Retrieved from www.lequipe.fr/Football/Actualites/Mondial-des-clubs-l-arbitrage-video-cree-de-la-confusion-selon-zinedine-zidane/760460

La Gazetta dello Sport (2016, December 15). Mondiale per Club, Var delle polemiche. Modric duro: "Non è calcio". Retrieved from www.gazzetta.it/Calcio/Estero/15-12-2016/mondiale-moviola-var-polemiche-modric-18027400177.shtml

Levy, D. (2010, June 22). Why Replay Would Not Work In World Cup Soccer. Retrieved from www.sbnation.com/2010/6/22/1647329/why-replay-would-not-work-in-world

Marsden, S. (2016, December 17). Barcelona boss Luis Enrique backs video technology. *Espn.co.uk*. Retrieved from www.espn.co.uk/football/barcelona/story/3021465/barcelona-boss-luis-enrique-backs-video-technology

Matthew, H. (2016, December 18). Sharp Antlers give Madrid real scare. *BBC*. Retrieved from www.bbc.com/sport/football/38338721

Mauri, C. (2017, November 23). Copa Libertadores: un desenlace polémico, pero esta vez sin el empleo del VAR Retrieved from www.lanacion.com.ar/2084871-copa-libertadores-un-desenlace-polemico-pero-esta-vez-sin-el-empleo-del-var

Murray, C. (2015, September 9). Can video replay work in soccer? MLS thinks so. Retrieved from www.theguardian.com/football/2015/sep/16/can-video-replay-work-in-soccer-mls-thinks-so

"No iba a protestar si cobraban penal" (2017, November 1). *Olé* Retrieved from www.ole.com.ar/lanus/marcone-mano-penal-iba-protestar_0_1898210224.html

Nueva polémica con el videoarbitraje con el Real Madrid de protagonista. (2016, December 15). *SPORT*. Retrieved from www.sport.es/es/noticias/real-madrid/nueva-polemica-con-videoarbitraje-con-real-madrid-protagonista-5693230

Oliver, A. (2017, December 13). "El gol de Casemiro, en LaLiga, se hubiera dado por válido". *Marca*. Retrieved from www.marca.com/futbol/mundial-de-clubes/2017/12/13/5a316c25468aeb93428b466c.html

Patterson, M. (2016, March 5). Video technology trials given go-ahead by IFAB as football laws are revised. *Skysports.com*. Retrieved from www.skysports.com/football/news/11095/10192598/video-technology-given-go-ahead-by-ifab-as-football-laws-are-changed

Players criticise VAR at Club World Cup, but opinion split as FIFA say it was 'seamless' (2017, December 18). *Inside World Football* Retrieved from www.insideworldfootball.com/2017/12/18/players-criticise-var-club-world-cup-opinion-split-fifa-say-seamless/

Relaño, A. (2016, December 16). VAR system clouds the Club World Cup. Retrieved from https://en.as.com/en/2016/12/16/football/1481868324_211898.html

Scott, M. (2016, November 16). VAR not ready for live testing, Infantino. Retrieved from www.goal.com/en/news/468/international-football/2016/11/16/29569982/var-not-ready-for-live-testing-infantino

Video referees may be introduced for World Cup 2018, IFAB secretary says. (2016, November 3). *ITV*. Retrieved from www.itv.com/news/2016-11-03/video-referees-may-be-introduced-for-world-cup-2018-ifab-secretary-says/

Warsaw, A. (2016, December 16). Video ref 'confusion' snags trials at Club World Cup in Japan. *Insideworldfootball*. Retrieved from www.insideworldfootball.com/2016/12/16/video-ref-confusion-snags-trials-club-world-cup-japan/

West, H. (2016, December 15). Technology must be clearer, say Real boss Zidane. *GOAL*. Retrieved from www.goal.com/en-us/news/88/spain/2016/12/15/30532132/technology-must-be-clearer-says-real-boss-zidane

Whaling, J. (2017, June 28). Portugal vs Chile brings more VAR controversy as clear penalty is NOT given by television officials. Retrieved from www.mirror.co.uk/sport/football/news/portugal-vs-chile-brings-more-10706486

12 VAR Experiments in the Bundesliga

Otto Kolbinger

Introduction

Right from the beginning of the VAR experiments in football, the Deutsche Fußball Liga (German Football League, hereinafter DFL) made sure to claim a leading role in the introduction process of this technological officiating aid. In fact, the DFL and the Deutscher Fußball Bund respectively (German Football Association, hereinafter DFB) were among the first group of six competition organizers to confirm their participation in the VAR experiments, alongside the Australian Football Federation, the Brazilian Football Association, the Royal Netherlands Football Association, Major League Soccer, and the Portuguese Football Association. Even before that, during the negotiations about the introduction of goal-line technology in German professional football, the DFL commissioned and funded a study of the Technical University of Munich, in which Kolbinger et al. (2015) discussed the potential use of video replay in football. Consequently, the DFL announced on 25 August 2016 to start with trial runs in two steps: an offline trial in the 2016/17 season, followed by live tests in the 2017/18 season of the German Bundesliga (DFL, 2016).

The Video Assistant Referee, in German often referred to as *video proof*, is the third kind of such interventions to support the referee in applying the laws of the game, besides the above-mentioned goal-line technology and vanishing spray, which both were approved by the DFL and the DFB in 2014. Besides those interventions, the referees are equipped with two further technological devices that help to improve the communication process between the referee and his or her assistants: radio-based headset systems and offside flags, which were both introduced in the 2009/10 season. Each of those introductions needs, beside the mandatory approval of the IFAB, to be permitted by the DFL as well as the DFB. Therefore, it is worthy to point out the relationship of those two organizations and its consequences for the German Bundesliga. The DFB is the overarching governing body of German football and member of FIFA. The DFL, which was founded in 2000, is responsible for the organization and marketing of the first and second division in Germany, called Bundesliga and 2 Bundesliga, respectively, and is a member of the DFB. The relationship between those two organizations is regulated by a basic contract, which inter alia defines that the DFB is responsible for jurisdiction and refereeing. Thus, referees are not employed by the DFL and therefore the DFL is not in full control of the intervention process, as for example the video assistants and the respective supervisors are also employed by the DFB, which led to some unique developments regarding the VAR experiments in Germany.

Thus, a significant part of this chapter will focus on this relationship and the problems and controversies caused. Further, the course of the experiments and the process of the introduction of the VAR in the Bundesliga will be reproduced through press releases of the DFL and DFB, prominent comments of important stakeholders, and case analysis of selected incidents. Besides those qualitative descriptions, quantitative data will be provided to describe the extent and patterns of the VAR in the Bundesliga.

Methods

The search of relevant published material was performed in two steps. First, all press releases of the DFL and DFB, published after the approval of the VAR by IFAB (5 March 2016), were screened carefully for VAR content. Second, as demanded by the editor, a Google keyword search was run on a week to week basis, to track the evolution of the VAR experiments in the German Bundesliga. The keywords included several common German expressions of VAR, the names of the heads of the DFB and the DFL, as well as the names of the assigned supervisors for the VAR experiments in Germany:

Video Assistant Referee, Video Assistant, video proof, Videobeweis, video referee, Videoschiedsrichter, Reinhard Grindel (head of DFB), Reinhard Rauball (head of DFL e.V.), Christian Seifert (CEO of the DFL GmbH), Ansgar Schwenken (director of football affairs and fans), Hellmut Krug (project manager VAR until 6 November 2017), Lutz Michael Fröhlich (project manager VAR for the rest of the 2017/18 season and head of the elite referees committee).

Only articles of the websites of the following nationwide magazines and newspapers were used: the three weekly magazines with the biggest turnover (*Spiegel*, *Focus* and *Stern*); the three bestselling football magazines (*Sportbild*, *Kicker* and *11Freunde*); and the five most read newspapers (*Bild*, *Sueddeutsche Zeitung*, *Frankfurter Allgemeine Zeitung*, *Welt* and *Handelsblatt*). Further, we screened articles we found on the websites of the main broadcasters of the German Bundesliga: *Sky*, *Eurosport* and *ARD* (which includes its sport platform *Sportschau* and the regional public broadcasting organizations *BR*, *HR*, *MDR*, *NDR*, *RB*, *RBB*, *SR*, *SWR* and *WDR*). In addition, exclusive interviews of the above-mentioned persons were used irrespective of the medium. The search was performed for the period between the date of approval of the VAR by IFAB until 31 July 2018.

The VAR Experiments in the Bundesliga – Trial and Error

From Approval to the First Live Test – A Story of Education and Expectations

As mentioned above, the German Bundesliga pursued a leading role in the introduction of the so-called VAR in football. Consequently, Ansgar Schwenken, director of football affairs and fans and member of the executive board of the DFL, appreciated the decision of the IFAB to approve the VAR experiments in March 2016 and announced the intent to start with the experiments at the beginning of the 2016/17 season. CEO Christian Seifert also pointed out that the Bundesliga will be able to realize those tests rapidly, as the DFL possesses its own production and broadcasting subsidiary (Sport-Informations-Dienst, 2016).

Consequently, on 24 August 2016, the assembly meeting of the Bundesliga approved offline tests for the 2016/17 season. Former referee Hellmut Krug was announced as project manager for the VAR experiments. Those experiments inter alia should serve to collect quantitative data about the frequency and duration of interventions, optimize the assistant's behaviour and test protocols, as well as to educate video assistant referees (DFL, 2016).

In January 2017, the DFL not only announced the start of pre-live tests in the first months of the year, but also reassured that live tests will be performed in the Bundesliga in the 2017/18 season (DFL, 2017a). The months leading up to the live implementation were dominated by prompting the new intervention and by educating the public as well as internal stakeholders, inter alia via a bilateral workshop in April 2017, what to expect from the VAR. The latter should also serve to raise the awareness of problems that could be caused by the intervention and to show the limits of the VAR.

As an example, in November 2016, Hellmut Krug pointed out that the way referees call a game will change, as from now on there is a chance to double-check goal calls – but not no-goal calls, if the play was interrupted before. This would also force defenders to change their behaviour (Schneider, 2016). However, as Krug stated two months in advance of the introduction, referees should not fundamentally change the way they umpire (Roth, 2017a). Further, the front-office employees of the DFB and DFL repeatedly explained that the VAR will lead to more correct decisions, but not to 100 per cent accuracy. The explicit goal should be to avoid "clear and obvious mistakes", a term that was also used as standard of review. In addition, it was asked for patience and indulgence, as the VAR experiments at this stage are still a work in progress (Roth, 2017b).

Nevertheless, especially after the shaky application at the Confed Cup, the German organizations never get tired to assure that they are well – and even much better – prepared for live tests. Hellmut Krug announced that the replays won't take as much time (Roth, 2017a) and that the assigned referees enjoyed a much more extensive preparation period (Sport-Informations-Dienst, 2017a). A behaviour, which was criticized by one of the German referees, Manuel Gräfe, who complained about the so effected national and international pressure and wondered if it is smart to make those statements before the live tests even started in Germany (Goldmann, 2017).

Krug repeated his claims, as on opening night of the 2017/18 Bundesliga season, two weeks after its first use at the Super Cup, the VAR was used smoothly to overturn a call: out of the sight of referee Tobias Stieler, Leverkusen midfielder Charles Aranguiz grabbed Munich striker Robert Lewandowski's shoulder, which was detected by the video assistant Felix Zwayer. After 36 seconds, Stieler changed his no-call and awarded a penalty for Bayern Munich: after 54 years, the first overturned call by replay review during a match of the German Bundesliga.

The Technical Crisis and First Expressions

After a good start, the VAR experiments went downhill the very next day, as the DFL had to release a statement that "despite the comprehensive preparation phase, [...] the use [of the VAR] was not possible or not possible during the whole time for three games. Moreover, for none of the games the virtual line to support offside calls has been available" (trans. from DFL, 2017b). The respective games in which the VAR could not be used were Hamburger SV vs. FC Augsburg (whole game), TSG 1899 Hoffenheim vs. SV Werder Bremen, and Hertha BSC vs. VfB Stuttgart.

The DFL further stated that they won't accept this situation and announced bilateral meetings with the executive board of Hawk-Eye, the VAR provider for the Bundesliga. A statement of Hellmut Krug even supported rumours, which mentioned a possible temporary suspension of the experiments, as he explicitly appreciated the decision to continue those. As a consequence, however, it was decided to suspend the application of the virtual offside lines, which wouldn't cause significant disadvantages, according to Krug, as the video assistants could rely on sufficient camera angles (DFB, 2017a).

Besides those technical issues, the first controversial incidents occurred. On the second match day, Hoffenheim's forward Mark Uth tried to run by Leverkusen's defender Benjamin Henrichs, which led to a slight collision and a tumble of the defender effected by Uth, who played on to tie the game. Neither the referee nor his video assistant saw enough evidence to call an offensive foul or to overturn the no-call respectively, which upset Leverkusen's manager Rudi Völler, who even wondered if the video assistant "has fallen asleep" and concluded that the experiments should be ended if the VAR is not working (Krull, 2017). In the match-up between

VfB Stuttgart and FSV Mainz 05, players and officials complained about the different standards that seemed to be applied by the VAR, as he overturned one disputable no-call in favour of the home team, but did not do so in a similar situation that would have resulted in a penalty for the away team.

Further, on the third match day, the German Bundesliga experienced its first on-field review, as Benjamin Cortus went to the review area and then came back on the pitch to send off Freiburg's Yoric Ravet for his violent tackle on Dortmund's Marcel Schmelzer – a correct decision and a successful intervention of the VAR, which, however, led to negative reactions by players of SC Freiburg. Midfielder Amir Abrashi even called the video proof a "giant shit" (trans. from Schröter, 2017).

Cologne vs Cologne and Universal Shortcomings

On the following two match days, the name of the city Cologne – home of the club 1. FC Köln as well as of the VAR headquarters – could be used almost interchangeably with any words to describe the problems or controversies caused by the VAR experiments. The 1. FC Köln even considered filing a protest after their match at Dortmund. The dispute was based on an overrule that led to the second goal for Borussia Dortmund, which was not performed according to the VAR protocols. Initially the referee awarded a free kick for Köln, after Dormtund's Sokratis Papastathopoulos pushed Köln defender Dominique Heintz into his teammate Timo Horn, before he kicked the ball into the net. After a short communication with video assistant Felix Brych, referee Patrick Ittrich changed his initial call and awarded a goal for Borussia Dortmund. Besides the argument of whether there was enough or even some evidence to overturn the initial call, disputes were caused by the fact that the referee already interrupted the match before the ball crossed the goal line. Consequently, according to the VAR protocols, the respective incident was not reviewable. Therefore, Köln manager Jörg Schmadtke not just raged about a "blatant blown call" (trans. from Meininghaus, 2017), but also about this violation of the protocol.

Three days later, Köln hosted Eintracht Frankfurt in a game that included three highly disputable penalty calls, two of them in favour of the away team. The video assistant referee used in this game, Wolfgang Stark, initially was also scheduled as VAR for Köln's next match against Hannover 96 but was changed to another game at short notice.

Two more incidents, not involving 1. FC Köln, provoked rather negative reactions, for different reasons. In the match between VfB Stuttgart and VfL Wolfsburg, the already booked goalkeeper of Wolfsburg, Koen Casteels, hit Stuttgart's Christian Gentner with his knee in the face, which was neither seen as an illegal contact by the referee nor the video assistant. In a first statement, Hellmut Krug talked about an unlucky collision – Gentner needed two surgeries and was on the disabled list for two months – but confirmed the initial call of referee Guido Winkmann as a border-line call, but acceptable (Sport-Informations-Dienst, 2017b) – a conclusion that was later revised by Krug, as he admitted that it was a blown call and should have been overturned to be a red card and a penalty (Wallrodt, 2017).

Concerning a penalty for Bayern München in their game at Gelsenkirchen, a controversial debate was initiated by the front office of Schalke 04. The incident leading to the penalty was a blocked pass by Naldo, in which the ball deflected from his foot to his arm, with both arms stretched out above his head during the whole tackle. Schalke manager Christian Heidel accused Krug of having announced in a meeting that players won't be called as deliberately handling the ball if a player deflects a ball with another part of their body before touching it with the hand. Of course, the allegations were dismissed by Krug (Rütten, 2017).

Irrespective of those accusations, the incident illustrated one problem, which was explicitly mentioned in a summarizing conclusion of the head of the elite referees committee, Lutz Michael Fröhlich, after the first five match days. A lot of players and team officials still struggle to differentiate between apparently similar situations, which are in fact different and therefore need to be treated differently by the VAR protocols. Fröhlich also pointed out that the German referees still appreciate the new technological officiating aid. Once again, he asked for patience and indulgence, especially in regard of the human fallibility that also applies to video assistants (Meuren, 2017) – certainly a legitimate reason for some of the disputable – maybe even wrong – decisions of the VAR, but less for the technical short-comings and violations of the VAR protocol. In addition, the head of the elite referees committee remained silent regarding a significant change made by the committee, which would lead to further controversies over the course of the following months.

The Fundamental Crisis

In the beginning of November, an internal letter of the committee got leaked in which a change of the standard of review was admitted: "For difficult calls, which can't be rated indisputably as 'obvious mistakes', but the video assistant doubts the initial decision by a high degree, he shall immediately contact the referee to initiate an on-field review" (trans. from Petersen and Roth, 2017) – a decision that was not supported by President Grindel, who even stated that he did not know about this correction (Deutsche Presse-Agentur, 2017a). The president himself later on caused new confusions as he erroneously claimed another change in the standard of review: the VAR should only be used if there has been an error of perception, not a judging error. He withdrew this claim the same day, rejecting the responsibility for this confusion and rather stated that it seems to be very difficult to explain the complex standard in a way that can be understood by everybody (DFB, 2017c).

In addition to the confusion about the application of the protocols, the project was shaken up by charges against Hellmut Krug that he used the VAR to fix matches in favour of Schalke 04. This was denied by the involved referee, Marco Fritz, and the DFB later added, after almost three months of internal investigations, that there are no indications for those allegations (DFB, 2018b).

In addition, one referee anonymously complained in an interview, broadcasted by the WDR on 29 October 2017, about an insufficient preparation phase. The referee mentioned problems of the communication system during the offline test, which weren't tackled, as well as that there haven't been tests with five games at the same time, which could have caused the problems on the first match day. Further, several referees complained about insufficient quality of the provided footage and poorly educated operators (Best, 2017). To protect the experiments, on 6 November 2017, the DFB decided to discharge Krug from his duties as head of the project. Lutz Michael Fröhlich was announced as successor (DFB, 2017b). In the same announcement, the German Football Federation also announced that there haven't been any changes of the standard of review.

The DFB was further attacked by the DFL, which published a statement in which they demanded the foundation of an independent referee association. Dr Reinhard Rauball, head of the DFL e.V., pointed out that the pilot phase of the VAR has mercilessly disclosed problems, which had already existed for a long time and needed to be addressed (DFL, 2017c). He also added that all stakeholders, explicitly including the clubs, need to collaborate in a disciplined and constructive way to successfully perform the VAR experiments. This part of the statement was similar to statements of the president of the DFB and the head of the DFL GmbH, with the latter also asking for a better clarification of the standard of review (Sport-Informations-Dienst, 2017c).

Those structural problems and controversies were further fuelled by a couple of poor applications of the VAR in practice. For two incidents, the referees later even admitted blown calls. On match day ten, Freiburg defender Çağlar Söyüncü touched the ball with his hand after he was slightly pushed by Stuttgart forward Daniel Ginczek. After a brief communication with the video assistant Felix Zwayer, referee Tobias Stieler went to the review area and decided to send off Söyüncü for the denial of an obvious scoring opportunity. After the game, Stieler mentioned that a yellow card probably would have been more appropriate (Tillenburg, 2017).

The other referee admitting a wrong decision was FIFA referee Felix Brych. In this case, during the FSV Mainz 05 vs. 1. FC Köln match on match day 12 (yes, again the 1. FC Köln), the referee stuck to his initial call after contacting the video assistant Tobias Welz. He awarded a penalty for Mainz, after Pablo de Blasis was supposedly hit at the knee by a defender. The video assistant, as well as de Blasis, confirmed the contact.

However, after the game Brych acknowledged that, after watching the respective play after the match, he could not identify any contact on the video footage (Wieserner, 2017).

In another incident, a seemingly overturned penalty call for FC Augsburg against Bayer Leverkusen led to confusion of members of Augsburg's front office. Manager Stefan Reuter not only claimed that it should have been a penalty, but also how there could have been enough doubt to contact the video assistant, as the initial call of the referee was a penalty. However, referee Christian Dingert alleged that his initial call was a goal kick for Leverkusen, not a penalty for Augsburg (Deutsche Presse-Agentur, 2017b).

Furthermore, executive board members of the DFB went on the record with suggestions that could lead to less confusion regarding such incidents and the VAR in general. Ronny Zimmermann, vice president of the DFB, expressed the need for ways to inform the people in the stadiums about ongoing reviews. DFB President Grindel became even more explicit, once again suggesting to use the video screens in the stadiums to inform the audience and teams (Knaack, 2017). In addition, it was suggested by, inter alia, Lutz Michael Fröhlich, to consider a challenge system, to shift the responsibility to the coaches and avoid discussions about an appropriate application of this part of the standard of review (Boßmann, 2017). Whereas those proposals weren't feasible without IFAB, this was not the case for a third suggestion: more transparency and better communication, which should, inter alia, be realized by a special workshop on 4 December 2017.

Happy Holidays and New Year's Resolutions

Retrospectively, the special workshop and a silent December – with regard to VAR – could be seen as a turning point in the VAR experiments in the Bundesliga. After the workshop, several front-office employees went on record with very positive statements about the conversations (e.g., Heidel and Völler in DFB, 2017d). This impression was confirmed by executive board members of the DFL and DFB. Ansgar Schwenken pointed out that the workshop improved the mutual understanding and, as did Lutz Michael Fröhlich, that the VAR experiments are back on track (DFB, 2017d).

The incidents leading to the biggest arguments in December were two incidents involving Bayern Leverkusen, one review that led to a red card for Leverkusen's Wendell after a bad tackle on Dortmund's Gonzalo Castro, and one no-call after Stuttgart's Santiago Ascacibar jumped studs-first on the ankle joint of Leverkusen's Julian Brandt. In a further incident, a penalty for Mönchengladbach was overturned, as a duel between Oscar Wendt and Schalke's Daniel Caligiuri, happening several seconds in advance of the tackle that led to the penalty, was rated as a foul after video review. The core of the discussion for those incidents was the standard of review.

Nevertheless, Lutz Michael Fröhlich concluded that for the last four match days before the Christmas break, the overall acceptance of the VAR increased, but added that work still needs to be done to clarify the standard of review. He also announced further improvements of the communication between the referee and video assistant, which should be brief and precise (Hermanns, 2018). President Grindel, who also praised the good application at the end of the first leg, even recommended less overall communication between the referee and the VAR, to ensure that the referees remain the ones in charge (Rudolf, 2017).

In a press conference on 11 January 2018, the DFB announced that during the first leg 1,041 calls and no-calls were double-checked, meaning 6.8 per game. 750 were performed without further communication with the referee (so-called "silent checks"), in 241 plays the initial decision of the referee was confirmed ("checks"), and in 50 situations the video assistant recommended to change the initial call or to perform an on-field review. Of those 50 situations, 37 were corrected and twice the referee stuck to the initial call correctly. Eleven reviews resulted in a blown call. Ansgar Schwenken demanded that this number has to decrease close to zero. Ronny Zimmermann, however, claimed that "we have been too German" in the sense of demanding a 100 per cent rate, which is not possible in his opinion (trans. from DFB, 2018a).

As explicit adaptations for the second leg, Lutz Michael Fröhlich pointed out that the video assistant referee should not act like a detective. Rather, the assistant should solely focus on obvious mistakes, which is the case if a wrong call can easily be identified via the video footage. He further reinforced his statements about the communication processes and prompted a quicker proceeding. Ansgar Schwenken reported the average durations of 30 seconds for checks and about one minute for interventions. Regarding the visualization of reviews on the video boards in the stadiums, Schwenken claimed that this is currently not possible due to technical reasons and the subjective nature of specific decisions (DFB, 2018a).

Smooth Start to 2018 (Except between the Lines) and Permanent Introduction

Especially during the first months of the second leg, there were astonishingly few discussions about the VAR. Consequently, it was no surprise that the DFL assembly meeting approved the permanent introduction of the Video Assistant Referee in the German Bundesliga on 22 March 2018, with 17 clubs voting for the introduction (one abstention). It was also announced to start an offline test period in the second Bundesliga in the 2018/19 season (16 yes, 1 no, 1 abstention) (DFL, 2018a).

In the first three months of 2018, the few controversies were almost exclusively caused by the lack of virtual offside lines. One of those incidents resulted in two letters of the ostensibly disadvantaged club

Hannover 96, to ask for a written statement of the DFB and the DFL. In a 2–3 loss against RB Leipzig, a game-tying goal of Niklas Füllkrug was overturned due to an offside position. While Hannover's coach even claimed, after seeing the footage, that it was not offside, manager Horst Heldt did not question the final call but wondered why the video assistant referee was allowed to overturn the call. He further criticized that another close call in Bayern München's game on the same match day was not overturned and ranted that with those two incidents, "injustice reached its peak" (trans. from Sport-Informations-Dienst, 2018a). As one response, Reinhard Grindel claimed that the clubs themselves declined the use of virtual offside lines (Richter, 2018), which was, however, highly doubted by Heldt (Krischel, 2018).

Not a Strong Finish after a Break Catches the VAR

The VAR did not finish strong in the 2017/18 Bundesliga season, with a couple of wrong decisions in the Bundesliga and a correct appeal, which arguably received the most attention. Shortly before halftime in the match between FSV Mainz 05 and SC Freiburg (match day 30), a cross of Mainz's Daniel Brosinsiki was deliberately deflected by Freiburg's Marc Oliver Kempf, using his left arm. Referee Gudio Winkmann, who did not perceive a deliberate handling of the ball, was informed by video assistant Bibiana Steinhaus that a penalty should be awarded. However, as Winkmann already blew the whistle to end the first half, a lot of the players, including all of SC Freiburg, had already left the field. Thus, all players had to return, and seven minutes after the misplay, Pablo de Blasis converted the penalty for the 1–0 lead. This successful shot at least prevented negative reactions by Mainz 05 players, as, according to the rules, for this penalty there would not have been the possibility to score on a rebound. Not the only odd detail that resulted from the fact that Winkmann had already blown the whistle to end the first half: in such cases, the referee is only allowed to restart the game if he hasn't left the pitch yet. The DFB later released a statement by Lutz Michael Fröhlich in which it was admitted that Winkmann had already left the pitch, but claimed that Steinhaus had already initiated the review process at this time. Thus, in his opinion, the review was in accordance with the laws of the game, but Fröhlich admitted that the process was rather unfortunate (DFB, 2018c).

At least just as unfortunate were two further incidents on the final match days of this season: a missed red card for Mönchengladbach's Rafael on match day 31 and one incident two match days later in the SV Werder Bremen vs Bayer Leverkusen match. This incident also contained the first time in Bundesliga that a case of mistaken identity was found by the video assistant. Referee Robert Hartmann revoked the yellow card for Lucas Alario, but did not award the yellow card to Kevin Volland, who committed the offence. Fortunate for the latter, as it would have resulted in a sending-off for the already booked Volland.

After the Bundesliga season, a use of the VAR in a German competition gained nationwide attention one more time. In the final of the DFB-Cup, with the heavy favourite Bayern München trailing by one goal, Frankfurt's Kevin Prince Boateng tripped Bayern's Javi Martinez. Although it was obvious that Boateng hit Martinez, which was even admitted by his coach Nico Kovac, referee Felix Zwayer stuck to his call after an on-field review. Lutz Michael Fröhlich later acknowledged that the committee has to consider whether those decisions are still reasonable, as there has been an overwhelming majority to call a penalty (Schröter-Lorenz, 2018).

Quantitative Analysis of the Live Tests 2017/18

On 6 June 2018, the DFB released summary results of the use of the VAR in its premiere season (DFB, 2018d). Naturally, the DFB and the DFL drew a positive conclusion and highlighted the 64 initially wrong calls that were overturned by the VAR. However, Lutz Michael Fröhlich admitted that there have been 14 incidents in which the intervention of the VAR did not lead to a correct decision. Nevertheless, he pointed out, as did Ansgar Schwenken, the positive development of the live tests, as the frequency of intervention as well as the number of incorrect overturns decreased over the course of the season. In this press release the DFB announced that the number of double-checked situations was 1,870, including 1,321 silent-checks (no communication with the referee), 461 checks (communication with the referee and confirmation of the initial call), and 88 interventions. On average each check was performed in 29 seconds and each intervention in 57 seconds (with an average of 53 seconds in the second leg).

Comparing the overall numbers with those of the first leg, the positive trend in terms of fewer interventions can only be confirmed for the absolute numbers. The total for interventions in which the VAR recommended to change the initial call decreased from 50 to 38. However, comparing it with the total number of double-checked situations by the VAR, an intervention occurred for 4.8 per cent of those incidents in the first leg (1,041 observed situations leading to 50 interventions) and 4.6 per cent in the second leg (829 observed situations, 38 interventions). It is at least questionable if the decrease in the overall number of observed situations of the VAR is due to a different standard of review.

Irrespective of the frequency, there was a positive development regarding the number of blown calls after review, which decreased from eleven to three. Further, Ansgar Schwenken pointed out that the DFL is working on ways to increase transparency. In May 2018, the DFL presented different ways to inform fans in the stadiums at the SportsInnovation event, hosted by the DFL and the Messe Düsseldorf.

Discussion – The VAR: A Story of Success?

The introduction of technological officiating aids must be seen as interventions in social settings. More than 50 years of evaluative research, using Suchmann's work as origin (1967), have taught us that those interventions affect myriad factors and are, vice versa, affected by them. And, consequently, it's not getting less complex if one needs to examine the success of an intervention. Kolbinger (2018) suggested an overarching framework for the evaluation of technological officiating aids that illustrates this complexity (Figure 12.1). The VAR experiments in the Bundesliga once again confirm the complexity of such interventions and the challenges to assess them.

The VAR helped to overturn 64 wrong initial calls of the referee. A success? There is no incontestable, objective answer. At most, it is possible to reach some kind of consensus between all stakeholders or a representative panel, as for example Stufflebeam (2001) suggests in his evaluation model. If such a step was performed in the evaluation process of the DFB and DFL, it was not communicated, which is why one can assume it did not happen. At least the delegates of the current Bundesliga clubs rated the intervention as a success, as they voted to permanently introduce the VAR in the Bundesliga. It remains unclear if they used information about the opinions of other stakeholders, but it is not an uncommon opinion in evaluation theory that the reactions of the public and the media influence the decision of (sport-) political decision-makers (Weiss, 1982), which are usually not collected in a scientific manner. According to our model, the stakeholders' opinions should be tracked (empirically) throughout the introduction and the application process.

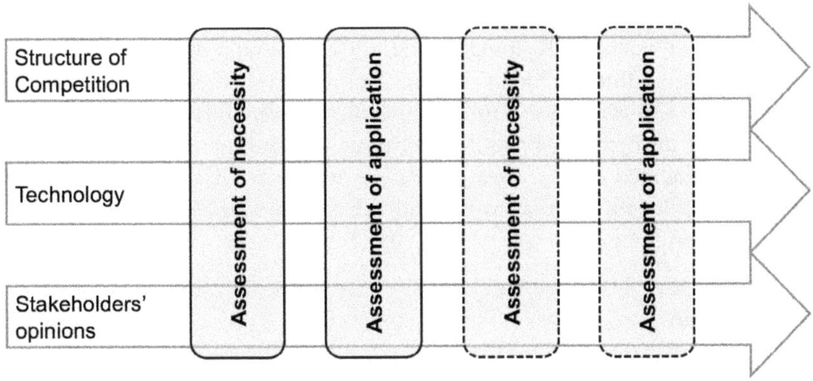

Figure 12.1 Overarching framework for the evaluation of technological officiating aids in game sports

The same applies to information about the technological progress, regarding alternative devices as well as the system used for the respective interventions itself, which includes, inter alia, two issues that occurred during the VAR experiments in the Bundesliga. First, discussions about the accuracy, especially regarding the virtual offside lines that should have been used but were dismissed after the first match day and not reintroduced due to the lack of a verification system. This raises the question: why should they have been used without verification in the first place? The second issue, which arguably led to the most discussions, was the standard of review. To only correct "obvious and clear mistakes" is a standard that should reduce the number of reviews, but is hard to define in a sufficient manner, which led to many discussions whether the VAR should have intervened or not. Berman (2011) discussed benefits and disadvantages of the current standard of review in the NFL, the so-called Indisputable Visual Evidence (that the initial call was wrong), to a "de novo standard", which would roughly mean that the initial call has no influence on the judgement of the video assistant referee. No one can answer if a different standard would have led to more or fewer overturned calls in the Bundesliga or correct decisions respectively. The same is true for side effects, such as the patterns of interruptions. This would demand tests of different standards in live settings and one has to wonder why this was not permitted by the IFAB.

The live experiments themselves demonstrated another challenge for the assessment of the application of such interventions: the choice of the respective competition for this part of the evaluation. Of course, the assessment of application can be performed in a different, less meaningful, competition. A strategy that was, for example, performed by the ATP, as the system for automatic line-calling was first tested at the Champions Tour, or the MLB, which tested the video replay system in the Arizona Fall League. However, competitions of different performance levels can differ in their structure of competition, which can influence the evaluation results. As an example, Kolbinger et al. (2015) found a different frequency of critical goal-line incidents in the Bundesliga compared with the 2. Bundesliga. On the other hand, taking the meaningful competition itself can lead to distortions of the competition, for example if the standard of review is changed during a season. At the very least, it can cause negative reactions and emotions, as we could see in the 2017/18 season of the German Bundesliga.

Conclusion

The DFL and the DFB experienced the above-mentioned challenges connected with the introduction of technological officiating aids. During the year, the associations were forced to perform some changes, like the dismissal of Hellmut Krug or the more defensive approach of the video assistant referees, but they also started to be proactive to enhance the situation. For the 2018/19 season the DFL announced to finally forward information about the ongoing replays to the people in the stadiums via the video boards, as well as that they would reintroduce virtual offside lines (DFL, 2018b). However, as illustrated in Kolbinger's framework, there will still be a need to evaluate those technological officiating aids in the future, as the influencing factors are dynamic and also because the VAR was far from perfect in its first season.

References

Berman, M. N. (2011). Replay. *California Law Review, 99*(6), 1683–1743. doi:10.2307/41345443.

Best, B. (2017, October 29). Zur Verschwiegenheit verpflichtet [Bound to be silent] [Video file]. Retrieved from www1.wdr.de/mediathek/video/sendungen/sport-inside/video-zur-verschwiegenheit-verpflichtet-100.html.

Boßmann, B. (2017, November 21). Schiri-Boss für Challenges und Videobeweis im Stadion [Head of referees in favour of challenge-system and VAR illustrations in the stadiums]. Retrieved from https://sportbild.bild.de/bundesliga/2017/1-bundesliga/schiedsrichter-boss-lutz-michael-froehlich-challenges-videobeweis-stadion-53931054.sport.html.

Deutsche Fußball Liga. (2016, August 25). Start der Testphase für den Video-Assistenten in der Bundesliga [Start of the test period for the VAR in the Bundesliga] [Press release]. Retrieved from www.dfl.de/dfl/files/pressemitteilungen/2016-pressemitteilungen/2016_Pressemitteilung_44-VAR.pdf.

Deutsche Fußball Liga. (2017a, January 23). Bundesliga plant ab 2017/18 mit Video-Assistent [Bundesliga introduces VAR for the 2017/18 season] [Press release]. Retrieved from www.bundesliga.com/de/bundesliga/news/bundesliga-plant-ab-2017-18-mit-video-assistent-noblmd17n.jsp.

Deutsche Fußball Liga. (2017b, August 19). Video-Assistent: Dienstleister Hawkeye mit massiven Problemen [VAR: Provider Hawkeye with massive problems] [Press release]. Retrieved from www.dfl.de/dfl/files/pressemitteilungen/2017-pressemitteilungen/30-Video-Assistent_Hawkeye.pdf.

Deutsche Fußball Liga. (2017c, November 7). DFL-Präsidium hält strukturelle Neuordnung des Schiedsrichterwesens für zwingend erforderlich [DFL-presidium sees mandatory need for a structural reorganization of the refereeing body] [Press release]. Retrieved from www.dfl.de/dfl/files/pressemitteilungen/2017-pressemitteilungen/40-DFL-Pr-c3-a4sidium-20zu-20Schiedsrichterwesen.pdf.

Deutsche Fußball Liga. (2018a, March 22). Bundesliga ab Sommer offiziell mit Video-Assistent – 2. Bundesliga mit Offline-Testphase ["Official use of the VAR in the Bundesliga starts this summer – 2. Bundesliga starts offline-tests"] [Press release]. Retrieved from www.dfl.de/dfl/files/pressemitteilungen/2018-pressemitteilungen/10-Video_Assistent.pdf.

Deutsche Fußball Liga. (2018b, August 10). Video-Assistent 2018/19: Mehr Transparenz für Zuschauer in Stadien und an Bildschirmen – Kalibrierte Abseitslinie wird eingesetzt – Offline-Tests in der 2. Bundesliga [VAR 2018/19: More transparency for live and TV audience – Application of the virtual offside line – Offline-Tests in the 2nd Bundesliga] [Press release]. Retrieved from www.dfl.de/dfl/files/pressemitteilungen/2018-pressemitteilungen/Video-Assistent-2018-19-Mehr-Transparenz-für-Zuschauer.pdf.

Deutsche Presse-Agentur. (2017a, November 3). Schiedsrichter übergehen DFB-Präsidenten Grindel [Referees pass over DFB-president Grindel]. Retrieved from www.faz.net/aktuell/sport/fussball/bundesliga/dfb-praesident-reinhard-grindel-kritisiert-modifizierung-des-videobeweises-15275625.html.

Deutsche Presse-Agentur. (2017b, November 4). „Irritation" um Aufregerszene in Augsburg: Referee im Fokus [Confusion about upsetting incident in Augsburg: referee in the spotlight] [Agency report]. Retrieved from www.focus.de/regional/augsburg/fussball-augsburg-spielt-1-1-gegen-leverkusen_id_7802551.html.

Deutscher Fußball Bund. (2017a, August 25). DFB und DFL setzen Video-Assistenten auch am 2. Spieltag ein [DFB and DFL will continue to use the VAR in week two] [Press release]. Retrieved from www.dfb.de/news/detail/dfb-und-dfl-setzen-video-assistenten-auch-am-2-spieltag-ein-172846/.

Deutscher Fußball Bund. (2017b, November 6). Fröhlich leitet Projekt Video-Assistent [Fröhlich leads video assistant project] [Press release]. Retrieved from www.dfb.de/news/detail/froehlich-leitet-projekt-video-assistent-177536/.

Deutscher Fußball Bund. (2017c, November 12). „Video-Assistent hilft Schiedsrichter bei Wahrnehmungsfehlern" ["VAR supports referee in case of errors of perception"]. Retrieved from www.dfb.de/news/detail/video-assistent-hilft-schiedsrichter-bei-wahrnehmungsfehlern-177962/.

Deutscher Fußball Bund. (2017d, December 4). Workshop zum Video-Assistenten: „Wir sitzen alle in einem Boot" [VAR-Workshop: "We're all in this together"]. Retrieved from www.dfb.de/news/detail/workshop-zum-video-assistenten-wir-sitzen-alle-in-einem-boot-179162/.

Deutscher Fußball Bund. (2018a, January 11). "Mut zur Entscheidung bleibt Qualitätsmerkmal des Referees" ["Courage to take a decision remains quality feature of the referee"] [Press release]. Retrieved from www.dfb.de/news/detail/mut-zur-entscheidung-bleibt-ein-qualitaetsmerkmal-des-schiedsrichters-180430/full/1/.

Deutscher Fußball Bund. (2018b, January 30). Externer Untersuchungsbericht: Kein Mobbing, keine Maniplationen [Report of external investigations: No bullying, no manipulations] [Press release]. Retrieved from www.dfb.de/news/detail/externer-untersuchungsbericht-kein-mobbing-keine-manipulation-181189/.

Deutscher Fußball Bund. (2018c, April 17). Fröhlich: So kam es zum Elfmeter für Mainz gegen Freiburg [Fröhlich: That's how the penalty call for Mainz against Freiburg happened] [Press release]. Retrieved from www.dfb.de/news/detail/winkmann-so-kam-es-zum-elfmeter-fuer-mainz-gegen-freiburg-185168/.

Deutscher Fußball Bund. (2018d, June 6). Saisonbilanz: Video-Assistent verhindert 64 Fehlentscheidungen [Resume of the season: VAR prevents 64 wrong decisions] [Press release]. Retrieved from www.dfb.de/die-mannschaft/news-detail/saisonbilanz-video-assistent-verhindert-64-fehlentscheidungen-188004/.

Goldmann, S. (2017, August 21). Was Schiedsrichter Manuel Gräfe von seinen Kollegen denkt [How referee Manuel Gräfe thinks about his colleagues]. Retrieved from www.11freunde.de/interview/was-schiedsrichter-manuel-graefe-von-seinen-kollegen-denkt.

Hermanns, S. (2018, January 1). Lutz Fröhlich: Es geht um einen behutsamen Umgang mit dem Videoassistenten" [Lutz Fröhlich:" It's about a gentle handling of the VAR"]. Retrieved from www.tagesspiegel.de/sport/videobeweis-in-der-fussball-bundesliga-lutz-froehlich-es-geht-um-einen-behutsamen-umgang-mit-dem-videoassistenten/20800496.html.

Knaack, B. (2017, November 8). „Absoluter Murks" ["Absolute bodge"]. Retrieved from www.spiegel.de/sport/fussball/videobeweis-in-der-bundesliga-absoluter-murks-a-1176969.html.

Kolbinger, O. (2018). *Innovative technische Hilfsmittel zur Unterstützung von Schiedsrichtern in Spielsportarten als Gegenstand von Evaluationsforschung [Innovative Technological Officiating Aids as object of Evaluative Research]* (Doctoral dissertation, Technical University of Munich). Retrieved from http://nbn-resolving.de/urn/resolver.pl?urn:nbn:de:bvb:91-diss-20180705-1368469-1-4.

Kolbinger, O., Linke, D., Link, D. & Lames, M. (2015). Do We Need Goal Line Technology in Soccer or Could Video Proof Be a More Suitable Choice: A Cost-Benefit-Analysis of Goal Line Technology in Soccer and Thoughts about an Introduction of Video Proof. In Cabri, J., Barreiros, J. & Pezarat-Correia, P. (eds.) *Sports Science Research and Technology Support* (107–118). Schweiz: Springer International Publishing. doi:10.1007/978-3-319-25249-0_8.

Krischel, M. (2018, April 5). „Setzen, sechs" – Heldt schießt scharf gegen Grindel ["Sit down, F" – Heldt fires against Grindel]. Retrieved from www.kicker.de/news/fussball/bundesliga/startseite/720961/artikel_setzen-sechs-heldt-schiesst-scharf-gegen-grindel.html.

Krull, P. (2017, August 27). Die Liga ist genervt und Rudi Völler auf der Palme [The league is annoyed and Rudi Völler driven nuts]. Retrieved from www.welt.de/sport/article168028564/Die-Liga-ist-genervt-und-Rudi-Voeller-auf-der-Palme.html.

Meininghaus, F. (2017, September 18). „Eine krasse Fehlentscheidung" ["A blatant blown call"]. Retrieved from www.sueddeutsche.de/sport/aerger-nach-videobeweis-eine-krasse-fehlentscheidung-1.3671302.

Meuren, D. (2017, September 23). Kommt jetzt der Lippenleserbeweis? [Lip-reading-assistant next?]. Retrieved from www.faz.net/aktuell/sport/fussball/bundesliga/rote-karte-fuer-vedad-ibisevic-von-hertha-bei-0-1-in-mainz-15213809.html.

Petersen, J. & Roth, T. (2017, November 2). DFB passt Videobeweis heimlich an [DFB secretly adjusts VAR protocols]. Retrieved from www.kicker.de/news/fussball/bundesliga/startseite/709808/artikel_dfb-passt-videobeweis-heimlich-an.html.

Richter, M. (2018, April 3). Grindel weist 96-Beschwerde zum Videobeweis zurück [Grindel rejects Hannover 96's appeal regarding the VAR]. Retrieved from www.kicker.de/news/fussball/bundesliga/startseite/720832/artikel_grindel-weist-96-beschwerde-zum-videobeweis-zurueck.html.

Roth, T. (2017a, July 9). Hellmut Krug: „Die Schiedsrichter sollen sich nicht ändern" [Hellmut Krug: "The referees should not change"]. Retrieved from www.kicker.de/news/fussball/bundesliga/startseite/701308/artikel_hellmut-krug_die-schiedsrichter-sollen-sich-nicht-aendern.html.

Roth, T. (2017b, August 17). Regeln, Schiris, Bezahlung: Das ist neu in der Bundesliga [Rules, referees, pay: That's new in the Bundesliga]. Retrieved from www.kicker.de/news/fussball/bundesliga/startseite/704104/artikel_regeln-schiris-bezahlung_das-ist-neu-in-der-bundesliga.html.

Rudolf, T. (2017, December 26). Grindel: „Weniger Kommunikation mit Kölner Keller" [Grindel: "Less communication with the cellar of cologne"]. Retrieved from www.kicker.de/news/fussball/bundesliga/startseite/713748/artikel_grindel_weniger-kommunikation-mit-koelner-keller.html.

Rütten, F. (2017, September 20). Bayern-Bonus? Retrieved from www.stern.de/sport/fussball/bundesliga/debatte-um-angeblichen-bayern-bonus-nach-handelf meter-gegen-schalke-7628870.html.

Schneider, M. (2016, November 15). Video-Schiri – wie die Revolution noch an der Praxis scheitert [VAR – How the revolution is denied by the real world]. Retrieved from www.sueddeutsche.de/sport/technische-hilfsmittel-im-fussball-video-schiri-wie-die-revolution-noch-an-der-praxis-scheitert-1.3250127.

Schröter, C. (2017, September 10). Abrashi: „Ein Riesenscheiß, das gefällt mir gar nicht" [Abrashi: "A giant shit, I don't like that"]. Retrieved from www.kicker.de/news/fussball/bundesliga/startseite/705845/artikel_abrashi_ein-riesenscheiss-das-gefaellt-mir-gar-nicht.html.

Schröter-Lorenz, C. (2018, May 21). Zwayer: „Ich stehe zu meiner Entscheidung" [Zwayer: "I stick with my call"]. Retrieved from www.kicker.de/news/fussball/dfbpokal/startseite/724328/artikel_zwayer_ich-stehe-zu-meiner-entscheidung.html.

Sport-Informations-Dienst. (2016, March 5). Videobeweis: IFAB segnet Testphase ab [VAR: IFAB permits experiments] [Agency report]. Retrieved from www.eurosport.de/fussball/videobeweis-ifab-segnet-testphase-ab_sto5293499/story.shtml.

Sport-Informations-Dienst. (2017a, June 26). Videoschiedsrichter: Krug rechnet in der Bundesliga mit weniger Diskussionen [VAR: Krug expects less discussions in the Bundesliga] [Agency report]. Retrieved from www.eurosport.de/fussball/videoschiedsrichter-krug-rechnet-in-der-bundesliga-mit-weniger-diskussionen_sto6231529/story.shtml.

Sport-Informations-Dienst. (2017b, September 16). Krug zur Gentner-Szene: "Grenzwertig, aber vertretbar" [Krug about the incident involving Gentner: "Borderline, but acceptable"] [Agency report]. Retrieved from www.welt.de/newsticker/sport-news/article168711735/Krug-zur-Gentner-Szene-Grenzwer tig-aber-vertretbar.html.

Sport-Informations-Dienst. (2017c, November 9). Abpfiff für den DFB, Anpfiff für die Klubs: DFL-Boss Christian Seifert macht klare Ansagen [DFL-Boss Christian Seifert sending clear messages to the DFB and the clubs] [Agency report]. Retrieved from www.eurosport.de/fussball/bundesliga/2017-2018/abpfiff-fur-den-dfb-anpfiff-fur-die-klubs-dfl-boss-christian-seifert-macht-klare-ansagen_sto6404267/story.shtml.

Sport-Informations-Dienst. (2018, April 2). „Es geht um Willkür" ["It's about arbitrariness"]. Retrieved from www.sueddeutsche.de/sport/videobeweis-es-geht-um-willkuer-1.3928548.

Stufflebeam, D. L. (2001). Interdisciplinary Ph.D. programming in evaluation. *The American Journal of Evaluation*, *22*(3), 445–455.

Suchman, E. A. (1967). *Evaluative Research: Principles and Practice in Public Service and Social Action Programs*. New York: Russell Sage Foundation.

Tillenburg, R. (2017, October 30). Kritik am Videobeweis wird (noch) lauter [(Even) more criticism for the VAR]. Retrieved from www.sportschau.de/fuss ball/bundesliga/videobeweis-stimmung-kippt-weiter-100.html.

Wallrodt, L. (2017, September 24). „Die Debatte um den Videobeweis ist scheinheilig" ["The VAR debate is hypocritical"]. Retrieved from www.welt.de/sport/fussball/article168957567/Die-Debatte-um-den-Videobeweis-ist-scheinheilig.html.

Weiss, C. H. (1982). Policy research in the context of diffuse decision making. *The Journal of Higher Education*, *53*(6), 619–639.

Wieserner, J. (2017, November 18). Videobeweis-Ärger: Brych gesteht Fehlentscheidung ein [VAR troubles: Brych admits wrong call]. Retrieved from www.kicker.de/news/fussball/bundesliga/vereine/710970/artikel_videobeweis-aerger_brych-gesteht-fehlentscheidung-ein.html

13 VAR Experiments in the Italian Serie A League

Juan Antonio Simón

The sources used in this research are the press notes produced by the principal governing bodies in Italian football, such as the FIGC, the AIA, and the Lega Serie A, and which appear on their corresponding websites, as well as information published by the main Italian newspapers along with the most relevant bibliographical documentation on this subject. Concerning the latter, the press sources used are those with the widest circulation in Italy, in terms of sports readership, such as *La Gazzetta dello Sport*, *Corriere dello Sport*, and *Tuttosport*, as well as mainstream dailies such as *Il Sole 24 Ore* or *Repubblica.* The VAR introduction process in Italy has shown the importance for success of the integration of this technology, the specific training of referees, and the role of the media as an ideal tool for football fans to understand the specificity of the VAR.

Introduction: The Arrival of Video-Assisted Refereeing in Italy: Will Technology Improve Football?

In few countries has football generated, and continues to generate, quite as much passion as in Italy. Four times winner of the FIFA World Cup, a huge number of Italian clubs, managers, and players form part of the world-wide history of this sport. It is impossible to explain Italian society without understanding what football has represented there. The Federazione Italiana Giuoco Calcio (FIGC), the governing body of the sport in Italy, has always been proactive when it comes to taking on board the guidelines and recommendations made by the international bodies that govern the sport, such as the International Football Association Board (IFAB), the Fédération Internationale de Football Association (FIFA), or the Union of European Football Associations (UEFA), concerning improvements to refereeing or the inclusion of new technologies. It should be recalled, regarding this question, that Italian football was among the first to back the introduction into its competitions of the Additional Assistant Referees (AARs) and of Goal-Line Technology. Concerning the VAR system, Italian football has been one of the first to offer its championship as a testing ground for the first trials.

Starting in April 2016, Italian football began the process by which the VAR system would be developed, and which would have as a culminating moment its introduction into the top flight of the national championship, the Lega Nazionale Professionisti Serie A (Lega Serie A). The Italian national team itself was one of the first to take part in the trials of this technology, in international friendly fixtures, such as those against France and Germany that same year. Both the FIGC and the Associazione Italiana Arbitri (AIA), the Referees' Association, as well as the Lega Serie A itself – the latter being the body that governs the most important club football tournaments in the country – threw themselves into the challenge that developing a technology as innovative as the VAR system meant for the game, in a sport that has traditionally been resistant to big changes in its regulations and to the introduction of new technologies.

After the first trial run of the VAR system in a friendly between Italy and France, the FIFA president, Gianni Infantino, told the press at the end of the game that the first experiment had been positive and that "we cannot, in 2016, allow refereeing mistakes that everyone watching on television can see but that the referee does not". Despite this, he did recognise that it would be necessary to continue testing to see whether the system could work or not. The delegate, and supervisor for the VAR system project in Italy, Roberto Rosetti, was from that moment on in charge of the planning that was designed for the introduction of the VAR system. This process had to adhere to FIFA's programming, which indicated that, as of the seventh match day of the 2016–17 season, there would be an "offline" phase for matches, to allow for the referees to familiarise themselves little by little with the technology. As of January 2017, a second, "online" phase would begin, in which the VAR system would be trialled in Coppa Italia games ("Moviola in campo", 2016).

In November 2016, the VAR system was tested again in an international between Italy and Germany. After that match, Massimo Busacca, in his role as Head of FIFA Refereeing, commented that the trial had been very positive and that the VAR had even been decisive at three points in the game. In the comments that he would make later to the Italian medium *Radio Sportiva*, Busacca pointed out that the figure of the two additional referees positioned beside each of the goals would disappear with the introduction of the VAR system, though he also made it clear that referees themselves would lose none of their authority when this technology was brought in: "Technology should not replace referees, because that would alter the game of football, it must be something that supports their work" ("Fifa, Busacca promuove", 2016).

Early in January 2017 it would be, once again, the Italian national team, the U-19 side this time, who would put the VAR system to the test, one that was run directly by the Lega Serie A, in a fixture with Spain at the Friuli stadium, in Udine. Roberto Rosetti commented that "FIGC and the Lega Serie A believe strongly in this project and are working with it seriously and with great professionalism to carry out a highly important experiment for the future of football" ("Nel giorno della Var", 2017). Later, tests were continued using the online version in two further games featuring the U-19 and the U-18 national sides respectively, friendlies against France at the Renato Dall'Ara stadium in Bologna, on 22 February, and against Denmark, on 8 March. In total, in Italy, there were 31 VAR offline trial games played, fixtures in which there was no direct communication with the referee, before trials were started in games in which there would be full communication. Italy was ready to include the VAR system in the Lega Serie A 2017–18 season as of the first game of the campaign.

The principal aim of this chapter is to analyse the impact, the most controversial problems, the critiques, and the praise that the introduction of the VAR system has brought with it since 2016. At the same time, special attention will be given to looking at how the Italian media, both the mainstream news media and the media specifically focused on sport, have analysed the introduction of VAR into the country, how they have reflected the comments and opinions of the main actors involved, and what role they have played in the spreading of information and the adequate understanding of the specific, differentiating features of this technology among the fans at large.

Methodology

In terms of methodology, the sources used in this research are the press notes produced by the principal governing bodies in Italian football, such as the FIGC, the AIA, and the Lega Serie A, and which appear on their corresponding websites, as well as information published by the main Italian newspapers along with the most relevant bibliographical documentation on this subject. Concerning the latter, the press sources used are those with the widest circulation in Italy, in terms of sports readership, such as the *Gazzetta dello Sport*, el *Corriere dello Sport*, and *Tuttosport*, as well as mainstream dailies such as *Il Sole 24 Ore* or *Repubblica*. Specific searches were made on each of their official websites using the simple search criteria *VAR*, *Video Assistant Referees*, *video*, as well as searches that combined criteria such as *technology* AND *var*, *var* AND *Marcello Nicchi* (the chief of refereeing in the AIA), *Roberto Rosetti* AND *var*, *video* AND *var*, *Italia* AND *var*, and *Video Assistant Referees* AND *var*. Based on those search results, each news item and its contents were analysed one by one, as was the information that was repeatedly published in different media with no new content being added.

The screening of the press was completed by Google searches, using the same criteria mentioned above, week by week, from the moment FIFA gave its approval, on 5 March 2016, up until 31 July 2018. What has been attempted, then, is to gather all information concerning the VAR system and its introduction into Italian football that fleshed out the headlines in the publications mentioned above.

Lastly, an analysis of the content covering the VAR system that appeared on the websites of media such as the *RAI* and *Mediaset*, as well as that of television programmes, such as those broadcast by *Sky Italia*, have been included as relevant information.

The Long Road to the Implementation of the VAR System in Italy: A Success Story with Controversy

The Early Steps of the VAR System in Italy

On 14 April 2016, the FIGC announced, via their president Carlo Tavecchio, that they had been chosen by IFAB, along with another group of international federations, to trial the VAR system. Tavecchio stated that he had called a meeting on 21 April in order to inform the Lega Serie A, Lega Serie B, and the AIA, and explain everything that would be necessary on a technical level for the trials to begin: "we have been the first to give backing to the use of on-field technology and we believe we have all the requisites to be able to make our contribution to this important experiment" (Federazione Italiana Giuoco Calcio [FIGC], 2016a). In that meeting, it was announced that the VAR trials would begin during the 2016–17 season, offline first of all, with there being no contact with the match referee in order to then, in the second phase (which would take place over the 2017–18 season), begin the online trial in which there would be contact between the on-field referee and the VAR room (FIGC, 2016b). The initial strategic plan for the introduction of the VAR in Italy foresaw the 2018–19 season as the one in which it would be effectively introduced into the Serie A championship, while for the 2017–18 season it was only planned to use the VAR in the TIM Cup round of 16.

Late in 2016, however, and after examining the data gathered from the different offline trials during the first phase, the AIA president, Marcello Nicchi, confirmed at the end of the FIGC federal council meeting that for the 2017–18 season that the VAR system would be used online as of the first match day of the Serie A championship. From the outset, the head of Italian refereeing tackled the doubters among the public, who were worried that the technological innovation would take power away from the referee. Nicchi was very clear about this, giving assurances that the VAR's function was to help referees, not to diminish their authority on the pitch, adding that even the players "will feel protected, while the final decision is always the referee's". To address the doubts that had been raised among journalists regarding the immediate introduction of VAR in Italian football, it was pointed out that, unlike volleyball or tennis, which use the Hawk-Eye system, the new technology would be applied as many times as was necessary throughout a match, but without forgetting that the idea was for it to intervene as little as possible in play. At that time it was still not known whether specialists in this type of refereeing would be used, it was mentioned that serving referees would be used, because whoever had to control the monitor on a Sunday had to have the same level of preparation and training as the one on the pitch and do everything that referees have been doing at Coverciano for years. By April 2017, some 80 per cent of the Serie A stadia had had VAR checks run in them. Marcello Nicchi also pointed out the importance of both managers and team captains attending as many of the meetings regularly organised by referees as possible ("Nicchi e la rivoluzione Var", 2017).

In June 2017, in the semi-final of the Final Eight of the Coppa Italia Primavera (Italian Spring Cup) between Juventus and Sampdoria, the VAR made its debut in an official club encounter. Roberto Rosetti, in charge of the VAR project in Italy, told *Sky Sport* journalists that it was going to be a hard summer owing to the fact that:

> it will be necessary to prepare for the job that we have already begun in the best way possible. In July there will be several important friendly fixtures which will be used to give the system a trial run-out at Sportilia and Coverciano.
>
> ("Primavera, Juventus-Sampdoria", 2017)

The man in charge of the development of the VAR system in Italy told the media that there were, by then, 14 countries testing the VAR system in their national competitions, as well as FIFA itself, in the Under-20 World Cup in South Korea, and all of them were producing interesting results. Rosetti reminded them that it was not a simple innovation and that "time and patience" would be necessary, although "this is the future, and will bring positive results". The spirit of this technology was to have minimum intervention and maximum benefit on the game, remarking that "it will be a sort of insurance for the referee, to help him restore justice if he has made a decision that is totally wrong" (Spignesi, 2017).

Two months later VAR would make its big debut in a series of pre-season friendly games. Milan went down 2–1 to the Spanish side Betis, and the manager of the Italian side, Vincenzo Montela, when asked about the refereeing decisions taken during the game, said that:

> so many people have requested the [introduction of] the VAR system, we cannot be complaining about it ahead of time: we have to get used to it, above all those who are there, on the spot, and those who are in the stadium. Then, the questions it raises will be many and we will have a lot of fun with them, but the referee will have the chance to choose with more footage at his disposal.
>
> ("Milan, André Silva", 2017)

In that game, the VAR changed two incorrect decisions that were of great importance: a goal that had initially been given was disallowed, and a penalty that the referee had not awarded eventually was. The media took on the debate concerning the suitability of the new technology, and journalists such as Luciano Cremona initially came out in favour of the VAR, pointing out that the system was still in the experimental stage and that its main function was to put right those calls that had been made incorrectly on the pitch. This same journalist also called for the footage of the plays in question to be replayed for all the fans in the stadium to see, along with the referees giving their explanations live, therefore coming out in favour of greater transparency in the taking of decisions (Cremona, 2017). The cost, in cash terms, for the FIGC and the Serie A of the VAR system being implemented would be somewhere between €1.5 and €2 million, including the €300,000 that each of the 20 top-flight clubs had had to contribute (Bellinazzo, 2017).

The VAR Is Presented in the Serie A Championship, and the Controversy Begins

After the second Serie A match day, some specialist journalists and commentators from the Italian digital television platform *Sky Sport* gave their opinions regarding the VAR system. The former Milan player, Alessandro Costacurta, pointed out quite clearly that over the two match days it had been obvious that it was:

> a great help and the very fact that there had been so little controversy was a sign that it had been successful. Clearly, referees will improve in how they evaluate their decision-making, but I think the VAR is an excellent instrument.

Journalist Paolo Condò also took a favourable position in this debate towards video-refereeing, pointing out that "many mistakes have been corrected, and this is a tremendously positive thing", also remarking that it would be interesting for the referee to give a brief explanation to the media concerning certain decisions that had been taken at the end of the game ("*Sky Calcio*", 2018).

Zvonimir Boban, the former Milan player, currently FIFA's Deputy Secretary General, told *La Gazzetta dello Sport* in March 2018 that his opinion, prior to the introduction of the VAR into European football, was that referees made big mistakes once every three games, and that now only once in every 19 would you see a similarly major mistake, one that would directly affect the result of a match. Another controversial element was the time that was taken up while decisions were being made. Boban argued that nine minutes were taken up before the ball was put back into play after a foul had been called, using data that showed that once every three games 55 seconds were lost while the referees corrected a serious mistake by means of the VAR. For these reasons, he considered that it was necessary for "one to think it through before speaking" and criticising this aspect of the VAR. In Italy, Boban continued, for every 100 serious mistakes, the VAR system has corrected 95, and while it is true that the way goal celebrations are carried out, with players' behaviour being changed from what everyone is used to, the FIFA representative was of the opinion that:

> if you look at the benefit of the corrected decision, and the fact that people will not be suffering for days on end as a result of those mistaken decisions, in that case do we have to let people carry on celebrating goals in the way they were used to doing? I don't think so. (Ceniti, 2018a)

Regarding this, the situation that came up in the Serie A game between Milan and Chievo on 18 March 2018 was also much discussed in the media. The Milan forward Patrick Cutrone scored the goal that was to make the score 2–2, but had to wait for the VAR to confirm definitively that the strike was legal and, once it was, ran over to the fans to celebrate with them, a scene that until now has not been common in football but which, maybe, both players and fans will have to get used to from now on ("Una celebración diferida", 2018).

Despite the controversies that arose during the 2017–18 season regarding the correct application of the VAR in certain situations, its success in terms of the significant reduction in the percentage of errors endorsed the project. FIFA's backing also let them think about how to improve the communication of referees' decisions to the crowd in the stadia. Roberto Rosetti himself told the media shortly before it was confirmed that the VAR would be used in the 2018 Russia World Cup that a possible change in the near future might be to do with the transmission of replays on the giant stadium screens:

> It will probably be possible to inform the spectators and the replays of the modified decisions will be shown on the giant screens, on this point of view it is a very positive impact on the public. VAR has become a social phenomenon: making the right decision has become the most important goal and everyone is seeing it in the most positive way.

Regarding certain controversial decisions taken in some matches, and the doubts raised about the need to review the protocol, Rosetti recognised that it was necessary to keep improving and that the aim was to achieve greater uniformity of information and intervention, "but in just six months a very good job is being done" ("Var, Rosetti", 2018). Early in August 2018, La Lega Serie A authorised the showing of the VAR videos on the giant stadium screens in Serie A. The authorisation was limited to the showing of footage from the game being played at the stadium, and after the refereeing decision had been taken, leaving out the possibility of showing fouls and offsides ("Le clip sugli schermi", 2018).

In March 2018, the European Club Association (ECA), a body that represents 160 European football clubs, held a meeting in Rome, presided over by Andrea Agnelli, the Juventus chairman. Among other topics, the introduction of the VAR system into European football was discussed, and Agnelli made his views on this technological innovation very clear:

> The VAR system has been introduced in several countries, and will also be used in the World Cup. It is an irreversible process. It is a subject that involves several federations. It is in Italy, it is in Germany and it will also be adopted in the United Kingdom. We must respect UEFA's timing on this issue. Staff, which means the referees, must be trained. This question involves everyone, but I think the system can be introduced as early as the 2019/20 season.
>
> ("Eca, Agnelli", 2018)

The First Results of the VAR System in Italian Football

Once the first half of the Serie A season was over, it was a good moment to take stock and draw the first conclusions regarding the VAR system. In January 2018, the AIA members and the Serie A managers met to analyse the results of the VAR system since its introduction into Italian football. One of the first conclusions reached was that nobody had any doubts that there had been mistakes made but, at the same time, there was no going back now. The meeting was chaired by the head of the AIA, Marcello Nicchi, as well as the person in charge of refereeing appointments, Nicola Rizzoli, and Roberto Rosetti. The statistical data being handled in Italy points to a level of error of only 1 per cent in those months, with 1,708 decisions being reviewed over 210 matches, which led to 60 decisions being corrected, 11 of those being wrong, including seven that influenced the outcome of the game. The statistics tell us that in Italy 5.1 "[silent] checks" are carried out per game, most of them silent and with no need to notify the match referee. These "checks" refer to 579 goals given, 282 penalties, and 214 possible sendings-off. The response time for decisions to be taken, one of the most controversial aspects in the early days, went down in the Italian championship from 1 minute 22 seconds over the first three match days, to 29 seconds by mid-season. On the other hand, the time taken for decisions to be reviewed and modified was of 1 minute 11 seconds, compared with 2 minutes 35 seconds over the first three match days. In the same vein, the referees commented in the meeting that in this championship, effective playing time had gone up by 40 seconds, the number of penalties awarded had increased by 5.5 per cent, and there had been reductions in the number of fouls, by 8 per cent, cards shown, by 18.8 per cent, and diving was down by 23 per cent. Lastly, regarding sendings-off, of the 40 that had occurred, not one had been owing to protests, compared with data from the previous season when, of 51 sendings-off, five had been for ongoing protests to the referee (Var, incontro arbitri-allenatori: "Impossibile tornare indietro", 2018).

Marcello Nicchi, the AIA president, claimed that "the VAR system is a product that works. There have been mistakes, and there will be more, but if we think that in just five months we have got to this point, then I am not at all troubled." The head of refereeing also wanted to differentiate between the model applied in Italy and that applied in German football, in an attempt to distance himself from the problems and controversies that the introduction of the VAR system had been immersed in in that country:

> we are an example for all the federations around the world, we have been getting requests from Asia to train their referees abroad. [...] In Germany it is working less well than here, because we are using it in a different way. We have a VAR room in each stadium, they have one centre from where they watch every stadium and that, evidently, works less well.

Nicchi also stated that in the following days a specialised centre would be ready in Coverciano, the headquarters of the FIGC, where all referees could be trained in the use of VAR technology in the best conditions. FIFA chose Coverciano as a pilot VAR training centre and all referees appointed to take part in the World Cup had to go there to be trained in the use of the technology ("Il Var e i suoi limiti", 2018).

In the same meeting, some team managers also gave their opinions after the first five months of the VAR system being introduced. The Inter Milan manager, Luciano Spalletti, pointed out that while some adjustments would be needed, the statistics were saying that in general things were working correctly and that "it would be dramatic to turn back now". One of the critical points, according to Spalletti, was when offsides are flagged and the wait for the response from the video is too long: "our linesmen are good, they know how to choose, they have a percentage of correct calls that means they should be given more responsibility than the VAR system allows them." Stefano Poli, the Fiorentina manager, also gave his backing to the VAR system:

> I believe the VAR system is an element that provides greater fairness. There are always mistakes made, but now the number is much lower. I think that in the future, some of the VAR decisions will be screened in the stadium, making them public.

Regarding the possibility that the managers themselves could request a VAR review during games, he commented that "the rules don't allow it and the referees have stressed this situation, but the referees are improving and are now better than they were at first, thanks to the efforts that have been made, so things are working out well" ("Il Var e i suoi limiti", 2018).

The Media and Their Pedagogical Function Regarding the VAR System

It was seen throughout the first season of the VAR system being applied in Italy that the media was a key player in the rapid and successful introduction and implementation of this technology in Italian football. A large part of the problems and controversies that arose during the early weeks of the VAR system's implementation were to do with the lack of precise knowledge on the part of journalists themselves, and fans, regarding the exact function of the VAR system and how it is applied on the pitch. Not knowing what actions can be reviewed by the VAR or ignorance of the protocol that rules how decisions are taken meant that players, managers, fans, and journalists all came out with a great many statements, above all in the early months, that showed a clear lack of awareness of the basic rules that were behind the VAR system. This fact points to the importance of an adequate training strategy, one that allows all the actors involved in what we might call "football society" in each country, to have sufficient knowledge of the application criteria of this new technology. This is one of the most important things to learn from the Italian experience and is, probably, one of the facets that helps us to understand how it became efficiently internalised in that country's footballing culture.

At the same time, however, this fact also tells us that it is necessary for both the national refereeing committees and the mainstream media in each country to work together and do the important work of educating the fans at large. Nor should we forget that this technology was still in the testing phase and, as had been seen in various countries where it was being trialled, the first months of the VAR system's application meant certain aspects of the protocol had to be reviewed as they were not absolutely clear. In this initial period, it was seen that the media could play a vital role in disseminating the fundamental concepts the application of the VAR system rest on and clearing up fans' doubts. The collaboration between the refereeing groups and the media was key in the Italian case. By means of the training process, the journalists themselves, and fans (with the help of special TV programmes in which referees and journalists took part), the controversies that arose in the early games, owing to the ignorance of certain aspects of the protocol, progressively died down, while at the same time, and in a parallel process, coordination among the refereeing team improved and the time needed to reach decisions was reduced.

One television experience that was an interesting example of good practice was that carried out by *Sky Sport Italia*. This digital TV platform, led by its chairman Federico Ferri, decided to offer its viewers a programme that explained, in a simple but rigorous way, the fundamental criteria that the use of the VAR system rests on. For this project, *Sky Italia* had the AIA as a collaborating body, particularly its main VAR representatives: Nicola Rizzoli and Roberto Rosetti. The programme was divided into two parts, during the first of which the way the VAR system was applied was explained to specialist football journalists and commentators, showing them data and stats gathered over the first six match days of the league championship. In the second part of the programme, there was an analysis of situations in which the technology had been applied, either correctly or mistakenly, in which the viewers were shown the conversations between the on-field referee and the video assistant referees (see Chapter 4), while at the same time both Rizzoli and Rosetti explained in a simple and educational way how the communication had gone and how it had led to the decisions finally taken.

On this programme, and given that six match days had already gone by, a series of highly relevant statistics were brought into play. It was highlighted that, out of a total of 59 matches, there was an average of one refereeing decision having been changed every 3.3. Equally, it was mentioned that over that period 264 incidents had been analysed, of which 59 per cent were goals, 27 per cent were penalties, and 15 per cent red cards. In some of those 264 incidents, there was not even any communication with the referee, with the "check" going no further. Of those 264 incidents, in 246 the referee's original decision was confirmed as correct, while only 18 were modified owing to having been wrong. Of those 18 modifications, in 13 of them the referee went in person to review the play on the monitor. These figures show that the central role of the referee on the field remains undiminished. Concerning the response time, it is pointed out that the analysis had been structured by splitting the statistics between the first three match days of the championship and the last three played, showing that in the early games the average time taken to reach a decision from the moment the incident occurred until the referee received communication was of 1 minute 22 seconds, while in the three most recent it had gone down to 40 seconds. Following these statistics, over 32 match days, some 130 mistakes might have been corrected.

Some media outlets had taken a stance in favour of the introduction of new technologies into football from the outset. The editor of *La Gazzetta dello Sport*, Sebastiano Vernazza, had no doubts in stating that the number of refereeing mistakes had been reduced considerably since the introduction of the VAR system into Italian football. In the match report of the fixture between Inter Milan and Lazio on 31 December 2017, this same journalist explained how referee Gianluca Rocchi's decision to award a penalty as a result of a suspected handball by an Inter defender was corrected by the VAR and by the referee himself when he saw the replay and realised that it was actually unintentional. Vernazza ironically dedicated this decision to those who were still ferociously critical of the VAR system in the name of a false romanticism. The new technology might have reduced the number of refereeing mistakes, although it should not be forgotten that "there is no such thing as perfection and the VAR can be used correctly and incorrectly, but the

VAR has reduced the number of imperfections, and anyone who denies that fact does so in bad faith" (Vernazza, 2017, p. 10).

The team managers involved in that game, however, were not both of the same opinion as the *La Gazzetta* journalist. On the one hand, Inter manager Luciano Spalletti, when asked for his assessment of the VAR system, commented that one had to "see the positive things that have been done. There is no comparison between the fair ones and the mistaken ones; it's not me that's saying it, everybody is saying it"[1] (Taidelli, 2017, p. 13). Simone Inzaghi, on the other hand, complained about how the VAR decisions had gone against his team:

> We are missing seven points because of the VAR. I think that on the question of Skriniar's handball the penalty could just as easily have been awarded as not. But I also think that in our games with Fiorentina and Torino, the intervention of the VAR was very questionable and was always resolved against us. In the first case we lost two points, in the second we should have been given a penalty but instead they sent off one of our players and we eventually lost all three points. And with this episode at San Siro, that now makes it seven points we are down in the standings because of the VAR.[2]
>
> (Cieri, 2017, p. 13)

That same match day, the VAR system also took centre stage in AS Roma's game against Sassuolo when it signalled offside in the goal that referee Orsato di Schio had initially given. Eusebio di Francesco, the AS Roma manager, was very critical in his post-match assessment of the new technology, saying that:

> I don't like it, there were mistakes before and there are mistakes now. As well as that, how are the decisions reached? An obstruction, like the one that was done to us in Juve's first goal in Turin, that's a foul isn't it? These things make you think, there is no uniformity.
>
> (Stoppini, 2017, p. 18)

Some newspapers as influential as *La Gazzetta dello Sport* did make some proposals for improving the VAR system, five months after it was introduced. Francesco Ceniti proposed five points which, for *La Gazzetta*, were necessary to reduce the number of mistakes and clear up the remaining doubts about the system. First, it was stressed that there was a problem of uniformity of refereeing decisions concerning handballs in the area. The newspaper recommended that all handballs in the area should be automatically reviewed by the referee on the monitor, wondering whether the "clearly unintentional" ones could be excluded when the player concerned has his arm against his body or is protecting his face. All other plays should be checked by the referee. A second innovation they suggested was to introduce effective playing time (two halves of 30 minutes) in order to, in this way, take the pressure off the VAR to make a hasty decision, thus getting rid of the controversies that stoppage time and unsporting timewasting caused. The third element they focused on was the so-called "clear mistake" or "serious mistake", a fundamental concept when it comes to understanding how the VAR should be applied. In this case, the leading journalist considered that between the current vagueness of the concept and the risk of the control of the game passing into the hands of the VAR, there was a compromise, which would consist of considering the application of the VAR system to all "mistakes", increasing the number of reviews, but always limiting them to the cases foreseen in the protocol (goal/no goal, penalty/no penalty, direct red cards, and mistaken identity) (see Chapter 10). This way, according to the newspaper, the grey area that comes from the uncertainty about whether to apply the VAR or not in given plays would be cleared up. Another thing they advocated was for total transparency regarding the footage the referee used to make decisions. *La Gazzetta* considered that this footage should be shown simultaneously to the fans at the stadium and those watching the game on television, so that everybody understands what is happening. They are also of the view that the coordinator of referees should publicly explain at least once a month what led the referees to not review certain plays and analyse the errors. Lastly, one of the most controversial aspects in the early stages of the application of the VAR system in Italy was the question of offsides and the millimetric precision of the technology to pinpoint the details. Until such time as 3D footage is available, it would be necessary to reform the rule to make it more objective in the case of the VAR assessing and to help the assistant referees in their work (Ceniti, 2018b).

It is interesting to analyse how the VAR system became integrated into Italian football in so few months, to the point where the absence of this technology from the Champions League gave rise to protests and requests for it to be brought in immediately. This was seen in May 2018, in the return leg of the semi-final in that competition between AS Roma and Liverpool. There were several controversial refereeing decisions in that game, above all a penalty that was not awarded for a very clear handball by the defender Alexander-Arnold following El Shaarawy's shot in the second half. In the post-match press conference, Di Francesco said, "today the VAR would have given us more than it would have taken away. They took something away from us in the most important moment of the match" ("La Roma sfiora", 2018). Even AS Roma's sporting director, Ramón Rodríguez, "Monchi", said "the VAR should be brought into the Champions League, what's happening is not normal. I cannot understand why it's not being used in the most important competition there is" ("La queja de Monchi", 2018).[3]

One Year of the VAR in Italia

The good management of the VAR project in Italy meant that Roberto Rosetti was given the job of VAR Refereeing Project Leader during the Russia World Cup, and is now UEFA's Chief Refereeing Officer, a post previously held by his compatriot, Pierluigi Collina. The recognition of Italian referees at international level also meant three more of them, Massimiliano Irrati, Daniele Orsato, and Paolo Valeri, were included on FIFA's VAR Specialist list for the World Cup, along with Gianluca Rocchi, who during the tournament carried out both the role of on-field referee and that of VAR (FIFA, 2018). Italy was the only country to have three representatives on the list of video match officials. FIFA also recognised Italian refereeing by choosing the International VAR Centre in Coverciano as the training centre for all referees appointed for the Russia World Cup.

A year on from the introduction of the VAR system in Italian football, it is very interesting to be able to look at the statistics thrown up by this process. Over the 2017–18 season, the VAR was used in 397 matches, 380 in Serie A and 16 in the TIM Cup.[4] Over these matches, there were a total of 2,023 checks (an average of 5.1 per game), of which 117 became overrules (that is, refereeing decisions that are changed), which means an average of 1 for every 3.29 matches. Of the 2,023 checks carried out, 1,060 concerned goals (53 per cent), 407 involved red cards (20 per cent), and 538 looked at penalties (27 per cent). Of the 117 overrules, 36 per cent were in the category of goals, 50 per cent in that of penalties, and 14 per cent were to do with red cards. The benefits seem to be evident in view of the reduction in the number of mistakes. With the VAR, the percentage of errors was 0.89 per cent as opposed to the 5.78 per cent it would have been without the use of the VAR system (FIGC, 2018, pp. 4–5).

The time the VAR and the on-field referee take to reach a decision is another subject that sparked most debate in those months. Data from the first season show that the average response time, the time the VAR spends analysing the play and making a decision, went down from 1 minute 22 seconds over the first three match days to 31.5 seconds by the end of the season. The time taken to overrule decisions also went down clearly from the early match days onwards: from the 2 minutes 35 seconds at the beginning of the season, the average time went down to 1 minute 22 seconds by the end, which is broken down into an average of 32 seconds for the VAR to look at the play and send the signal to the referee and the 50 seconds that he himself takes to look at the replay on the screen beside the pitch. If we analyse the effective playing time, we also see significant improvements over previous seasons. From the average of 49 minutes 10 seconds of effective playing time in the 2015–16 season, this went up to 51 minutes 13 seconds in the 2017–18 season, an increase of 2 minutes 3 seconds (FIGC, 2018, p. 6).

The introduction of the VAR system into Italian football has also had an impact on the way players behave on the field. Unlawful conduct was noticeably reduced over the course of the season. The number of yellow cards shown went down by 12.3 per cent, from 1,719 in the 2016–17 season to 1,508 in the last campaign. Protests also dropped, (–17.5 per cent), as did diving (–35.3 per cent) and unsporting conduct (–6.1 per cent). Lastly, the number of red cards was reduced from 97 to 91, with the number of sendings-off for protesting dropping from 11 to just 1 (FIGC, 2018, p. 7).[5]

Conclusions

Despite all the controversies that arose during the opening games, we can consider the introduction of the VAR system into Italian football as an experiment that has been highly successful. It is evident that it took several weeks for the referees to perfect the techniques in communication between the on-field referee and the VAR room (see Chapter 10) in order to make decision-making speedier and more effective. In some cases, the problems came about as a result of technical problems with the audio equipment the referees were using. At the same time, the whole process of internalising and learning of the main elements of the VAR protocol on the part of the mainstream media and the fans should not be forgotten. It was shown how the process of coordination and technical improvement on the part of referees ran parallel to a drop in the number of controversies that, in the early weeks, had been the result of a lack of awareness of the key elements of the protocol. This learning process allowed the media themselves, over the course of the first year of the system's implementation, the possibility to offer interesting analyses on aspects of the system that may be reflected on in order to propose possible changes and improvements in the coming seasons.

The Italian case has become something of an international reference point, and this is largely attributable to the good level of training the country's referees have. This has led us to reflect on the importance that the development of new technologies specifically aimed at the training of referees in matters that, until now, have not had any connection with their role, such as the use of new technologies adapted to audio-visual environments and television production, will have in the near future. It is necessary for centres that specialise in training referees in the use of the VAR system, such as the International VAR Centre in Coverciano, to be set up in every country, thus allowing all of them access to training of the same quality and to the same standard, and to take part in courses aimed at improvement and updating all year round. Another aspect worth highlighting is the benefit, or harm, that could come from having the management of the VAR system in one place for all games or, in contrast, every stadium having its own Video Operator Room (VOR). On the one hand, each stadium having its own VOR and the on-field referee and his assistants forming a team with the VAR and the AVAR themselves, forging personal bonds, could be a factor that helps improve communication between those in the VOR and those on the pitch, thus reducing the response time and avoiding errors. At the same time, it seems that the tendency recently is to centralise the handling of all the games in one VOR, thus making it more viable and cutting costs. Some time will need to pass before it will be possible to say with any degree of accuracy what benefits and what harm may come from this choice of where to locate the VAR.

Lastly, it is still too early to be able to say whether the introduction of the VAR system into football may lead to big changes in behaviour on the part of fans at the stadium and how they react to refereeing decisions. Certain decisive questions are still very much up in the air. One of them is whether the use of the VAR system will encourage a change in the attitude of the fans when it comes to accepting refereeing decisions, changing their way of behaving and making it normal, as has been the case for several years in rugby (see Chapter 8), for the decisions to be explained directly to the fans. The march towards greater fairness in football by means of the use of new technologies now seems unstoppable, but it is still too early to say how this will change the way the sport is practised and the experience of the fans, both those present in the stadiums and the ones watching in their own homes.

Notes

1 Original: "vedere le cose positive che sono state fatte. E non c'è paragone tra quelle giuste e quelle sbagliate: non lo dico io, lo dicono tutti".
2 Original: "A noi mancano sette punti per colpa della Var. Penso che sul mani di Skriniar il rigore si poteva dare come se poteva non dare. Ma penso anche che nelle nostre gare con Fiorentina e Torino l'intervento dell Var è stato molto discutibile e si è sempre risolto a nostro danno. Nel primo caso ci abbiamo rimesso due punti, nel secondo ci dovevamo dare un rigore ed invece hanno espulso un nostro giocatore e alla fine abbiano perso tre punti. E con questo episodio di San Siro diventano sette i punti che ci mancano in classifica per colpa della Var".
3 UEFA's Executive Committee decided early in December 2018 that the VAR system would be used from the Round of 16 onwards in the 2018–19 UEFA Champions League (Moñino, 2018).
4 According to the official document, 397, but 380 + 16 is 396.
5 For greater quantitative information, see Chapter 1.

References

Bellinazzo, M. (2017, September 27). Cina, via libera al Var dal 2018: investimento da 2,5 milioni di euro. *Il Sole 24 Ore*. [Blog post]. Retrieved from http://marcobellinazzo.blog.ilsole24ore.com/2017/09/27/cina-via-libera-al-var-dal-2018-investimento-da-25-milioni-di-euro/?refresh_ce=1

Ceniti, F. (2018a, March 3). Var, Boban: "Tanti errori evitati. E il gioco non ne soffre...". [Interview. Video file] *La Gazzetta dello sport*. Retrieved from http://video.gazzetta.it/var-boban-tanti-errori-evitati-gioco-non-ne-soffre/c2a04b04-1efe-11e8-9fe7-79e3758b0cea

Ceniti, F. (2018b, January 8). Dai 'mani' in area al tempo effettivo: le 5 proposte Gazzetta per migliorare la Var. *La Gazzetta dello Sport*. Retrieved from www.gazzetta.it/Calcio/Serie-A/07-01-2018/dai-mani-area-tempo-effettivo-5-proposte-gazzetta-migliorare-var-240754036344.shtml.

Cieri, S. (2017, December 31). Ira Inzaghi sulla Var: "Sfortunati ci mancano sette punti". *La Gazzetta dello Sport*, 13.

Cremona, L. (2017, August 10). Il Var funziona, basta solo abituarsi. E trasmettere i repaly sui tabelloni. *TGCOM24*. Retrieved from www.sportmediaset.mediaset.it/calcio/calcio/il-var-funziona-basta-solo-abituarsi-e-trasmettere-i-replay-sui-tabelloni_1166513-201702a.shtml.

Eca, Agnelli: "Ridurre numero gare. Var? Avanti così…". (2018, March 27). *La Gazzetta dello Sport*. Retrieved from www.gazzetta.it/Calcio/Serie-A/Juventus/27-03-2018/eca-agnelli-ridurre-numero-gare-var-avanti-cosi-26081090563.shtml.

Fédération Internationale de Football Association. (2018). FIFA World Cup Russia 2018: List of match officials. [online]. Retrieved from https://img.fifa.com/image/upload/w7qdlidnrq4lah2pvvc3.pdf

Federazione Italiana Giuoco Calcio. (2016a, April 14). La FIGC selezionata dall'I-FAB per testare il sistema "video assistant referees". Retrieved from www.figc.it/it/federazione/news/la-figc-selezionata-dall-ifab-per-testare-il-sistema-video-assistant-referees/.

Federazione Italiana Giuoco Calcio. (2016b, April 21). Video assitenza agli arbitri, in FIGC incontro positivo con AIA, LEGA A e B. Retrieved from www.figc.it/it/federazione/news/video-assistenza-agli-arbitri-in-figc-incontro-positivo-con-aia-lega-a-e-b/.

Federazione Italiana Giuoco Calcio. (2018, June 6). Report finale "Video Assistant Referee Project". Retrieved from www.aia-figc.it/download/news/09_06_2018media_var_report.pdf.

Fifa, Busacca promuove la Var: "Test positivo in Italia-Germania". (2016, November 17). *Repubblica*. Retrieved from www.repubblica.it/sport/calcio/2016/11/17/news/fifa_busacca_promuove_la_var_test_positivo_in_italia-germania_-152222794/?ref=search.

Il Var e i suoi limiti, arbitri e tecnici a confronto. Spalletti: "Ma tornare indietro sarebbe devastante". (2018, January 15). *Repubblica*. Retrieved from www.repubblica.it/sport/calcio/2018/01/15/news/var_errori_allenatori_arbitri_spalletti_inzaghi-186537837/?ref=search.

La queja de Monchi: "No es normal que la Champions no tenga el VAR". (2018, May 5). *La Vanguardia*. Retrieved from www.lavanguardia.com/deportes/futbol/20180502/443196134964/monchi-var-champions-roma.html.

La Roma sfiora l'impresa: Liverpool battuto 4-2, ma a Kiev vanno gli inglesi. (2018, May 2). *La Stampa*. Retrieved from www.lastampa.it/2018/05/02/sport/la-roma-sfiora-limpresa-liverpool-battuto-ma-a-kiev-vanno-gli-inglesi-WRvd3HsaIhANHyfL444siL/pagina.html.

Le clip sugli schermi degli stadi di A. (2018, August 7). *Repubblica*. p 38.

Milan, André Silva non basta: rossoneri ko in amichevole con il Betis, Var decisiva. (2017, August 9). *Repubblica*. Retrieved from www.repubblica.it/sport/calcio/serie-a/milan/2017/08/09/news/milan_andre_silva_non_basta_rossoneri_ko_in_amichevole_con_il_betis_var_decisiva-172746202/?ref=search.

Moñino, L. J. (2018, December 4). La Champions utilizará el VAR a partir de octavos de final. *El País*. Retrieved from https://elpais.com/deportes/2018/12/03/actualidad/1543829873_637394.html

Moviola in campo, l'arbitro: "Ho preso tempo, chiesto agli assistenti e poi ho deciso". (2016, July 2). *TGCOM24*. Retrieved from www.sportmediaset.mediaset.it/calcio/calcio/moviola-in-campo-l-arbitro-160-ho-preso-tempo-chiesto-agli-assistenti-e-poi-ho-deciso-_1116432-201602a.shtml.

Nel giorno della Var, Italia under 19 sconfitta dalla Spagna. (2017, January 18). *Repubblica*. Retrieved from www.repubblica.it/sport/calcio/2017/01/18/news/a_udine_esordio_della_var_on_line_nella_sfida_tra_italia_e_spagna_under_19-156324451/.

Nicchi e la rivoluzione Var: "Siamo pronti, l'arbitro d'area diventa arbitro video". (2017, April 24). *Repubblica*. Retrieved from www.repubblica.it/sport/calcio/serie-a/2017/04/24/news/nicchi_e_la_rivoluzione_var_siamo_pronti_l_arbitro_d_area_diventa_arbitro_di_porta_-163806155/?ref=search.

Primavera, Juventus-Sampdoria: debutto storico per la Var. (2017, May 4). *Repubblica*. Retrieved from www.repubblica.it/sport/calcio/2017/06/04/news/primavera_juventus-sampdoria_1-0_bianconeri_in_semifinale_debutto_storico_ma_anonimo_per_la_var-167241908/?ref=search.

Sky Calcio Show Countdown (2018, August 28). [Television broadcast]. Sky Italia.

Spignesi, R. (2017, May 29). Campionato Primavera, ufficiale l'utilizzo del VAR nelle fasi finali. Retrieved from www.spaziocalcio.it/calcio-giovanile/campionato-primavera-ufficiale-lutilizzo-del-var-nelle-fasi-finali/.

Stoppini, D. (2017, December 31). Per il titolo non ci siamo. Var? Non c'è uniformità. *La Gazzetta dello Sport*, 18.

Taidelli, L. (2017, December 31). Spalletti applaude: "Equilibrio e carattere Dietro ok". *La Gazzetta dello Sport*, 13.

Una celebración diferida por el VAR. (2018, March 19). *La Vanguardia*. Retrieved from: www.lavanguardia.com/deportes/futbol/20180319/441710786754/milan-patrick-cutrone-celebra-gol-diferido-var.html.

Var, incontro arbitri-allenatori: "Impossibile tornare indietro". (2018, January 15). *TGCOM24*. Retrieved from www.sportmediaset.mediaset.it/calcio/calcio/var-incontro-arbitri-allenatori-impossibile-tornare-indietro-_1191891-201802a.shtml.

Var, Rosetti: "Moviola in Russia? Fiducioso. Presto immagini sui maxi schermi". (2018, February 19). *La Gazzetta dello Sport*. Retrieved from www.gazzetta.it/Calcio/19-02-2018/var-rosetti-moviola-russia-fiducioso-presto-immagini-maxi-schermi-250389562430.shtml.

Vernazza, S. (2017, December 31). Il bicchiere mezzo pieno. *La Gazzetta dello Sport*, 10.

14 VAR Experiments in Major League Soccer (MLS)

Clinton Warren

Introduction

In 2014, Major League Soccer (MLS) Commissioner Don Garber approached FIFA with an interest in becoming the first professional football league in the world to adopt a video review system in which managers on the sideline would have the opportunity to challenge referee decisions in the interest of improving the accuracy of league referees (Wahl, 2014). Garber's initial comments were supported by then FIFA President Sepp Blatter. However, as the International Football Association Board's (IFAB) rules on the implementation of a video assistant referee (VAR) have become clearer, MLS's initial push for a coach's challenge system is no longer a viable request. IFAB's principles related to video review of referee calls established the external referee position of the VAR. Specifically, the VAR is a match official who has independent access to match footage and is allowed to assist the on-field referee in the event the VAR witnesses a "clear and obvious error" or "serious missed incident" (IFAB, 2018). Notably, the IFAB protocol also states that only the on-field referee can initiate a review with the VAR and other match officials being allowed to recommend reviews. As this relates to Garber's comments, there is no room in the current IFAB VAR protocol for a manager or coach challenge to a referee decision. While the system originally envisioned by the MLS commissioner is not realistic within the IFAB protocol, MLS is a league that sought to be at the forefront of the adoption of a video review system. As a result, IFAB named MLS, along with Australia's A-League, the German Bundesliga, Brazil's Campeonato Brasileiro, and select competitions in the Netherlands and Portugal as the original six participants in experimentally testing VAR.

To fully understand the manner by which initial VAR experiments were conducted in the United States in accordance with IFAB's approval, a brief primer on the organizational and competitive structure of soccer in the United States is needed. MLS has been the top division in North American soccer since it began play in 1996. MLS was launched as part of the United States' initiative to host the 1994 FIFA Men's World Cup.

From a business perspective, MLS established a series of strict capital structure, marketing and management, player acquisition, and roster management policies that were designed to keep the league from overspending early in its existence. This was done to avoid the fates of early soccer leagues in America that folded as expenses began to far outstrip revenues. In creating its league structure, MLS and USSF agreed to operate a closed system in which no promotion and relegation of clubs is possible through on-field accomplishments. This closed system is popular in North American professional sports. To further accomplish its goals of conservative growth and long-term viability, MLS operates as a single-entity business in which franchise owner-operators are granted a licence to run their teams rather than complete ownership autonomy (Warren and Agyemang, 2019).

In addition to the unique, closed nature by which MLS operates its league, MLS and the USSF work with a marketing and management company called Soccer United Marketing (SUM) that handles all commercial negotiations with respect to the MLS and US National Team television and sponsorship rights. Individual MLS teams are still afforded some autonomy with regard to generating revenue, but larger, league-level agreements are handled through SUM. Interestingly, SUM also serves as the marketing partner for the Mexican National Team. While SUM initially worked to serve the USSF and MLS, in 2013 MLS agreed to a formal partnership with a league called USL Pro, today's USL (Warren and Agyemang, 2019). At the time, USL Pro was considered the third tier of soccer in the United States and Canada. The initial agreement was formed to facilitate play between USL Pro teams and MLS Reserve teams as well as a seamless loan system for MLS clubs in similar geographic areas as USL Pro clubs (Doyle, 2013). In 2018, the USSF granted the USL second division status in the United States soccer system (Associated Press, 2018). Since that time, the commercial relationship between MLS, SUM, and the USL has grown. MLS has accepted expansion fees from USL owners, who have then earned a spot in MLS. A prime example of this is FC Cincinnati's move from USL to MLS in 2019. Further, USL has re-branded using an English blueprint and expanded its league system to include three tiers. At present, the United States soccer system includes MLS as the first division, the USL Championship as the second division, USL League One as the third division, and USL League Two as the fourth division. While promotion and relegation based on team performance is still not possible, this restructuring has led to significant growth in the professional game in North America.

As a result of the league's formal, commercial partnership with the USL, MLS staged its first official VAR experiment on 12 August 2016 in a USL match between New York Red Bulls II and Orlando City B, hosted at Red Bull Arena in Harrison, New Jersey (Borg, 2016). During the match, head referee Ismail Elfath made history as VAR was utilized for the first time in a competitive football match (Dyer, 2016). Following the initial experimental use of the technology in this match, USL tested the technology in a total of four other New York Red Bulls II home matches during the 2016 season. This chapter will discuss these initial experiments in detail.

Following the USL experiments in 2016, Garber further advanced the league's video review initiatives in his annual "State of the League Address" on the eve of the 2016 MLS Cup Final. He stated MLS would use the first half of the 2017 season as an opportunity to test the implementation of VAR in league matches with the intent of fully adopting the system for competitive league matches following the MLS All-Star game on 2 August 2017 (Borg, 2016).

Culturally, Garber and MLS's positive position on the implementation of a video review system for MLS is not surprising. The usage of a video review system is commonplace in other North American sports. The National Football League (NFL), National Basketball Association (NBA), Major League Baseball (MLB), and the National Hockey League (NHL) all have video review systems that allow officials and team personnel to review referee calls to ensure accuracy. Garber himself has prior sport industry experience working in the NFL. As such, the usage of video review in the United States is a generally accepted practice. However, video review, or instant replay as it is called in other American sports, suffers from the typical criticisms of the technology. The most popular criticism of instant replay in the NFL, NBA, and MLB is that officials take too long to review calls, and, as a result, gameplay is slowed down and the overall game experience suffers. Compounding this issue, instant replay in the four major North American sports is not solely left to the discretion of the game's referee crew. In each of these leagues, coaches are allowed to challenge calls of game officials. This challenge system represents an added layer to the typical reviews of scoring and other high-impact plays during games. While MLS is considered a video review pioneer as a result of its work with IFAB, video review is not a new phenomenon in North American professional sport.

Methods

This chapter provides a historical and descriptive assessment of the early implementation and usage of VAR in MLS and its affiliated league, the USL. A historical account of the initial USL and MLS experiences with VAR is provided after reviewing official match reports. Additionally, interview statements recorded by the Professional Referee Organization (PRO) are analysed to provide the referee perspective on the implementation of VAR in North America. Further, initial quantitative evidence assessing the technology's effectiveness is reported upon. Future research is then suggested that would help assess the overall implementation of VAR in MLS.

Initial Experiments in the USL

During the 2016 league season, USL experimented with the use of VAR. Five matches that took place at Red Bull Arena in Harrison, New Jersey, the home of MLS side New York Red Bulls and USL side New York Red Bulls II, were the subject of these initial experiments. New York Red Bulls II hosted Orlando City B (12 August), Louisville City FC (19 August), FC Cincinnati (28 August), Harrisburg City Islanders (7 September), and FC Montreal (17 September) in these five initial matches. The officiating crew on 12 August 2016, led by head referee Elfath, made history by using VAR to review two calls in the match. First, in the 36th minute of the match, Orlando City B's Conor Donovan fouled Junior Flemmings of New York Red Bulls II as Flemmings was heading in on goal. Elfath called the foul near the edge of the 18-yard box but did not award a penalty, and approximately one minute after the initial foul was called, the VAR, Allen Chapman, recommended the play be reviewed. The review confirmed the foul had occurred outside the penalty area, but upon review, Elfath determined Donovan's foul had denied an obvious goal-scoring opportunity and he was shown a red card. It took Elfath less than one minute at the video review monitor to make his decision. Later in the match, Elfath issued a red card to another Orlando City B player, Kyle McFadden, for his challenge on a 50–50 ball. The VAR communicated to the head referee that he should consider reviewing the play. Upon Elfath's review, he determined McFadden should only receive a yellow card for his challenge. Finally, in stoppage time of the match, a goal was scored by Zoumana Simpara of New York Red Bulls II. The goal-scoring play was automatically reviewed, during which time it was determined the player who was potentially offside was, in fact, onside and the goal was good (Dyer, 2016). This match served as a watershed moment for the implementation of VAR. Following the match, David Elleray, the Technical Director of IFAB, stated this initial experiment with VAR should receive "a very high grade" (Dyer, 2016).

In all, a VAR review was only used one more time in the five initial matches that were used for the experiment. The head referee of the New York Red Bulls II v Louisville City FC match on 19 August 2016, Hilario Grajeda, needed to make a very similar decision as Elfath had the previous week. Once again, Flemmings was fouled at the edge of the 18-yard box, and the ball was placed just outside the penalty area. The VAR, Chapman, radioed to Grajeda indicating he should review the play for a potential clear and obvious error in the placement of the free kick. Upon review, Grajeda made the determination the foul had occurred inside the penalty area, and he awarded New York Red Bulls II a penalty kick. The Red Bulls converted the penalty and went on to win the match 1–0 (Gass, 2016). Following this brief, initial success of VAR in the USL, MLS and the USL agreed to continue conducting tests over the course of 25 early-season matches in 2017 (MLS Soccer, 2017). Results of these initial experiments confirmed the league's rationale to fully adopt VAR for all competitive matches following the 2017 MLS All-Star Game.

PRO Referees and the Video Review System

The Professional Referees Organization (PRO) is an independent company that is responsible for the administration of soccer refereeing programmes in North America. Elfath and Chapman, PRO referees, made history in being the first pair to utilize video review, and the position of the VAR (Chapman), in a competitive league match. Each referee had positive things to say of their initial experiences in implementing the video review system in USL. Chapman stated:

> The VAR concept is a beneficial tool to the game and provides an opportunity to get more precise accuracy in decisions; particularly game critical decisions. Initially I had reservations about the impact to the game but after significant testing I now believe it's a valuable tool and concept.
> (PRO Media, 2016)

Chapman also highlighted the challenges of serving as a VAR. He indicated feeling:

> the pressure of checking every decision and staying within the protocol set by IFAB was equally challenging in addition to working with the replay operator and the need to communicate with the center referee. I felt like an integral part of the crew and even though I wasn't on the field … .
> (PRO Media, 2016)

Elfath reflected on his experience to the PRO media staff after the match by indicating how humbled he was to play a role in refereeing history. Further, he added his opinions on the addition of video review and the position of VAR:

> This is a major milestone in the ever-evolving profession that we are in; I see nothing but a bright future for refereeing now that we have such a valuable tool at our disposal. Congratulations must go to all involved in making it a reality … . As referees, we share the same passion for the game as any avid fan, so any tool that helps in correcting clear errors of game changing decisions is very much welcomed. The testing so far has eased the concerns regarding game flow, so we are happy.
> (PRO Media, 2016)

While Chapman and Elfath were complimentary of the system and its implementation in this initial experiment, they both indicated room for improvement in their post-match interviews. Chapman said the "first live testing wasn't flawless" and Elfath expressed a unique experience in knowing the VAR was "checking" his calls for any clear errors. As an organization, PRO has continued to be a pioneer in the training of referees to serve as VARs and to implement a video review system in North American soccer matches.

In 2017, PRO announced it had hired Premier League and Champions League veteran Howard Webb to serve as its Manager of VAR Operations (ESPN, 2017). After joining PRO, Webb introduced training programmes for referees that included classroom sessions, hands-on training with the video review system, as well as additional training matches at the Generation Adidas Cup, a youth tournament hosted by MLS academy teams. This training for PRO referees focused on communication between the VAR and head referee as well as the identification of "clear and obvious errors" so as to better define these crucial moments in matches (PRO Media, 2017).

While MLS and PRO have been at the forefront of education related to the usage of VAR, Webb has admitted the system's implementation has not been flawless in North America. Webb stated:

> We've had a couple of situations that have not quite gone to plan, but that's inevitable with a system like this that is new and challenging … The consistency is critical and keeping the threshold high, making sure we don't intervene on things we don't need to get involved with.
>
> (Rivett, 2018)

It is apparent the position of PRO is that the implantation of VAR is a work in progress and that continued training and education is critical to develop the type of consistency Webb expects.

Initial Quantitative Evidence from 2017 and 2018

Initial data reported by ESPN and PRO Referees suggests the VAR experiments in the United States are proving the technology to be successful. Table 14.1 includes a summary of data from the first 452 official league matches in which VAR was implemented in the United States. The data represented in Table 14.1 includes all MLS matches in 2018 from the start of the season through 5 September. The 2017 data includes 137 MLS matches and 17 USL matches. In total, 452 matches utilized VAR, with 140 official VAR reviews occurring. This represents an average of 0.31 reviews per match. Of the 140 plays that were reviewed, 104 calls were overturned. This indicates 74.29 per cent of reviewed calls were initially deemed to be incorrect, with VAR resolving the refereeing mistakes. Following post-match reviews by the leagues, it was determined that of the 140 calls that were reviewed using VAR, 91.43 per cent led to the correct decision.

Table 14.1 Summary of data from the first 452 official league matches in which VAR was implemented in the United States

MLS Season	*Matches*	*Reviews*	*Overturned*	*Total Correct*	*% Overturned*	*% Correct*
2018	298	95	68	84	71.58%	88.42%
2017	154	50	36	44	72.00%	88.00%
Total	452	140	104	128	74.29%	91.43%

Sources: ESPN (2018) and Rivett (2018)

The fact that a total of 74.29 per cent of reviewed calls are overturned suggests PRO referees are making mistakes on almost three out of every four calls they make as it relates to goals, penalties, direct red cards, and instances of mistaken identity. However, it is more likely referees are altering the manner by which they make these calls as a result of the understanding they will be able to review their decision using video technology. For example, a head referee may allow a play in which a player appears to be offside to proceed to conclusion. If that play leads to a goal, the head referee can request a video review and determine the player was offside, thus removing the goal from the score record. No data yet exists to assess the frequency with which PRO referees are using VAR in this strategic manner.

Notable Early Uses of VAR in MLS

During Week 22 of the 2017 MLS season, video review was utilized for the first time in the American soccer top flight. Philadelphia Union was hosting FC Dallas, and Hilario Grajeda served as the VAR with Ricardo Salazar as the head referee. In the 79th minute, Union goalkeeper John McCarthy collided with Dallas forward Cristian Colman while attempting to collect the ball in his penalty area. As the players made contact, the ball rolled to onrushing Dallas forward Maximiliano Urruti, who shot the ball into the goal. Following the goal, Grajeda and Salazar immediately communicated as, per league rules, all scoring plays are reviewed. Salazar then signalled for an official video review of the play, and after 46 seconds, he determined the goal should be disallowed as a result of a foul committed by Colman on McCarthy. The entire review process, from the time the goal was scored to the time Salazar overturned the call, took slightly more than two minutes (Brisendine and Boehm, 2017). Following the match, Union forward C.J. Sapong provided a player's perspective on the first use of VAR in MLS when he stated, "It's nothing that interrupted the flow of the game too much. It's good they got it right" (Brisendine and Boehm, 2017). During the first week in which MLS utilized VAR, one other incident was officially reviewed. Gyasi Zardes of the LA Galaxy had a goal disallowed and was shown a yellow card for purposely using his hand to play the ball in the penalty area of the opposing Portland Timbers. In all, the first week of MLS usage of VAR featured nine matches and the above mentioned instances were the only times when the refereeing crew used VAR to complete a full review of a play. During 2017, VAR was used in a total of 137 MLS matches, and the system averaged 9.6 call checks and 0.32 full video reviews occurring per match (Rivett, 2018). Notably, the 2017 MLS Cup Final utilized VAR; however, no calls were made in which a full video review occurred.

In general, reports of the early usage of VAR in MLS, and other domestic leagues in the United States, were positive. However, as the implementation of the technology gained wider usage throughout the league, it became a focal point of criticism levied at MLS and PRO. Interestingly, the criticism VAR has typically received in MLS is unrelated to the technology slowing down play or disrupting the flow of matches. Rather, the use of VAR has been criticized with respect to judgement calls made by the head referee after an official review has been instituted. Specifically, questions have been raised as to whether the overturning of a goal that was originally allowed by calling the play offside after video review is consistent with IFAB's protocol of review calls that are "clear and obvious errors". For example, in a match in May 2018, Atlanta United's Josef Martinez scored a goal in the 6th minute of play against Sporting Kansas City. Martinez's goal was reviewed as the player appeared as though he may have been offside prior to scoring. Ultimately, the head referee of the match, Mark Geiger, disallowed the goal after video review. By rule, Geiger's decision was correct, as such, the usage of VAR in this instance should be considered a success. However, the overturned goal was largely criticized as fans saw the usage of VAR in that instance as helping the referee make a judgement call rather than reviewing a "clear and obvious error". The usage of video review to assist in refereeing judgement calls is also being questioned by players.

Two examples from 2018 matches illustrate MLS player frustrations with the argued inconsistent and judgement-driven usage of VAR. Zlatan Ibrahimovic of LA Galaxy was sent off in a Week 12 match versus Montreal Impact after video review indicated he slapped the back of the head of an opposing player. Ibrahimovic was shown a straight red card after Elfath completed a full video review after consulting with the VAR (Creditor, 2018). Then, during Week 13 of the same season, video showed Columbus Crew's Cristian Martinez slap Seth Sinovic of Sporting Kansas City in retaliation of Sinovic's contact with him after the whistle (Creditor, 2018). In this instance, the head referee, Robert Sibiga, was communicated with by the VAR, but Sibiga did not conduct a full video review of the play. Following the match, MLS retroactively issued a red card and two-match ban to Martinez. Multiple MLS players, including

Benny Feilhaber (Los Angeles FC), A.J. DeLaGarza (Houston Dynamo), Michael Parkhurst (Atlanta United), and Justin Meram (Columbus Crew) criticized MLS, PRO, and VAR in general via social media following the inconsistent calls in Weeks 12 and 13 of the 2018 MLS season (Creditor, 2018). While no system of officiating can be executed perfectly, the inconsistency of calls like these have been an issue for MLS and PRO. Further, the growing frustrations of players and their vocal criticism via social media in conjunction with widespread fan criticism of a similar nature creates a significant public relations problem for MLS and PRO. Importantly, this issue is not lost on PRO. According to Howard Webb, the biggest challenge PRO faces in its continued implementation of VAR is gaining consistency in understanding when the VAR should intervene. Webb has cited the subjectivity of many calls in the sport as an ongoing challenge to the review system (Rivett, 2018).

Conclusion

This chapter provides a descriptive, historical account of the role MLS played after IFAB determined a course of action in experimenting with and ultimately adopting its VAR protocol. As a league, MLS expressed early interest and intent in adopting a video review policy. The MLS affiliate league, the USL, used the first VAR in a competitive match, and early results of VAR in North America were viewed as positive. While the initial data suggests the usage of VAR is having a generally positive effect on refereeing in MLS and USL, there is growing concern among players, technical staffs, and fans that PRO referees are struggling to seamlessly integrate the technology into their match officiating. The challenge for PRO has been the negative reaction has not been related to the expected criticism of game play being disrupted, but rather, the inability of referees to make the appropriate judgement calls when undergoing a video review. As the implementation of VAR continues to evolve, this will be a key area of consideration for MLS, USL, and PRO.

While this chapter outlines the early implementation of VAR in North American soccer, future research should be considered to fully assess the success, or failure, of video review in MLS and other American domestic leagues. First, MLS and PRO should consider utilizing an independent, external review board to assess the accuracy of VAR usage. MLS often suffers from the perception that it is a closed league in which decisions are made in secrecy. An independent, and objective, review of VAR procedures would assist the league in combating its issues with respect to perceptions of poor judgement calls and a lack of transparency. Continuing to track the number of checks, reviews, and overall call accuracy will be important as VAR moves forward. Second, MLS should consider collecting data from players and fans. While these two constituents have very different roles within the game, both groups have reported frustrations with the VAR protocol and its implementation by PRO. Developing a better understanding of what is at the core of these concerns will help MLS and PRO move forward in positively implementing IFAB's video review protocol.

References

Associated Press. (2018). US soccer grants USL 2nd division status. *Chicago Tribune.* [online] Retrieved from: www.chicagotribune.com/90minutes/mls/ct-90mins-us-soccer-grants-usl-2nd-division-status-20180117-story.html.

Borg, S. (2016). MLS will seek to introduce Video Assistant Referees (VAR) during 2017. *MLSsoccer.com.* [online] Retrieved from: www.mlssoccer.com/post/2016/12/10/mls-will-seek-introduce-video-assistant-referees-var-during-2017.

Brisendine, S. and Boehm, C. (2017). Breaking down the first weekend of Video Review in MLS. *MLSsoccer.com.* Retrieved from: www.mlssoccer.com/post/2017/08/06/breaking-down-first-weekend-video-review-mls.

Creditor, A. (2018). The MLS XI, Week 13: The players have had enough of VAR. *Sports Illustrated.* [online] Retrieved from: www.si.com/soccer/2018/05/28/mls-week-13-var-sporting-kc-columbus-alan-gordon-armenteros-cristian-techera.

Doyle, M. (2013). MLS, USL Pro reach deal on restructured reserve league. *MLSsoccer.com.* [online] Retrieved from: www.mlssoccer.com/post/2013/01/23/mls-usl-pro-reach-deal-restructured-reserve-league.

Dyer, K. (2016). History is made as video replay technology is used in an official USL match. *MLSsoccer.com.* [online] Retrieved from: www.mlssoccer.com/post/2016/08/13/history-made-video-replay-technology-used-official-usl-match.

ESPN. (2017). PRO hires Howard Webb to oversee video assistant referee program in MLS. *ESPN.com.* Retrieved from: www.espn.com/soccer/major-league-soccer/story/3055827/pro-hires-howard-webb-to-oversee-video-assistant-referee-program-in-mls.

ESPN. (2018). VAR review in La Liga, Serie A, Bundesliga, Ligue 1 and MLS this season. *ESPN.com.* [online] Retrieved from: www.espn.com/soccer/blog/espn-fc-united/68/post/3624367/how-var-has-performed-in-la-liga-serie-a-bundesliga-ligue-1-and-mls-this-season.

Gass, D. (2016). Complete recap of 2nd Video Assistant Referee experiment at Red Bull Arena. *MLSsoccer.com.* [online] Retrieved from: www.mlssoccer.com/post/2016/08/20/complete-recap-2nd-video-assistant-referee-experiment-red-bull-arena.

IFAB. (2018). Video Assistant Referees principles and practicalities. [online] Retrieved from: http://theifab.com/projects/vars/principles-practicalities-protocol.

MLS Soccer. (2017). MLS and USL partner to conduct 25 live Video Assistant Referee experiments. *MLSsoccer.com.* [online] Retrieved from: www.mlssoccer.com/post/2017/03/20/mls-and-usl-partner-conduct-25-live-video-assistant-referee-experiments.

PRO Media. (2016). PRO officials make history with use of video replay technology in USL. *PROreferees.com.* [online] Retrieved from: http://proreferees.com/2016/08/15/pro-officials-make-history-with-use-of-video-replay-technology-in-usl/.

PRO Media. (2017). Generation Adidas Cup a big step in VAR training – Howard Webb. *PROreferees.com*. [online] Retrieved from: http://proreferees.com/2017/04/12/generation-adidas-cup-a-big-step-in-var-training-howard-webb/.

Rivett, C. (2018). Video review: How we got here. *PROreferees.com*. [online] Retrieved from: http://proreferees.com/2018/01/12/video-review-how-we-got-here/.

Wahl, G. (2014). MLS tells FIFA it wants to be first to experiment with in-game video review. *Sports Illustrated*. [online] Retrieved from: www.si.com/soccer/planet-futbol/2014/09/12/mls-video-review-challenges-replay-don-garber-sepp-blatter-fifa.

Warren, C.J. and Agyemang, K.J. (2019). Soccer in the United States. In S. Chadwick, D. Parnell, P. Widdop, and C. Anagnostopoulos (Eds.), *Routledge handbook of football business and management* (pp. 591–601). New York: Routledge.

15 How Twitter Has Portrayed the VAR

Raquel Gallego

Twitter as a Communication Tool

In the same way that the VAR may be changing football, one technology that has changed the future of the human race is the Internet. More than half of the world's population now uses the Internet. In fact, the number of Internet users worldwide had gone up by 9.1 per cent over last year, according to the latest report on digital development from We Are Social (We Are Social, 2019, p.8). This tool is used principally for searching for information, which has made the Internet the largest communication platform, ahead of television, radio, and the printed press. In this study, a substantial growth in the use of social networks around the world is also observed, with users growing to as many as 3.26 billion, which means a jump of 10 per cent in users in just a year.

Since the emergence of social networks, Twitter, which started in 2006, has managed to place itself among the front runners in terms of users. One of the reasons for its success lies in the fact that it is not only an open microblogging network on which other members' content can be seen and "retweeted", but that the text used in each message is limited and may have photographs, videos, GIFs, and so forth, as well as "hashtags", indices or tags that allow the user to locate what they are interested in in a matter of seconds. As well as this, content may go viral all around the world, meaning people may give their opinion on a given subject from any point on the planet.

Its creator, Jack Dorsey, considers that Twitter "is not a social network, rather it is a communication tool" (Muñoz & Riveiro, 2009). In fact, "governments and private businesses have included it on the list of communication media for disseminating actions and public information" (Meunier, 2012, p.1). The fact is:

> Twitter has changed the net and has completed the social turnaround that blogs started at the end of the nineties. Now, the whole planet is talking, and we can all listen to it. There is also a chance that the whole planet will listen to us. At least once. We have 140 characters to try it. Many times.[1]
>
> (Orihuela, 2011, p.21)

For all of these reasons, Twitter has become the perfect communication channel for spreading any information, by means of messages that allow other users to be informed, to be educated, to debate, and even to get mobilised.

Methodology

The main aim of this chapter is to portray a reality map of VAR in Twitter. It is centred on the message given out/emitted by the official FIFA and IFAB channels, as well the accounts of the Spanish, Italian, French, German, and English football leagues. In addition, an attempt will be made to show how the media in these countries have informed the public at large.

The research includes a quantitative section with the aim of verifying how many tweets have been sent from the accounts analysed, for what purpose, what intention they had, and the visibility they have achieved.

In order to define indices that are suitable and eligible for inclusion, an analysis of content is carried out, as this is "a valid research technique that allows for reproducible and valid inferences that can be applied to its context to be formulated from the data obtained from Twitter" (Bustos Díaz & Ruiz Del Olmo, 2016, p.279).

The choice of this technique lies in the fact that:

> the analysis of content is based on reading (textual or visual) as a means of information gathering, reading that, unlike common reading, must be done following scientific method, that is, it must be systematic, objective, replicable and valid... The characteristic thing about content analysis, and what distinguishes it from other sociological research techniques, is that it is a technique that intrinsically combines, and that is where its complexity lies, observation and production of data, and the interpretation or analysis of the data.
>
> (En Andreu, 2002, p.2)

The methodology applied in this study, understood by one of its great defenders as "a research technique for the objective, systematic and quantitative description of the manifest content of the communication" (Berelson, cited in En Andreu, 2002, p.2), has been successfully used in different research projects in which the content published by several Twitter accounts over a specific period is analysed. See cases like "How Do Sports Organizations Use Social Media to Build Relationships? A Content Analysis of NBA Clubs' Twitter Use" (Wang & Zhou, 2014), "Pandemics in the Age of Twitter: Content Analysis of Tweets during the 2009 H1N1 Outbreak" (Chew & Eysenbach, 2010), "Understanding Professional Athletes' Use of Twitter: A Content Analysis of Athlete Tweets" (Hambrick et al., 2010), "Social Listening: A Content Analysis of E-Cigarette Discussions on Twitter" (Cole-Lewis et al., 2015), or "La campaña virtual en Twitter: análisis de las cuentas de Rajoy y de Rubalcaba en las elecciones generales de 2011" (García & Zugasti, 2014), among others. Researchers, recognising the usefulness of content analysis – which arose during the 1930s and 1940s thanks to Harold D. Lasswell – have adapted it over the years, in order to be able to apply it to the analysis of new technologies.

As Klaus Krippendorff once said, "empirical interest in symbolic facts can no longer be applied to the study of messages in an isolated way", rather that they should bear in mind "the channels, the limitations of information flows, the processes of communication and its functions in and effects on society, the systems that include advanced technology and modern social institutions" (Krippendorff, 1980, p.11). To which may be added:

> the analysis of content may have as much descriptive aim as an inference (deductive) aim and may use both quantitative analysis techniques as it does qualitative analysis techniques; there is also agreement in that the analysis is not limited to the manifest content of the messages, but may be extended to their latent content, and in the fact that the analyses made must be subjected, like all analysis, to tests of validity and reliability.
>
> (López-Aranguren, 2010, p.595)[2]

The object of study of this analysis comprises the tweets posted from 5 March 2016, the day on which FIFA approved a two-year trial period for "live experiments with video assistance for clear errors in match changing situations approved" (IFAB, 2016, p.17), until 31 August 2018. It has been carried out by taking into account the publications made on the official FIFA and IFAB Twitter accounts, which are @FIFAcom and @TheIFAB.

At the same time, the Twitter accounts of Europe's strongest football leagues, namely @LaLiga, @SerieA_TIM, @Bundesliga_de, @LFPfr, and @premierleague, have been analysed. As well as this, one communication medium from each of the countries mentioned has also been analysed, the chosen ones being those which had the most followers at the time the research was undertaken, they are: @marca, @Gazzetta_it, @lequipe, @SportBILD, and @guardian_sport, respectively.

Over the course of the research, different tools such as Twitter Advanced and Twittonmy were used. The quantitative results thrown up by the survey, for their part, are handled and analysed by the SPSS data analysis program. The search criteria used in Twitter Advanced were: "VAR, videoarbitraje, l'arbitrage vídeo y videobeweis". The methodology design, for its part, has a series of contextual units: (a) sender; (b) date; (c) purpose of the tweet; (d) mention; (e) hashtag; (f) whether it includes keywords such as hashtag; (g) internal links; (h) external links; (i) multimedia elements; (j) retweet; (k) comments; (l) favourites; and (m) vision of the message, with which, as well as going more deeply into the study, a rapid description of the way in which the accounts have been using this tool may be outlined.

In this point of the methodology, the codification is fixed by using nominal and interval-type scales applied to the data resulting from the research, in line with the Tito model (Tito, 2013, p.165).

Initially, a pretest was done, which was published in the chapter "The visibility of the VAR on Twitter" (Gallego, 2017, pp.25–29) and the main aim of which was to ensure the methodological efficacy of the study, by testing the codification plan (Heinemann, 2003, p.169). It spanned all tweets published @TheIFAB, @FIFAcom, @marca, and @LaLiga, from 5 March 2016 until 1 November 2017.

Information via Twitter

The Features of a Good Content Strategy

According to Twitter's business web page, "there's no right or wrong number of times you should tweet each day, or when" (Twitter, n.d.). What is clear is that it is important to "focus on creating a regular cadence of content that's relevant to your target audience" (Twitter, n.d.). That said, tweets must comply with certain characteristics, such as:

- Keep it short, including a link to a blog post or website if you have a longer message to convey.
- Include forceful visual elements because they add personality and, above all, a greater number of interaction rates (retweets, comments, and favourites).
- Incorporate relevant hashtags that allow you to expand your reach. Best practices recommend using no more than two hashtags per tweet.
- Ask questions to interact with your audience and you can even run polls.
- Finally, retweeting relevant content and replying to tweets are great ways to maintain a robust Twitter presence (Twitter, n.d.).

Pretest on FIFA, IFAB, Marca, and LaLiga

The pretest was carried out by analysing the IFAB and FIFA accounts, as well as that of one of the media (in this case Marca) and one of the leagues (in this case Spain's La Liga), and taking into account only information published concerning the VAR. This preliminary study reflected that both FIFA and IFAB had met some of the requisites for a good strategy, if not all of them, as can be seen in Table 15.1. As for @marca and @LaLiga, the figures obtained are those that appear in Table 15.2.

In the evaluation the messages concerning the VAR offered based on the data gathered, certain surprising partial conclusions can be noticed, such as the fact that @TheIFAB showed itself to be completely neutral to

Table 15.1 Data from the @TheIFAB and @FIFAcom accounts obtained in the pretest

	Tweets	*Link*	*Multimedia*	*Tags*	*Interaction*	*Message type*	*Purpose*
@TheIFAB	48	Internal: 54% External: 10% No link: 36%	Photo: 33% Video: 4% GIF: 21% No MM: 42%	Hashtag: 52% Mention: 17%	Retweet: 0% Reply: 0%	Positive: 0% Negative: 0% Neutral: 100%	Inform: 71% Educate: 27% Engage users: 0%
@FIFAcom	25	Internal: 72% External: 4% No link: 24%	Photo: 68% Video: 4% GIF: 0% No MM: 28%	Hashtag: 56% Mention: 28%	Retweet: 0% Reply: 6%	Positive: 40% Negative: 0% Neutral: 60%	Inform: 88% Educate: 22% Engage users: 0%

Source: Compiled by the author.

the use of the VAR, despite it being one of the organisations that was impelled to promote the introduction of the system. In fact, in all of its tweets, there was no word used to indicate that the VAR is "good", "suitable", "necessary", or any other positive adjective. Meanwhile, @LaLiga was the one that took up the strongest stance in favour, with 100 per cent of its tweets being positive regarding the use of the new tool. About half way between @TheIFAB and @marca was @FIFA, with some 40 per cent of its tweets being in favour of the introduction of the new technology as opposed to 60 per cent in which it gave out a neutral message. As well as this, only @marca offered tweets with negative contents concerning video refereeing, which made up 26 per cent of the messages. This figure is a lot lower than the positive ones, which came to 53 per cent of all posts, but closer to the neutral tweets, which were 21 per cent.

If we take into account the tweets posted by these accounts, it is clear that their main goal was to provide information, with 88.25 per cent of the tweets (average of 71 per cent IFAB; 88 per cent FIFA; 94 per cent Marca, and 100 per cent La Liga), followed by educating the rest of the users, with 12.75 per cent (average of 27 per cent IFAB; 22 per cent FIFA; 2 per cent Marca and 0 per cent La Liga).

Despite the fact that interaction is one of the main characteristics of Twitter, only 1 per cent of tweets were looking for other users to participate.

Table 15.2 Data from the @marca and @LaLiga accounts obtained in the pretest

	Tweets	*Link*	*Multimedia*	*Tags*	*Interaction*	*Message type*	*Purpose*
@marca	47	Internal: 98% External: 2% No link: 0%	Photo: 89% Video: 2% GIF: 0% No MM: 9%	Hashtag: 52% Mention: 17%	Retweet: 0% Reply: 0%	Positive: 40% Negative: 0% Neutral: 60%	Inform: 94% Educate: 2% Engage users: 4%
@LaLiga	6	Internal: 50% External: 0% No link: 50%	Photo: 50% Video: 17% GIF: 0% No MM: 33%	Hashtag: 56% Mention: 28%	Retweet: 0% Reply: 0%	Positive: 53% Negative: 26% Neutral: 21%	Inform: 100% Educate: 0% Engage users: 0%

Source: Compiled by the author.

The pretest allowed us to verify that the variables to be studied were the right ones, even if at least one more should have been included, to identify the adjectives used. This is an open question, to see what index of positive and negative words are the ones most used in posts.

Final Results

With the aim of seeing more clearly the shifts that took place in the communication strategies followed by the accounts analysed, as well as the repercussion the contents have had on other users of this social network, the tweets were divided into four time periods:

A. Those published from 5 March 2016 to 2 March 2018.
B. Those posted from 3 March 2018 – the day on which FIFA announced that the VAR system was to be used in the FWC 2018 – to 13 June 2018.
C. Those posted during the FWC 2018.
D. Those written between the end of the FWC 2018 and 31 August 2018.

Thus, the research reflects how, in the accounts analysed, and referring always to information posted concerning the VAR, there have been shifts.

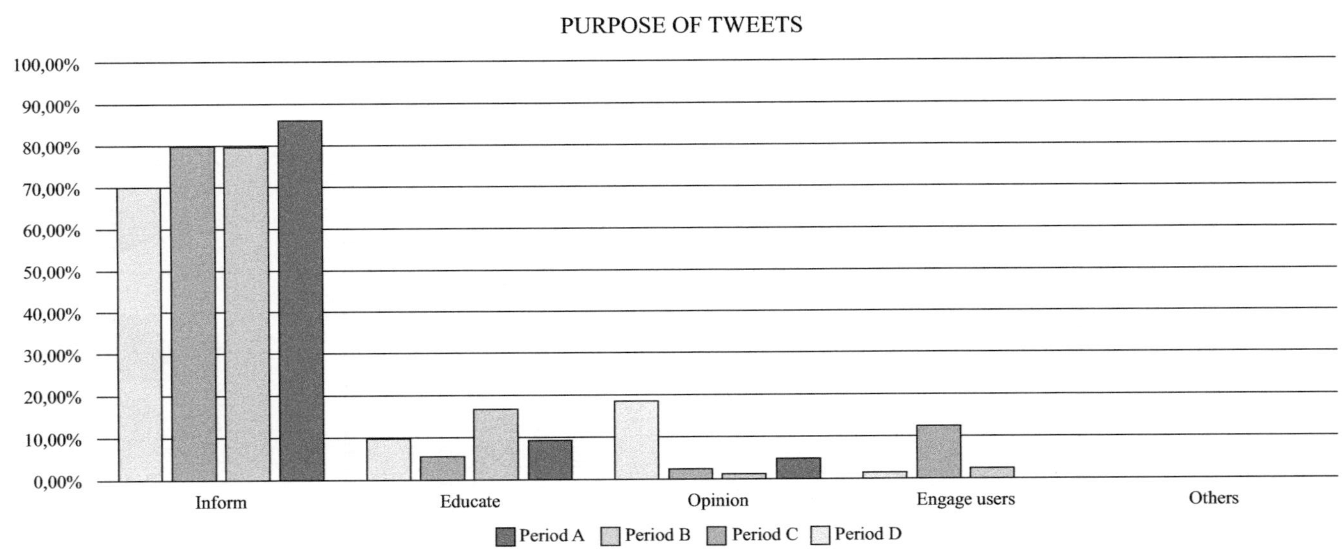

Figure 15.1 The objective of the tweets analysed throughout the four periods

Source: Author.

Important among them, for example, is the fact that if at the beginning in order to find tweets in which video refereeing appeared, it was necessary to use different words, such as "VAR", "videoarbitraje", "l'arbitrage vidéo", and "videobeweis" among others, as the search criteria, now almost all of the tweets can be located simply by searching for "VAR" (some 93.52 per cent in the last period as opposed to 61.51 per cent in the first).

Another fact to bear in mind relates to *feedback*,[3] which is very common on Twitter, but which the accounts analysed seemed not to want to know anything about initially. In fact, they went from writing just 1.48 per cent of their tweets with the aim of engaging with other users in period A, to publishing more than 12 per cent of their messages with this intention after the announcement that VAR was to be used in the FWC 2018. During the tournament this figure dropped considerably (to 2.38 per cent), even if the number of tweets aimed at educating the user did go up, reaching 16.67 per cent, as can be seen in Figure 15.1.

What barely changed at all, as is shown in Figure 15.2, was the use of photography in the majority of tweets, it being, at around 80 per cent, the multimedia file most used. That said, the use of video did increase over the duration of the tournament, rising from 2.39 per cent in tweets posted in period A to 12.96 per cent by the end. Meanwhile, the number of messages posted with no multimedia file – which is unusual on Twitter – went down from 17.43 per cent in the early days to 1.85 per cent in the last period.

Regarding the number of *comments*, *retweets* and *favourites*, they went up as soon as it was known that video-assisted refereeing was to be used at the FWC 2018. In fact, the number of messages with more than 60 *favourites* doubled as of that moment.

Lastly, throughout this research an increase in the number of tweets in favour of the VAR can be appreciated (from 24.86 per cent in the first period to 32.41 per cent in the last), despite there being a slight fall during the tournament (it remained at 22.62 per cent) which was to the benefit of

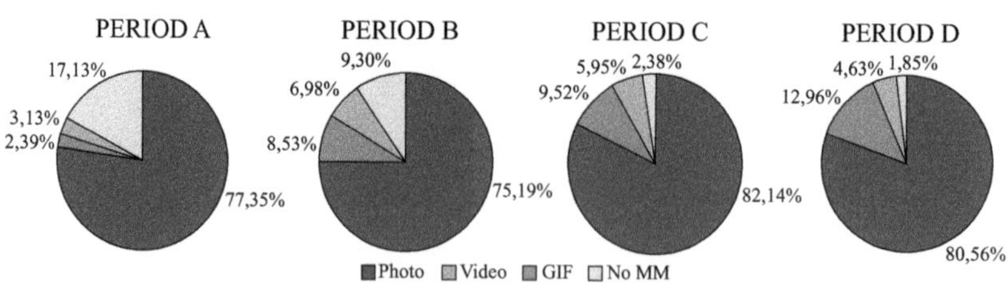

Figure 15.2 The use of multimedia files in the analysed tweets

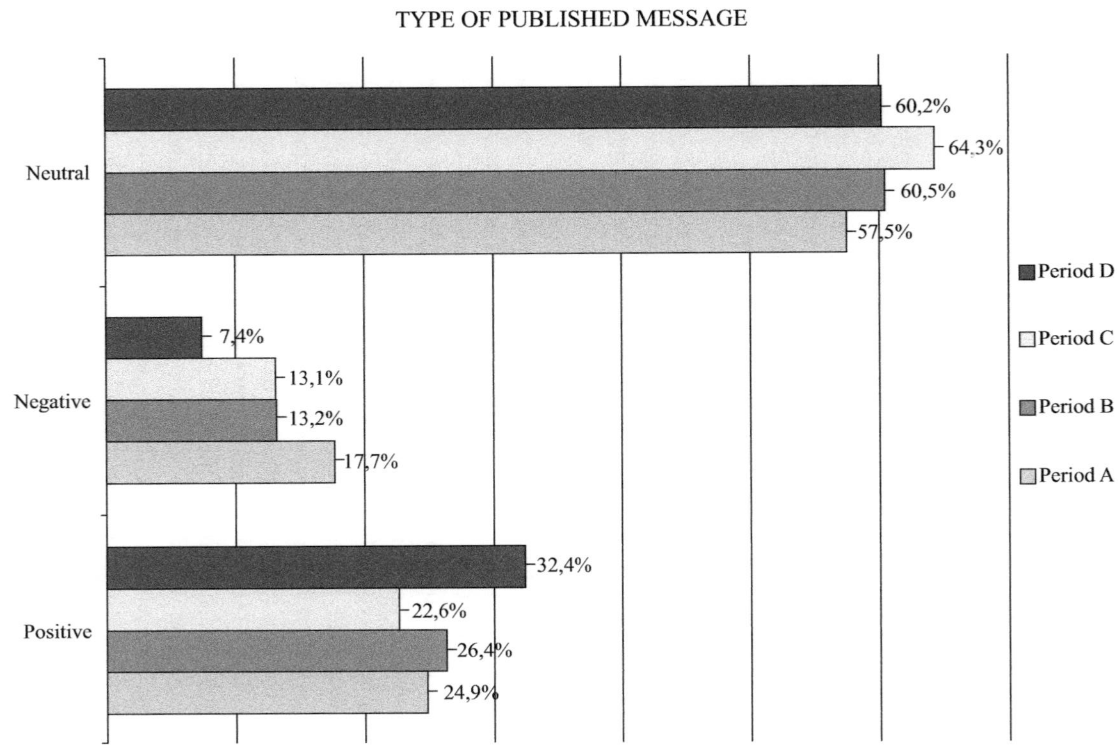

Figure 15.3 The type of published message in different accounts analysed

the neutral messages, which have always been around 60 per cent. All of this is reflected in Figure 15.3.

IFAB and FIFA's Line of Communication

The IFAB and FIFA were two of football's governing bodies which were obliged to advocate the suitable use of the VAR: in the early days, as could be seen in the pretest, they did not take full advantage of Twitter as a tool to publicise the system of video-refereeing successfully. The IFAB did, however, improve their communication strategy greatly as the months went by, as the data in Table 15.3 illustrates.

This fact becomes evident, above all, as regards the feedback with other users. One of the most striking conclusions to be drawn from the pretest was the fact that the aim of their tweets was to inform and, at times, to educate, but they left out interaction with other participants. This mistake was put right, in the second period especially, when the number of tweets whose purpose was to engage with other users rose drastically, from 3.30 per cent to 35.14 per cent.

IFAB looked for, and found, the perfect way of educating and getting other users engaged. How? They did it by posting messages which "in the form of a game" offered information that was relevant to the VAR and provided users with a link to their website page which was dedicated to explaining the way video-refereeing was to be used.

The redesign of their communications plan did not stop there, however. IFAB, who had barely taken advantage of statements made by some of the football world's leading figures, realised that by using them, they could bring about a positive shift among users, who would be able to see for themselves how their idols were in favour of the VAR. Some of them spoke out in favour of the system even though they were aware that if this technology had existed in their playing days, they might not have gone down in history for certain actions they did which are known worldwide. This is the case of Diego Armando Maradona and his "Hand of God". To do this, IFAB took advantage of news stories, interviews, and reports that were published in different media to post them on Twitter.

Following on with this line of positive messages, they started to post tweets showing, for example, the results of surveys in which fans openly supported the use of video-refereeing in football, as was the case of the tweet posted in August 2018 in which English fans are shown saying they want the VAR to be introduced even though the Premier League continued to reject it in its internal voting. As well as this, they started posting tweets that used positive language. There were also tweets showing the tournaments in which the VAR had featured, and, during the FWC 2018, one showing what actions the VAR had been key in resolving.

In this way, although the IFAB's line of communications was clearly marked by neutrality, the positive image that, a priori, it was supposed the

Table 15.3 Relevant data from @TheIFAB and @FIFAcom accounts

		Tweets	*Purpose*					*Multimedia*				*Hashtag*	*Interaction*		*Type of published message*		
			Inform	*Educate*	*Give opinion*	*Engage users*	*Other*	*Photos*	*Videos*	*GIFs*	*No MM*	#	*>60 Retweets*	*>60 Favourites*	*Positive*	*Negative*	*Neutral*
@The IFAB	Period A	91	59.34%	36.26%	0%	3.30%	1.10%	46.15%	6.59%	18.68%	28.57%	37.36%	3.30%	5.49%	18.68%	0%	81.32%
	Period B	37	56.76%	8.11%	0%	35.14%	0%	62.16%	10.81%	24.32%	2.70%	0%	8.11%	10.81%	24.32%	0%	75.68%
	Period C	11	63.64%	36.36%	0%	0%	0%	90.91%	0%	0%	9.09%	27.27%	18.18%	18.18%	36.36%	0%	63.64%
	Period D	15	100.00%	0%	0%	0%	0%	93.33%	0%	0%	6.67%	6.67%	13.33%	26.67%	33.33%	0%	66.67%
	Averages	**38.5**	**69.93%**	**20.18%**	**0%**	**9.61%**	**0.27%**	**73.14%**	**4.35%**	**10.75%**	**11.76%**	**17.83%**	**10.73%**	**15.29%**	**28.18%**	**0%**	**71.82%**
@FIFA com	Period A	31	61.29%	16.13%	16.13%	6.45%	0%	61.29%	6.45%	0%	32.26%	48.39%	35.48%	96.77%	35.48%	0%	64.52%
	Period B	7	100.00%	0%	0%	0%	0%	14.29%	85.71%	0%	0%	100.00%	71.43%	100.00%	28.57%	0%	71.43%
	Period C	0	-	-	-	-	-	-	-	-	-		-	-	-	-	-
	Period D	1	100.00%	0%	0%	0%	0%	100.00%	0%	0%	0%	100.00%	0%	100.00%	0%	0%	100.00%
	Averages	**9.75**	**87.10%**	**5.38%**	**5.38%**	**2.15%**	**0%**	**58.53%**	**30.72%**	**0%**	**10.75%**	**82.80%**	**35.64%**	**98.92%**	**21.35%**	**0%**	**78.65%**

Source: Compiled by the author.

body was to project but which it did not, at least in the first period, grew in importance over time.

Where evolution has also been noticed over the course of the study was in the power of repercussion of the messages posted from the @TheIFAB account since, even though the number of retweets and favourites was not especially high, the percentage of the rank of more than 60 improved in both cases, going up from 3.30 per cent of tweets with more than 60 retweets in the first period to 18.8 per cent in the third period and from a figure of 5.49 per cent of tweets with more than 60 favourites in the beginning to 26.67 per cent at the end of the study (see Table 15.3).

FIFA, for their part, always banked on keeping the same philosophy, at least as far as VAR-related tweets went. In them, they made it clear from the start that their main objective was, and continued to be, that of providing information (an average of 87.10 per cent of their tweets were for this purpose). To achieve that, they relied on photographs (58.53 per cent) and videos (30.72 per cent). As well as that, they reduced the lack of multimedia support to the extent that they reached 0 per cent of tweets with no multimedia, despite starting off by posting some 32.26 per cent of all messages with no images, videos, or GIFs.

Like the @The IFAB, the @FIFAcom account uses mentions in around 50 per cent of its messages. That said, it makes much greater use of hashtags. In the @FIFAcom account, these appear in over 82.80 per cent of their tweets, a much higher percentage than the 17.83 per cent of messages posted by @TheIFAB.

They also outstrip them in the percentage of tweets with more than 60 retweets, reaching an average of 35.64 per cent over the four periods, and in messages with more than 60 favourites, which went as high as 98.92 per cent in the same period.

As an anecdote, it is worth noting that the official FIFA account, @FIFA-com, did not publish any message during the whole FWC 2018. This was due to the fact that they created a specific profile for the tournament, from which they tweeted everything that happened as the championship unfolded.

The Media Take the Floor

"Today, you can take a look at the latest news with just a click, consume entertainment … or simply turn on the radio, television or social networking commentary some news you see in the newspaper every morning" (Vivian, 2012). Society is continually consuming information, and doing so by means of the traditional media, but also through applications and social networks. In the end, "the eruption of the Internet as an information platform has brought about deep changes in the whole newspaper industry over the last decade, affecting both its traditional economic structure and its systems of production and distribution of contents" (López & Murciano, 2012, p.173).

Social networks, as well, allow people to consume traditional media, as the press, radio, and television are, in a different way. Twitter, for example, "operates as a space for the acquisition of information" (Mitchelstein & Boczkowski, 2018, p.135). The reason for this? It is because 140 characters initially, and 280 since 2017, are enough to give you the latest update, to know what a news story is about and if it is worth clicking on the link to find out more. It also lets the user give their opinion and see what others have commented on the subject.

All of that without mentioning that the GIFs and videos that are added to messages known as tweets can be seen without downloading anything and without having to wait for the page to load. It is quite the invention for a century in which time is money and people want immediacy.

The traditional media are aware of the importance this social network has taken on. For that reason, they have also banked on having a profile on the platform and it is rare nowadays to find a newspaper, a radio station, or a television channel that does not have an official Twitter account.

For this study, as mentioned above, the accounts of five media have been chosen, namely those of *Marca*, *La Gazzetta*, *L'Équipe*, *The Guardian*, and *Sport Bild*. The choice to study these rather than others was based on the number of followers they have. In other words, they are the media with most followers in Spain, Italy, France, England, and Germany. In figures, @Marca has 4.95 million followers, @Gazzetta_it has 1.67 million, @lequipe has close on 5 million, @guardian_sport has 779,000, and @SportBILD has 388,000.

All five of these media reported over the four periods of the study on how the implementation of the VAR was turning out. Between 5 March 2016 and 31 August 2018, almost 600 tweets mentioned the VAR, and it was the Italian daily, with 257 messages posted, that wrote the most times about this technology.

In Table 15.4 the data gathered in the study can be seen, which will be broken down and analysed in the following paragraphs.

The terms used as search criteria varied widely: from "VAR" to "Videobeweis", but also included "Videoarbitraje", "l'arbitrage vidéo", "Video Assistent", "Arbitro Video Asistant", or "Video Assistant Referee".

As communication media, their main goal is to provide information, with @SportBILD being the one that dedicated most tweets to this mission, some 93.33 per cent of the total. Meanwhile, the @marca account worked most in the area of educating users, with 5.74 per cent of its messages having that purpose, and @lequipe is the one that sought to engage users most insistently, as the figure of 3.44 per cent shows.

Each one of the accounts relied differently on the multimedia elements that can be used on Twitter. Photography was the big star, and @SportBILD was the one that used it most, with some 98.10 per cent of its tweets containing photographs. On the other hand, and although it may seem strange, given that we are talking about the technological format that

Table 15.4 Data obtained from the accounts of analysed communication media

	Tweets		Purpose				Multimedia				Link	Hashtag	Mention	Interaction			Type of published message		
			Inform	*Educate*	*Give opinion*	*Engage users*	*Photos*	*Videos*	*GIFS*	*No MM*	*Yes*	*#*	*@*	*>60 Comments*	*>60 Retweets*	*>60 Favourites*	*Positive*	*Negative*	*Neutral*
@marca	Period A	78	65.38%	2.56%	29.49%	2.56%	93.59%	1.28%	0%	5.13%	98.72%	34.62%	23.08%	8.97%	38.46%	64.10%	43.59%	21.79%	34.62%
	Period B	13	76.92%	7.69%	15.38%	0%	100.00%	0%	0%	0%	100.00%	7.69%	15.38%	23.08%	46.15%	92.31%	23.08%	30.77%	46.15%
	Period C	29	96.55%	3.45%	0%	0%	86.21%	0%	10.34%	3.45%	86.21%	72.41%	10.34%	13.79%	44.83%	100.00%	31.03%	6.90%	62.07%
	Period D	54	83.33%	9.26%	7.41%	0%	81.48%	7.41%	9.26%	1.85%	92.59%	61.11%	14.81%	11.11%	14.81%	72.22%	33.33%	9.26%	57.41%
	Average	**43.5**	**80.55%**	**5.74%**	**13.07%**	**0.64%**	**90.32%**	**2.17%**	**4.90%**	**2.61%**	**94.38%**	**43.96%**	**15.91%**	**14.24%**	**36.06%**	**82.16%**	**32.76%**	**17.18%**	**50.06%**
@lequipe	Period A	50	84.00%	2.00%	12.00%	2.00%	90.00%	0%	0%	10.00%	92.00%	0%	2.00%	2.00%	20.00%	54.00%	22.00%	20.00%	58.00%
	Period B	17	88.24%	0%	0%	11.76%	88.24%	0%	0%	11.76%	88.24%	0%	0%	5.88%	11.76%	58.82%	52.94%	11.76%	35.29%
	Period C	4	100.00%	0%	0%	0%	100.00%	0%	0%	0%	100.00%	0%	0%	0%	50.00%	75.00%	0%	75.00%	25.00%
	Period D	4	100.00%	0%	0%	0%	100.00%	0%	0%	0%	100.00%	0%	0%	0%	25.00%	100.00%	25.00%	0%	75.00%
	Average	**18.75**	**93.06%**	**0.50%**	**3.00%**	**3.44%**	**94.56%**	**0%**	**0%**	**5.44%**	**95.06%**	**0%**	**0.50%**	**1.97%**	**26.69%**	**71.96%**	**24.99%**	**26.69%**	**48.32%**
@Gazzeta_it	Period A	200	72.00%	3.00%	25.00%	0%	84.00%	0%	0%	16.00%	90.50%	87.00%	5.50%	3.50%	1.50%	6.00%	19.00%	25.00%	56.00%
	Period B	36	100.00%	0%	0%	0%	75.00%	0%	0%	25.00%	91.67%	72.22%	2.78%	13.89%	8.33%	13.89%	27.78%	11.11%	61.11%
	Period C	5	100.00%	0%	0%	0%	60.00%	0%	40.00%	0%	60.00%	100.00%	0%	0%	0%	0%	20.00%	20.00%	60.00%
	Period D	16	100.00%	0%	0%	0%	100.00%	0%	0%	0%	93.75%	81.25%	12.50%	0%	0%	12.50%	31.25%	6.25%	62.50%
	Average	**64.25**	**93.00%**	**0.75%**	**6.25%**	**0%**	**79.75%**	**0%**	**10.00%**	**10.25%**	**83.98%**	**85.12%**	**5.19%**	**4.35%**	**2.46%**	**8.10%**	**24.51%**	**15.59%**	**59.90%**
@guardian_ sport	Period A	32	84.38%	0.00%	15.63%	0.00%	62.50%	3.13%	0%	34.38%	90.63%	3.13%	56.25%	0%	0%	6.25%	15.63%	25.00%	59.38%
	Period B	14	64.29%	21.43%	7.14%	7.14%	100.00%	0%	0%	0%	100.00%	7.14%	57.14%	0%	0%	0%	7.14%	35.71%	57.14%
	Period C	23	95.65%	0%	4.35%	0%	95.65%	4.35%	0%	0%	95.65%	69.57%	60.87%	0%	0%	4.35%	17.39%	21.74%	60.87%
	Period D	4	75.00%	0%	25.00%	0%	75.00%	25.00%	0%	0%	75.00%	25.00%	50.00%	0%	0%	25.00%	50.00%	25.00%	25.00%
	Average	**18.25**	**79.83%**	**5.36%**	**13.03%**	**1.79%**	**83.29%**	**8.12%**	**0%**	**8.59%**	**90.32%**	**26.21%**	**56.07%**	**0%**	**0%**	**8.90%**	**22.54%**	**26.86%**	**50.60%**
@SportBILD	Period A	35	80.00%	2.86%	17.14%	0%	94.29%	0%	0%	5.71%	94.29%	97.14%	20.00%	0%	0%	0%	20.00%	31.43%	48.57%
	Period B	4	100.00%	0%	0%	0%	100.00%	0%	0%	0%	100.00%	100.00%	50.00%	0%	0%	0%	0%	50.00%	50.00%
	Period C	0	-	-	-	-	-	-	-	-	-	-	-	-	-	-	-	-	-
	Period D	3	100.00%	0%	0%	0%	100.00%	0%	0%	0%	100.00%	100.00%	0%	0%	0%	0%	66.67%	33.33%	0%
	Average	**14**	**93.33%**	**0.95%**	**5.71%**	**0.00%**	**98.10%**	**0%**	**0%**	**1.90%**	**98.10%**	**99.05%**	**23.33%**	**0%**	**0%**	**0%**	**28.89%**	**38.25%**	**32.86%**

video-refereeing is supported by, video was hardly used at all, and only @guardian_sport (in 8.12 per cent of its tweets) and @marca (in 2.17 per cent) actually used it. Meanwhile, @Gazzetta_it was the medium that took greatest advantage of GIFs (in 10 per cent of its tweets).

In the vast majority of the tweets analysed, more than 92 per cent, the user could access further information by clicking on a link that was provided, which would take them to the relevant page of the newspaper and the news story in question.

The use of hashtags was quite generalised throughout the media analysed here, if not always in the same percentage. Thus, @SportBILD, 99.05 per cent of the time, was the one that made most use of them, followed by @Gazzetta_it, with 83.98 per cent (almost double that of @marca and three times that of @guardian_sport). At the other end of the scale was @lequipe, which never used them. What the French medium did do, although only in 0.5 per cent of its tweets, was put mentions. In this case, @guardian_sport topped the ranking, with 56.07 per cent.

As for the repercussions the 621 VAR-related tweets posted by the five accounts had, the most usual thing was for them to be *commented*, *retweeted*, and chosen as *favourites* by a maximum of 20 people. The Spanish sports daily was the one that managed the highest number of messages with over 60 *comments* (in 14.24 per cent of all tweets sent out), more than 60 *retweets* (36.06 per cent), and more than 60 *favourites* (82.16 per cent)

Regarding the type of message, they were valued as positive, negative, or neutral. @marca was the medium that has been more in favour about introducing the VAR (32.76 per cent). This is quite the opposite of @SportBILD, of whose messages some 38.25 per cent were classed as negative.

It has not, though, always been this way. In the period between 5 March 2016 and 2 March 2018, the Spanish daily posted a large number of tweets, some 33, in which it made it plain that VAR was good for football. It also published some, 17, in which it revealed some of the errors that were still being committed despite the VAR being used, as was the case on 18 December 2017, when it published a tweet with the most outrageous mistakes that had been made.

After it was announced that this technology was to be applied in the FWC 2018, @marca began a campaign that rather went against video-refereeing, publishing as many as 30.77 per cent of its messages with a hint of negativity. Some of these messages were, moreover, statements made by footballers and other personalities who, in principle, should have been supporting the VAR, as was the case of the former referee Pierluigi Collina, all of which contributed to an atmosphere of mistrust towards the new technology.

Everything changed as of the FWC 2018, during which the percentage of contrary tweets dropped to 6.9 per cent, while the positive ones rose, with messages that included terms such as “justice” or even “Long Live the

VAR". Later on, with the VAR now functioning in Spain's La Liga, tweets that were more critical grew again in number, rising to 9.26 per cent.

In the rest of the media there were also shifts in the type of message that was put out most. In the case of @Gazzetta_it, it was in the first period when they published a greater number of negative tweets, some 25 per cent, together with the third period, which was the duration of the FWC 2018 (20 per cent). Even still, after the tournament, the figure dropped to 6.25 per cent.

The period that saw most messages against the VAR was just when it was announced that it was to be used in the FWC 2018, with @SportBILD, @guardian_sport, and @marca having the highest percentages: 50 per cent, 35.71 per cent, and 30.77 per cent, respectively.

During the tournament, and in view of the good performances of referees and the VAR, all the media published a lower number of negative tweets, except @lequipe, which sent out a message against the use of the VAR in some 75 per cent of their tweets. That said, after the FWC 2018, they did not publish a single tweet that mentioned anything negative about the technology.

What Do the Top Leagues Have to Say?

The VAR intervenes to help the match official when he has to take decisions in which there is a strong element of doubt, thus avoiding mistakes and avoiding the protests of players and team managers during matches. Its introduction into various leagues and tournaments has been by no means uncontroversial. Despite everything, and having been used in the 2017/18 season in the German Bundesliga and Italy's Serie A, among others, Spain's La Liga and France's Ligue 1 also decided to embrace the VAR. They did so after the technology had been employed as well in the FWC 2018, held in Russia. So, of all of the so-called "top leagues", only the Premier League has remained on the margins, having decided by a vote that they did not want VAR for the next season (MacInnes, 2018).

Of the accounts analysed for this research, and which belong to the associations in charge of organising the national championships in Italy, France, England, Spain, and Germany, only the latter two show any real movement on Twitter, while the other three only managed six tweets between them in the period analysed.

This means that only the @Bundesliga_de and @LaLiga accounts took advantage of this tool to keep users informed and updated on the use of the VAR in their competitions. In fact, it is the strategy that the two of them followed in view of the, eventually, successful introduction of the technology.

Thus, in the case of @Bundesliga_de, it was done in the following way: first, they gave information about the first VAR trials (2 June 2016); afterwards, it was announced that there would be an assistant VAR in all

Bundesliga games, as well as in the relegation play-offs and the Super Cup (23 January 2017) and, later, a tweet was published in which they spoke of how on 28 March a refereeing team would receive VAR support for the first time (16 February 2017). In the opening month of the competition, information was provided concerning how communication between the match official and the VAR worked, as well as four examples of the use of the VAR (1 August 2017; see also Chapter 12). On the opening day of the competition, they put out a reminder of the changes that had happened (10 August 2017), while once video-refereeing had actually been put into practice, the two times it had been called upon were mentioned and its success was stressed (20 August 2017). This was a simple line of communication, one that was very clear, which showed the progress of the strategy.

Employing a similar strategy, @LaLiga first made its opinion known regarding the technology (29 March 2017) in order to then, almost a year later, announce, by means of an official notice, that the VAR was to be introduced the following season (23 January 2018). From then on, @LaLiga pursued two different lines of strategy: on the one hand, one that provided information concerning the use of video-assisted refereeing and, on the other, one that educated users, in a fun way, by means of tweets that posed questions with multiple-choice answers and whose aim was to build a connection with fans who would later be the ones to be supportive or critical of the VAR.

In total, they published as many as 12 tweets in two months (June and July 2018) with the sole aim of making clear when and how the technology was to be used. Once the fans had been educated, it was now the turn of referees, players, and press. The La Liga account showed the training work that had been undertaken with each of the agents involved with the beautiful game, and then slipped in between them new feedback messages from users.

One of the big differences between the @Bundesliga_de and @LaLiga is to be found in the fact that the latter used a lot of videos, which let the user obtain sufficient information without having to be redirected to another link. As well as this, they make use of the FIFA VAR slogan: "Minimum interference, maximum benefit". The fact of being able to "put" the knowledge users have concerning this technology "to the test" implies learning more about the VAR, and learning better.

Did the FWC 2018 Mark a Before and an After in the Use of the VAR?

After the FWC 2018, FIFA stated that the VAR had "contributed to the all-round success of the referees in the tournament" (FIFA, 2018). With that statement in mind, the question that then arose was if the tournament had marked a before and an after in the way VAR was viewed. To find an accurate answer, it was decided to carry out a cross-sectional study in order to be able to analyse if fans, players, club directors, team managers, and media were happy to accept the use of the technology in their leagues.

To look into this question, two accounts were chosen: @Gazzetta_it and @marca, leaving the other three aside. The reason for choosing only these two was the fact that they were the most active when it came to VAR references. As well as this, they belong to communication media from, respectively, Italy (which started to use the technology before the FWC 2018, see Chapter 1) and from Spain (where it was introduced after that tournament). The messages analysed in each of these newspapers' accounts, therefore, are from the first month and a half of competition of the seasons in which VAR was implemented in the two leagues, Italy in 2017/18, and Spain in 2018/19.

The results, which can be seen in Table 15.5, show that the number of negative messages concerning the use of the VAR was greater in Serie A than it was in La Liga (by 8 per cent). Thus, in the @Gazzetta_it account, the tweets concerning the VAR that were negative in tone reached 32 per cent of all messages posted from 19 August 2017 until 30 October 2017, a figure that was greater than that for positive messages, which were 22 per cent. That said, they were both lower than the percentage of neutral messages, which was 46 per cent.

In this new study, an attempt was made to see patterns and find out whether there was any relation between the type of message (negative, positive, and neutral), and the format (news story, match report, minute by minute, statements, etc.).

Following the case of the Italian medium, it was clear that the greatest number of negative tweets reflected statements made by players, team managers, and club directors (44 per cent), which were almost double the positive messages. In the case of the neutral messages, meanwhile, the majority concern tweets containing videos.

In the @marca account, for its part, the negative messages concerning the VAR made up only 8 per cent of all those posted over the first month and a half of the system being implemented in Spain's La Liga, between 17 August 2018 and 29 October 2018. Forty-two per cent were positive, while the other 50 per cent were neutral.

In the Spanish medium, most of the positive messages came from tweets from the well-known "minute-by-minute" feature during games, being 12 out of the 48 analysed (25 per cent of the total). In addition, as in the Italian case, statements were those which made up the greater part of the negative tweets, even though in @marca they were in fact only two messages out of 48. The news stories that appeared between those dates tried to be as neutral as possible (55 per cent) or positive (36 per cent), with only one of the tweets analysed within the category of information written by journalists being negative.

Table 15.5 Crossed data to better illustrate the relationship between the format and the type of message

	Format							*Tweets*		
	News	*Match reports*	*Statements*	*Opinions*	*Videos*	*MBM*	*Interviews*			
	36%	60%	0%	0%	–	46%	100%	42%	Positive	
@marca	9%	0%	67%	50%	–	0%	0%	8%	Negative	
48 tweets	55%	18%	33%	50%	–	54%	0%	50%	Neutral	
	17%	–	25%	–	29%	–	–	22%	Positive	**Type of published message**
@Gazzeta_it	28%	–	44%	–	14%	–	–	32%	Negative	
41 tweets	56%	–	31%	–	57%	–	–	46%	Neutral	

Source: Compiled by the author.

Conclusions

The VAR is changing the way football is watched. Although in the initial stages it had practically no visibility on the social networks, now it can be said that the users of platforms such as Twitter are beginning to see it as almost the perfect solution for avoiding big refereeing mistakes. It is clear that, as the system is still young, it must be improved. For the moment, however, and bearing in mind the road trodden so far, it can be said that it is progressing adequately and, on occasions, actually shines.

There are ever fewer voices speaking against its being implemented, although there are protests about the way it is used. One of the most controversial aspects is that the time needed for decisions to be taken has increased, which is not to the liking of either players or fans. In fact, right now, it is precisely the controversy that has been raised by this "loss of time" that occupies most space within media information that is against the VAR. That said, and interestingly, if players, team managers, and fans protested before about the arrival of the technology, nowadays their complaints are more to do with the fact that it is not used when, from their point of view, there is a controversial play. This means, in fact, that complaints are heard now, when similar incidents occur, because the referee does not always turn to the VAR for help.

In the analysis of the messages on the IFAB's official Twitter account, @TheIFAB, the neutrality that can be observed regarding the use of the technology contrasts with the positive tone of the messages put out by FIFA on theirs, @FIFAcom. The use of Twitter to spread the positive value of the tool coincides with the moment that FIFA chose to push for it and, more specifically, when the president, Gianni Infantino, decided to make the VAR something of a personal project (see Chapter 10). More evident still is the way La Liga used Twitter to create positive opinion, with 100 per cent of their messages being in favour of the introduction of the technology. Later research ought to look into showing a possible interest in transmitting transparency, in the case of FIFA, and an economic one, in the case of La Liga.

The use of photography in tweets has practically been a constant in the official accounts of FIFA (73.14 per cent) and IFAB (58.53 per cent) as well as mass media (*Marca*, 90.32 per cent; *L'Équipe*, 94.56 per cent; *La Gazzeta*, 79.75 per cent; *The Guardian*, 83.29 per cent, and *Sport Bild*, 98.10 per cent). Whereas the use of video (4.35 per cent in the case of FIFA; 30.72 per cent in IFAB; 2.17 per cent in *Marca*; 8.12 per cent in *The Guardian*; and 0 per cent in *L'Équipe*, *La Gazzetta* and *Sport Bild*) stays far away.

Another of the resources used principally by the IFAB was the game strategy in order to attract the attention of participants, posing questions that could be answered by looking on their dedicated VAR page.

The last communication device to attract attention on the IFAB's Twitter account was to tweet statements made by personalities from the world of football, and the use of statistics to promote a positive image of the VAR.

The image of the VAR that the media have been projecting has improved progressively, despite a great many protests about it. Until now, the tendency has been for the number of tweets in favour of the system to rise, to the detriment of the negative ones, even if those with a neutral message are around the 60 per cent mark.

There are several reasons for this shift, one of the main ones being the fact that the VAR was introduced in the FWC 2018 in Russia, where it was possible to see its crucial role; as well as this being the formative strategy carried out by FIFA, some leagues, and media. The Twitter platform has become a fundamental communication tool for helping users better understand how this technology works. It has been done by means of the use of videos, talks, games, and even instructional campaigns carried out in football stadiums.

All of this has a clear objective: to help the VAR become a reality with which, in line with the FIFA philosophy, "minimum interference, maximum benefit" is gained.

Notes

1 Original: "Twitter ha cambiado la red y ha completado el giro social que iniciaron los blogs a finales de los noventa. Ahora el planeta entero está conversando y todos podemos escucharlo. También existe la posibilidad de que el planeta entero nos escuche. Al menos una vez. Tenemos 140 caracteres para intentarlo. Muchas veces" (Orihuela, 2011, p.21).

2 Original: "El interés empírico por los hechos simbólicos ya no puede aplicarse al estudio de los mensajes de forma aislada", sino que deben tener en cuenta "los canales, las limitaciones de los flujos de información, los procesos de comunicación y sus funciones y efectos en la sociedad, los sistemas que incluyen tecnología avanzada y las modernas instituciones sociales". A lo que puede añadirse que "el análisis de contenido puede tener tanto un fin descriptivo como un fin inferencia (deductivo) y puede utilizar tanto técnicas de análisis cuantitativo como técnicas de análisis cualitativo; también hay un acuerdo en que el análisis no está limitado al contenido manifiesto de los mensajes, sino que puede extenderse a su contenido latente, y en que los análisis realizados deben someterse, como todo análisis, a pruebas de validez y fiabilidad" (López-Aranguren, 2010, p.595).

3 Feedback is the response a receiver makes to the sender, bearing in mind the message issued by the sender.

References

Andreu, J. (2002). *Las técnicas de análisis de contenido. Una revisión actualizada.* Sevilla, España: Editorial Fundación Centro de Estudios Andaluces.

Bustos Díaz, J., & Ruiz Del Olmo, F. J. (2016). La imagen en Twitter como nuevo eje de la comunicación política. Retrieved from: www.redalyc.org/articulo.oa?id=31048480018

Chew, C., & Eysenbach, G. (2010). Pandemics in the Age of Twitter: Content Analysis of Tweets during the 2009 H1N1 Outbreak. *PLoS ONE*, 5(11), p. e14118. doi:10.1371/journal.pone.0014118

Cole-Lewis, H., Pugatch, J., Sanders, A., Varghese, A., Posada, S., Yun, C., Schwarz, M., & Augustson, E. (2015). Social Listening: A Content Analysis of E-Cigarette Discussions on Twitter. *Journal of Medical Internet Research*, 17(10). Retrieved from: www.jmir.org/2015/10/e243/

FIFA. (2018). Arbitraje y VAR en Rusia 2018™: la nueva era del fútbol. fifa.com. Retrieved from: https://es.fifa.com/worldcup/news/arbitraje-y-var-en-rusia-2018tm-la-nueva-era-del-futbol

Gallego, R. (2017). La visibilidad del VAR en Twitter. In Sánchez Cid, M., Armenteros, M., & Benítez, A. J. (Eds.), *Tecnología Audiovisual e Innovación en el Deporte de Alto Rendimiento* (pp. 25–29). Madrid: Dyckinson.

García, C., & Zugasti, R. (2014). La campaña virtual en Twitter: análisis de las cuentas de Rajoy y de Rubalcaba en las elecciones generales de 2011. *Historia y Comunicación Social*, 19 February, pp. 299–311.

Hambrick, M., Simmons, J., Greenhalgh, G., & Greenwell, T. (2010). Understanding Professional Athletes' Use of Twitter: A Content Analysis of Athlete Tweets. *International Journal of Sport Communication*, 3, pp. 454–471.

Heinemann, K. (2003). *Introducción a la metodología de la investigación empírica en las Ciencias del Deporte*. Barcelona: Paidotribo.

IFAB. (2016). Minutes. *130th Annual General Meeting of the International Football Association Board*. Retrieved from: http://static-3eb8.kxcdn.com/documents/644/124828_140218_2016AGM_Minutes_v1.0.pdf

Krippendorff, K. (1980). *Content Analysis: An Introduction to its Methodology*. Beverly Hills, CA: Sage Publications.

López, P., & Murciano, M. (2012). Retos del periodismo ante el nuevo escenario digital. En adComunica. *Revista de Estrategias, Tendencias e Innovación en Comunicación*, 4, pp. 17–21. Retrieved from: http://repositori.uji.es/xmlui/bitstream/handle/10234/53599/83-113-1-PB.pdf?sequence=1

López-Aranguren, E. (2010). El análisis de contenido tradicional. In García Ferrando, M., Ibáñez, J., & Alvira, F. (Eds.), *El análisis de la realidad social: métodos y técnicas de investigación* (pp. 594–616). Madrid: Alianza Editorial.

MacInnes, P. (2018). Premier League clubs vote against introducing VAR next season. *The Guardian*. Retrieved from: www.theguardian.com/football/2018/apr/13/premier-league-video-assistant-referees-var-vote-against

Meunier, F. G. (2012). El uso de la Red Social Twitter como herramienta para la difusión de información pública. *Razón y palabra*, 81. Retrieved from: www.razonypalabra.org.mx/N/N81/V81/27_Meunier_V81.pdf

Mitchelstein, E., & Boczkowski, P. (2018). Juventud, estatus y conexiones. Explicación del consumo incidental de noticias en redes sociales. *Revista Mexicana de Opinión Pública*, 13(24), pp. 131–145. Retrieved from: www.revistas.unam.mx/index.php/rmop/article/view/61647/55593

Muñoz, R., & Riveiro, A. (2009). Twitter no es una red social sino una herramienta de comunicación. *El País*. Retrieved from: https://elpais.com/tecnologia/2009/03/25/actualidad/1237973279_850215.html

Orihuela, J. L. (2011). *Mundo Twitter: Una guía para comprender y dominar la plataforma que cambió la red*. Barcelona: Alienta.

Tito, J. A. (2013). El análisis de contenido como herramienta de utilidad para la realización de una investigación descriptiva. Un ejemplo de aplicación práctica para conocer las investigaciones realizadas sobre la imagen de marca de España y el efecto país de origen. *Publicación Provincia*, 29, pp. 135–173.

Twitter. (n.d.). What to tweet. *Twitter.com*. Retrieved from: https://business.twitter.com/es/basics/what-to-tweet.html

Vivian, J. (2012). *Media of Mass Communication*. Edinburgh: Pearson.

Wang, Y., & Zhou, S. (2014). How do sports organizations use osical media to build relationships? A content analysis of NBA clubs' Twitter use. *International Journal of Sports Communication*, 8, pp. 133–148.

We Are Social. (2019). Digital in 2019: Global Internet Use Accelerates. Retrieved from: https://wearesocial.com/blog/2019/01/digital-2019-global-internet-use-accelerates

Part V

New Challenges in Refereeing Performance

16 Educating International Football Referees

The Importance of Uniformity

Manuel Armenteros and Tom Webb

Introduction

The career of an elite football referee demands continuous development and progression in order to advance to the FIFA list and officiate international matches and competitions (Webb, 2017). Whilst referees at all levels of competition (local, national, international) are associated with FIFA, only those who have achieved the highest qualification level can officiate in international FIFA matches. All referees affiliated with FIFA have a designated grade and must advance in their grade in order to referee higher-level matches. Refereeing is also at the centre of FIFA's training and further education programme. Regional and international courses and seminars for instructors, referees, and assistant referees seek to ensure that basic requirements and criteria are met. FIFA identify that the rules of football should be uniformly interpreted and applied all over the world, with the continuous professional development of referees designed to contribute to this coordinated approach (FIFA, 2012).

To assist with the development of referees globally, FIFA created the Refereeing Assistance Programme (RAP) in December 2007, focused on the education of top referees, with the overarching objective to professionalise the environment in which referees develop and work, at both national and international level (FIFA, 2007). Following the construction of the RAP, which ran from 2007 to 2017, the training of FIFA referees' instructors has been coordinated by the FIFA Senior Manager of Refereeing Development, Fernando Tresaco Gracia, and a network of FIFA referee instructors. These individuals have actively participated in the innovation of new interactive multimedia teaching resources and methodologies based on a diverse range of new digital and video-based technologies around all FIFA confederations. The introduction of these innovations has meant an evolution in the teaching and instruction of FIFA referees. Courses designed around work, with the handbook version of the Laws of the Game, have been modified and learning and teaching activities have been introduced with interactive versions of the Laws of the Game, including 2D and 3D animations, match analysis, and online evaluation based on video.

As an associated extension of the introduction of technological advancements in football, the International Football Association Board (IFAB) approved the VAR experiments in March 2016 (see Chapter 11). The implementation of the VAR system has shown that elite referees, like any human being, make mistakes in their job. Sometimes these errors may be due to physical limitations (position, visual perception), or the limits of the technology that referees are using (not enough cameras to capture the action, or limited spatial or temporal resolution (see Chapters 5 and 8)), but some other limitations can be inherent to the subjective interpretation of referees (Mascarenhas, O'Hare, & Plessner, 2006; Webb & Thelwell, 2015).

The official data published regarding the use of VAR in different domestic leagues permits academic access to a valuable information resource that prior to VAR was unavailable; namely, the ability to understand and measure correct and incorrect referee decisions during a match. Because the TV video footage is also used by the VAR team to check referee decisions, the data published by leagues and federations allows academics to analyse the different referee interpretations concerning the same situations during a match. However, interrogation of the introduction of VAR and the decisions made during the course of a match by referees has led to further questions regarding the training and pre-match preparation of FIFA referees. Therefore, this chapter seeks to consider the following questions to provide new insight into the training and preparation of FIFA referees:

1. How are FIFA referees prepared for matches?
2. How are the confederations contributing to their training?
3. What does FIFA do to keep this elite body of referees updated and achieve consistency when applying the Laws of the Game across confederations and transnational boundaries?
4. What courses do the elite referees receive and who organises and manages their training?
5. And crucially, are the same criteria of uniformity and standardisation maintained?

Previous research has investigated the education of football referees. Studies have reviewed the homogeneity of referee decision-making, and the influence of culture on the performance of both referees and players and the behaviour of players (Webb, 2017; Webb & Thelwell, 2015). Findings have suggested that some factors can influence the homogeneity of decisions during the course of a match (Webb & Thelwell, 2015).

The developing and improving support systems provided to referees at both federation and confederation level, as well as through FIFA, have led to increases in performance. For example, there has been an observable and notable decline in home advantage since 1945 in both English and

Scottish professional leagues, with the improvements in referee support and training at elite levels postulated as an important contributory factor (Nevill, Webb, & Watts, 2013; Webb, Dicks, Thelwell, & Nevill, 2018). Moreover, we can also ascertain that the organisation, management, and training of referees within federations has evolved considerably, with referees acknowledged historically as an under-represented, underfunded, and forgotten group in football as the game evolved (Webb, 2014a).

The preparation and training of referees has become a subject that has received academic attention, with the importance of this group of individuals, particularly in professional football with the monetary rewards available for success, scrutinised and evaluated (Mascarenhas et al., 2006; Slack, Maynard, Butt, & Olusoga, 2013). The impact of culture on training and performance has also been considered in relation to referees. Webb and Thelwell (2015) suggest that culture can affect decision-making during a match, the interpretation of the Laws of the Game, and consequently how referees apply these rules. More specifically, this gives insight into observable differences between leagues, such as the acceptance that simulation and certain types of behaviour are a more inherent aspect of "Latin European" football, whilst in "Anglo" nations there is a perception that there is a greater acceptance of physical challenges.

The associated gaps in referee law interpretations demonstrate a disassociation with the IFAB and FIFA's statement that "football is a global sport and its rules must be interpreted and applied with absolute consistency wherever the game is played" (FIFA, n.d.). Consequently, this has led to attempts made by FIFA to homogenise referee interpretations of the Laws of the Game through intensive courses for elite referees, to counteract the differing interpretations and implementations of the laws across different countries (Giulianotti & Robertson, 2004, 2009). Indeed, strategies to manage homogeneity across federations and competitions have been considered in academic studies (Webb, 2017; Webb & Thelwell, 2015; Webb, Wagstaff, Rayner, & Thelwell, 2016). Furthermore, the challenges faced by managers and leaders in terms of the geographical and cultural differences, evident when leading referees, as well as the requirement for the provision of remote training, and the associated demands have also received attention (Webb, 2014b; Webb et al., 2016).

Given the above, and the changes to the operational role of the referee, which have also been identified, it is imperative to consider the role of FIFA and the confederations in the provision of the main refereeing programmes utilised to train international referees between 2011 and 2017. The structure and the principal objectives of these courses are described, as well as the role of FIFA and its regional technical instructors in the teaching chain.

In order to obtain information concerning how elite referees perceive their training and what they do to prepare themselves for international matches, more than 30 international referees from different confederations were invited to answer a detailed, structured survey. Between 2011 and 2017 direct and informed observations have been made during FIFA development courses and seminars for FIFA's referees and referee instructors. To be accepted as a natural part of the culture, it has been crucial to observe some aspects of this academic work and to become a participant in the design of methodologies and resources for training in class and on the pitch. In order to contribute to this research regarding the training of international referees around the world, current and recently retired referees were asked to participate voluntarily and anonymously in an open survey.

A pre-test pilot with three ex-elite referees, now FIFA instructors, was conducted prior to the launch of the survey. Following this process, five referees per confederation were invited to participate and a total of 18 officials answered the open survey, comprising 51 questions grouped in four sections: refereeing experience, comparative standards in refereeing training, culture and cultural values, and leadership and ICT.

Technical Training for Elite Referees and Technical Instructors

For a considerable period of time, the training of elite instructors, referees, and assistant referees was mainly based around on-field practical training, and on theoretical training focused on the use of the Laws of the Game handbook (IFAB, 2018). The social visibility of this group, and the need to homogenise interpretation and application of the Laws of the Game around the world, has multiplied efforts in FIFA and its confederations to optimise their training activities (Webb, 2017). In order to do this FIFA has developed and launched an initiative (Referee Assistance Programme, RAP) designed to address the challenges that referees and instructors face when interpreting and implementing the Laws of the Game during training and match situations.

The FIFA Referee Assistance Programme (RAP)

The RAP has two main activities, the FUTURO III courses for FIFA instructors and for member association (MA) referee instructors, and courses for MA top referees and assistant referees.

FUTURO III Courses for Referee Instructors

Two different types of Futuro III courses existed: one, focused on teaching and updating the knowledge of FIFA instructors; and another, conducted by FIFA instructors and aimed at MA referee instructors.

FUTURO III Course for FIFA Instructors

Once a year, FIFA organised a FUTURO III course for FIFA referee technical instructors. These FIFA referee instructors taught and conducted

FUTURO III courses in their regions for technical instructors. Table 16.1 presents an example of a FUTURO III course for FIFA referee technical instructors, hosted in Zurich, in February 2013.

This course has been one of the most important activities in the chain of teaching between FIFA instructors and local instructors to homogenise and standardise delivery across federations. It covered subjects concerning technical and fitness instructing, updating instructors on new teaching resources, and methodologies to enhance the level of referee instructors. Since the creation of the RAP programme, this annual course has been a showcase to promote new teaching resources and learning methodologies (Armenteros

Table 16.1 List of participants in the FIFA Technical instructors FUTURO III course, held in Switzerland, in 2013

AFC		*CONCACAF*	
ABDULLAEV Farkhad	Uzbekistan	BAHARMAST Esfandiar	USA
AL HAFI Ismail	Jordan	BENNETT Jennifer	USA
AL TRAIFI Ali Ahmed	Saudi Arabia	DARVILLE Stanley	Bahamas
ARIFF Mohamed Muhiseen	Sri Lanka	GUTIERREZ CASTRO Ronald	Costa Rica
BIN MOHD SALLEH Subkhiddin	Malaysia	PRENDERGAST Peter	Jamaica
ISHIYAMA Noburu	Japan	SIBRIAN Rodolfo Alexander	El Salvador
KOLPAKOV Viktor	Kyrgyzstan	VERGARA Hector	Canada
MAIDIN Shamsul	Singapore	*CONMEBOL*	
OIWA Mayunu	Japan	COLMAN MC LEOD Carolina	Paraguay
SAFAR Hany Taleb Ballan	Qatar	DE OLIVEIRA Silvia Regina	Brazil
SZOKOLAI Krystyna	Australia	LARRIONDA PIETRAFESA Jorge	Uruguay
YACOB Mohamad Rodzali	Malaysia	PEREZ Ana	Peru
BOKO Thomas	Benin	RUIZ ACOSTA Oscar	Colombia
ELSHENAWI Ahmed	Egypt	*OFC*	
GAYE Fatou	Senegal	Intaz HUSSAIN SHAH	Fiji
HADQA Yahya	Morocco	*UEFA*	
HENRIQUES Carlos Manuel	South Africa	BENNETT Steve	England
Joseph WELLINGTON	Ghana	CIRCHETTA Claudio	Switzerland
JOUINI Neji	Tunisia	DALLAS Hugh	Scotland
LIM KEE CHONG An Yan	Mauritius	DAMKOVA Dagmar	Czech Republic

(*Continued*)

Table 16.1 (Cont.)

AFC		*CONCACAF*	
MAGASSA Sidi Békaye	Mali	FERNANDEZ MARIN Juan Antonio	Spain
MAPHOSA Sabelo	Zimbabwe	HOLM Gitte	Denmark
MOUNGUENGUI Pierre Alain	Gabon	IVANOV Valentin	Russia
NABINA SEBUTU Dawili	Congo DR	SAJN Vladimir	Slovenia
TANGAWARIMA Felix Onias	Zimbabwe	TOMS Wendy	England
WAWERU Gladys	Kenya		

Source: Authors' own elaboration based on FIFA and confederations publications.

et al., 2019; Armenteros, Benítez, & Sillero, 2009; Armenteros & Curca, 2008) in order to improve the knowledge and uniformity of the Laws of the Game in the FIFA referee instructors panel.

The use of new learning and teaching materials has been a key consideration for FIFA in these courses. Since the launch of the RAP, in 2007, the development of video-based learning and teaching materials has been an intrinsic part of the FUTURO III courses for FIFA referee instructors. The objective for the FIFA instructors was to disseminate the new and emerging technologies and developments to their federations and domestic leagues, in order to demonstrate the advancements being made in referee training.

The first teaching resource developed for FIFA instructors was a match analysis in an interactive multimedia version of the analysis of match situations. A new interactive desktop version design for computers replaced the use of the classic DVD format and allowed the integration of other media such as animations (Armenteros & Curca, 2008). These new teaching materials were more interactive and allowed instructors to analyse videos with additional visual media, improving the understanding of all new changes made by the IFAB and FIFA.

The first new multimedia version of the Laws of the Game was created in 2008, complete with 2D and 3D animations to enhance the learning experience. Subsequently an interactive video quiz about the Laws of the Game and an interactive trivia test (Armenteros, Benítez, & Curca, 2010; Armenteros et al., 2009) were introduced. These new teaching materials started being used by FIFA referee instructors as a new tool to self-assess, and also to evaluate other regional referee instructors and referees. Finally, new learning activities were designed to enhance practice-based activities by groups or individuals, using the interactive video training modules and the online video database (Armenteros, Domínguez, Fernández, & Benítez, 2016). As part of these courses, experts in pedagogy and technologies were invited to teach FIFA instructors about "learning by doing" methods and pedagogical strategies to improve the teaching experience when using collaborative methodologies in FIFA FUTURO III courses for referee instructors (Armenteros et al., 2019).

These educational resources and teaching methodologies have been produced in the framework of collaboration between FIFA and the Carlos III University of Madrid, and they have been well received with extremely positive feedback in the different courses for referee instructors (Armenteros, Liaw, Fernández, Díaz, & Sánchez, 2013) and referees (Armenteros, Liaw, Sánchez-Franco, Fernández, & Sánchez, 2017). The successful delivery of these teaching and learning resources has been exported to FIFA confederations who have implemented the new technologies to produce their own teaching materials following the same FIFA standards.

In 2016, the withdrawal of the FIFA Senior Manager of Referee Development Fernando Tresaco Gracia and FIFA Technical Instructor Steve Bennett represented a turning point in the training of referee instructors, which overlaps with the arrival of VAR. Head of Department Massimo Busacca and new FIFA Chairman of the Referees Committee Pierluigi Collina focused their resources on preparing referees and assistant referees for competitions, specifically focused on the arrival of VAR experiments and the need to teach referees and instructors to use this technology. No innovations in the education of referee instructors and teaching resources have been made since these changes.

FUTURO III Courses for MA Referee Instructors

The FUTURO III courses for MA referee instructors focused on the education of member association instructors. It used a regional format, with a standard duration of five days, assembling and educating instructors selected by the member associations from neighbouring countries in a centralised venue. Up to ten countries benefitted from a single FUTURO III course every year in each confederation (FIFA, 2014b, p.2), although some confederations, because of their size or language, have more than one region. The final list of course participants were approved by FIFA and the confederation.

The organisation of regions was made according to the size of the confederation and the language(s) within the confederation. Eleven Refereeing Development Officers (RDOs) were in charge to coordinate these courses in their regions, together with the FIFA Refereeing Department. These RDOs were usually proposed by their confederations and approved by the FIFA Refereeing Department.

As an example, Table 16.2 shows the areas and the RDOs in FUTURO III MA referee instructor courses during 2015.

On each course, a maximum of two instructors per member association were permitted to participate in any one course, depending on the qualifications of the applicants and size of the country. In addition, a female instructor could be appointed if she was an instructor and actively participated in education programmes for top women referees and assistant referees, both before and after the FIFA course. The FUTURO III course was conducted by a FIFA coordinator and one local coordinator, with FIFA referee instructors also invited to attend the courses. These regional referee technical instructors were appointed by member associations and no public criteria to be appointed as a FIFA referee technical instructor has been found or is in the public domain at the time of writing.

Member Association Refereeing Courses

Member association refereeing courses focused on the education and instruction of member associations' top referees and assistant referees. These courses were conducted and supervised by two instructors who previously took part in the FIFA FUTURO III MA courses for referee instructors. A maximum of 30 top male and/or female referees and assistant referees from the association, plus five local instructors were

Table 16.2 FIFA Refereeing Development Officers in 2015

Name	*Confederation*	*Area*	*Language*
Ahmed ELSHENAWI	CAF	North Africa	English and Arabic
Carlos HENRIQUES	CAF	Central and South Africa	English and Portuguese
Sidi Békaye MAGASSA	CAF	Central Africa	French
Farkhad ABDULLAEV	AFC	Central Asia	English and Russian
Ismail AL HAFI	AFC	West and Central Asia	Arabic and English
Mohamad Rodzali YACOB	AFC	East and South Asia	English
Mark HESTER	OFC	Oceania	English and French
Peter PRENDERGAST	CONCACAF	North America and Caribbean	English
Ronald GUTIERREZ	CONCACAF	Central America	Spanish
Carolina Colman	CONMEBOL	South America	Spanish
Steve Bennett	UEFA	Europe	English

Source: Authors' own elaboration based on FIFA and confederations publications.

selected by the member associations to participate in the refereeing courses. The five local instructors had to be actively involved in education programmes, and in the instruction of the different categories of referees and assistant referees in the association before and after the FIFA course.

The standard duration of the courses was five days, but did vary according to the topics and financial resources of the organising member association. FIFA appointed a FIFA technical instructor, a FIFA physical instructor, and an RPO whose duties were to support the member association instructors participating in the FIFA FUTURO III refereeing courses, and the organisation of the refereeing and development programmes. FIFA was responsible for providing the teaching material and equipment for these courses.

MA technical instructors have tried to follow the FIFA refereeing standards to continue and improve the correct interpretation of the Laws of the Game, as defined and guided by FIFA. All teaching materials created by FIFA were used together with some other video-based teaching resources from member association competitions (such as the Gold Cup and Champions League). For instance, in the Rwanda Football Federation (FERWAFA) course, held in Remera, Kigali City in 2017, 25 Rwandan elite FIFA and top local referees, including nine women, who were all included on the FIFA List of Referees, attended. The course was conducted by two FIFA referee instructors, namely Rwanda's Ntagungira Celestin and Mohamed Houssien Ali (Djibouti). Both FIFA instructors were assisted by Abdulkhaliq Sekamana and Michel Gasingwa, local technical instructors, and Serge Mwambali, a local physical instructor.

Seminars Previous to Competitions

Apart from RAP courses, and prior to each competition, international referees and assistant referees met in FIFA seminars. Whether international, regional, or national competitions, the elite referees attended seminars where they received preparation courses that could range from five to ten days. These preparation courses concentrated on practice associated with the technical and tactical approach of referees. In order to address these areas, the course included reviews of important incidents in previous matches, utilised to move towards consistency and uniformity in referee decision-making.

Prior to the FIFA World Cup, the FIFA Confederations Cup, or the Olympic Games, elite referees received a series of specific and coordinated training sessions, over a period of time. This training was administered through a number of seminars. In the case of the FIFA World Cup, referee candidates attended at least one course in the months leading up to the World Cup, and after the final list of referees was identified, another course in the week before the start of the World Cup was conducted.

For instance, to prepare the referees for the 2014 FIFA World Cup Brazil,[1] the referees attended the following courses (FIFA, 2014b):

- 10–14 February 2014: Canary Islands (only referees)
- 24–28 March 2014: Home of FIFA Zurich (referees and assistant referees) (Confederations: UEFA, AFC, OFC)
- 7–11 April 2014: Home of FIFA Zurich (referees and assistant referees) (Confederations: CONMEBOL, CAF, CONCACAF)
- 1–10 June 2014: Rio de Janeiro (referees and assistant referees)

Because there were participants from different locations, courses were usually split into different time periods and locations, in order to make it easier for participants to attend the courses. In addition, the FIFA Confederations Cup, which was held one year before the FIFA World Cup in the same location, served as a training vehicle for referees who attended the FIFA World Cup the following year. The training activities were divided into physical and technical preparation, as it was argued that this work deals only with the technical aspects that can be developed in both the classroom and on the field.

In the classroom, and before the start of the competition, referees attended classes about the most important topics of the Laws of the Game. Prior to the seminar, a team of technical instructors, coordinated by the Senior Development Manager, Fernando Tresaco, watched all clips from previous competitions and proposed a pre-selection of clips that were analysed and discussed during the FUTURO III courses for FIFA referee technical instructors. The results were several multimedia presentations that were used in seminars before and during competitions. These presentations were: Hot Topics; the Team Arrival; and the Analysis of Match Situations multimedia presentations.

All the video clips in these multimedia presentations, used for elite referees during FIFA competitions seminars, had consensual decisions between all FIFA referee technical instructors and FIFA Head of Refereeing, although different interpretations were evident and therefore Massimo Busacca, Head of the Refereeing Department, had the final decision, and at times, he decided to remove some video clips from the seminars. Therefore, some divergent opinions related to the analysis of clips was observed between referee technical instructors and FIFA's Head of Refereeing.

On the pitch, new offside exercises based on players' tactics were implemented during seminars with a new digital technology that allowed a faster instant replay for assistant referees, providing a more instantaneous feedback mechanism.

The aspect of training which has not developed as quickly in the federations is that of psychological training for referees (Webb, 2017). Before 2011, the ex-Head of the Refereeing Department, José María

García-Aranda, introduced psychological training for FIFA referees. However, the new Head of Department, Massimo Busacca, decided by the end of 2011 that this training area was no longer required, attributing much more emphasis on other areas of training such as positioning and exercises with counterattacks during the training sessions on the pitch.

Seminars and Training Activities for Regional Competitions

Officials also received on-field-of-play experience as part of their training. This experience consisted of the participation in their own national leagues, in FIFA competitions, and in regional competitions such as the AFC Asian Cup, UEFA Champions League, CONCACAF Gold Cup, la Copa de América de la Conmebol, or the CAF Champions League, as well as off-field training, attending specific training courses in the months before the competition and during seminars, in the days leading up to the competition. For instance, the selected trios from AFC member associations based on their performance in FIFA and AFC Competitions (AFC, 2015, p. 14), were appointed to two AFC tournaments, namely the AFC U-16 Championship Thailand 2014 and the AFC U-19 Championship Myanmar 2014, where the main AFC instructors monitored their performance during the competitions.

The recent development of online solutions, such as the online video platforms (Armenteros et al., 2016), were used for pre-match technical preparation, and both FIFA and the confederations prepared online video questions that were used as self-assessment for referees and a tool for referee instructors to follow up with the elite referees. One of the key elements was ensuring that the match officials had the required technical knowledge about the Laws of the Game. For instance, it was essential for Asia's flagship national team tournament to do a series of online video questions that were conducted weekly from October through to the end of December 2014, before the 2015 AFC Asian Cup. Following the preparation for the 2015 AFC Asian Cup, international AFC match officials were also tested on match incidents, having also been quizzed on the Laws of the Game, and were provided with feedback and advice from the AFC main referee instructors after they had assessed their answers, with positive feedback received related to the construction of the activities:

> These weekly activities tremendously improved the match officials' technical knowledge and understanding of the Laws of the Game. Participation in the tests helped improve their knowledge base, which included how to recognise and differentiate the severity of fouls and misconduct and to be consistent with other match officials.
>
> (AFC, 2015, p. 20)

2017: Special Meeting with the Confederation Referees Department Heads

On 25 April 2017, one year after the election of the new FIFA president Gianni Infantino, a panel with the Referees Department heads and representatives from all six confederations, as well as AFC Referees Committee Deputy Chairman Hany Taleb B S Al-Raeesi, FIFA Head of Refereeing Massimo Busacca, and FIFA Referees Committee Chairman Pierluigi Collina, met in Kuala Lumpur. The aim of the meeting was to understand the various refereeing strategies currently employed by all six confederations and FIFA, as well as to identify opportunities for closer collaboration (AFC, 2017). Discussions revolved around departmental structures, annual budgets, referees, instructors and assessor panels, development programmes, special projects, supportive systems, and exchange programmes. One of the additional points of discussion was the structure and selection process of the FIFA international referee lists.

Among other recommendations, there was agreement on:

1. Referees' committees should be based on refereeing experts.
2. There should be increased involvement of the confederations in the selection process of referees for the FIFA list.
3. There should be reinforced supervision by confederations of the frequency of referees' appointments at the national level.
4. Confederations should ensure that minimum standards are met in order to establish a professional environment for referees.
5. Quality over quantity should be prioritised in FIFA competitions and on the FIFA list.

Another issue was that the annual budget that confederations received from FIFA for referees' training included "financial restraints and insufficient budgets to plan and carry out development activities" (AFC, 2017, p. 9). This highlighted issues in terms of the training provision and support that could be provided.

In terms of their development programmes, all confederations declared to have designed:

> tailor-made education programmes for referees and assistant referees, and benefited from the legacy of the FIFA Referee Assistance Programme (RAP), the course contents and supporting resources uniform and consistent throughout the globe … and express recognition to the previous support of the FIFA refereeing department in the development of refereeing.
>
> (AFC, 2017, p. 9)

In fact, they declared "to have, or plan to have, online training platforms and produce their own teaching materials following similar (FIFA) procedures: video analysis of competitions, matches, collection and database" (AFC, 2017, p. 9). The delivery of these aspects of the development programme are at different stages of progression and implementation, as well as differing in status between confederations.

Implementation of VAR: New Needs and Special Courses for MA Representatives

During 2017, there was a growing interest in the member associations to participate in VAR experiments (see Chapter 11). The IFAB underlined the importance of creating specific training programmes to prepare the referees for this new technology. Furthermore, FIFA promoted all kinds of activities focused on preparing the referees for VAR experiments and in the acquisition of new technical equipment to help and teach the new skills for VAR operations. As a result, more than 20 MA representatives were given guidance on how to train VARs in their respective competitions. The training was fed into the ongoing FIFA/IFAB Implementation Assessment and Approval Programme (IAAP), with refereeing development as part of FIFA's objectives (FIFA, 2018).

Referee Education Programmes Developed by FIFA Confederations

As mentioned previously, apart from FIFA training support through the RAP programme, and in seminars prior to FIFA competitions, each confederation designed and launched their own programmes to prepare elite referees and instructors around the world. For clarity, a summary of the programmes provided by each confederation are outlined below.

AFC

The AFC Future Referees Project was a two-year initiative that began in 2009, as a development programme for young referees. The candidates were selected from different regions of Asia, and the project was divided into four courses, with referees required to make the cut-off grade with each passing course. The referees participated in practical and classroom theoretical sessions to enhance their fitness condition and technical knowledge on key areas of the Laws of the Game, with the participants' knowledge of the Laws of the Game gauged through quizzes and video tests (AFC, 2016b). Referees also had to make group and individual presentations after analysing various match situations. A report-writing test was conducted to check their ability to convey accurate information on match incidents, as well as their ability to communicate in English. In the last stage of this programme, referees' performance was evaluated in real matches.

On 6 January 2016, Shamsul Maidin, a previous FIFA instructor, was appointed as AFC Director of Referees. In 2012, he declared that he was concerned about the need for a professionalisation of officials in AFC (AFC, 2012). The two main projects presented by the new Director of AFC Referees in 2017 were the MA Refereeing Administration Project and the AFC Referee Academy.

The AFC Elite Academy, discussed in previous meetings, and well received by the committee members in 2016 (AFC, 2016a), started in 2017. The Elite Academy focused on increasing the number of AFC referees at world-class level, as well as extending development support to the AFC member associations, and identifying and preparing potential referees for the AFC Elite Referees Panel, replacing the previous AFC Project Future Referees Programme. All AFC member associations could nominate two candidates to study at the academy during a four-year programme, conducted at three levels: introductory, intermediate, and advanced.

Despite the introduction of the Future Referees Project and the Elite Academy, referee responses to the development process were mixed. One referee believed that the support and training offered was satisfactory: "the referees committee were very accurate in terms of following up and correcting the mistakes. They still support and advise me." However, another referee believed that, despite the advancements, referees were still the part of the game that received the least attention from federations in particular: "referees are always the low priority in the domestic organisation … training sessions are always a challenge that we face. I'm well supported now as an elite referee compared to an ordinary level referee."

Referees in AFC also believe that training referees uniformly across confederations and internationally at FIFA level is a challenging exercise. Referees argued that it should be something that FIFA are aiming to achieve: "footballing authorities should educate to the same level and teaching materials at FIFA and confederation level should be the same." It was also recognised that there are cultural differences between countries, which should be taken into account when training referees and disseminating information: "the way we are trained should be different depending on the country and its environment." Nevertheless, referees identified that the changes made by FIFA and AFC are correct, even if they would like to see further training and development opportunities:

> Though football level varies in each member association, referees still should have the same level of teaching and understanding at all levels. I strongly feel that classroom education on match analysis and understanding match situations is crucial. More classroom analysis should be done at all levels, and then the application of positioning and movement should be executed in practical training with quality players on the field to enhance the training methodology.

Referees believed that the application of these concepts could be delivered through further investment in "quality instructors, teaching materials from FIFA for consistency, understanding and application, and on field training methods".

CAF

CAF develops different courses for officials at different stages of the development process. For example, CAF Elite "A" courses are aimed at elite female and male referees, whilst CAF Elite "B" courses cover the intermediary between the top-ranked Elite "A" and Young Talents categories, with match officials between the ages of 30 and 35.

Officials selected from amongst the best young referees in the continent, participated in five-days courses, with the help of CAF and FIFA technical instructors (CAF, 2018). The objective of these courses was to update the participants on the Laws of the Game (theoretical topics included severity of offences, match management, penalty area incidents, amongst others), as well as evaluate their technical abilities for major competitions, such as the Africa Cup of Nations and the African Nations Championship (CHAN).

CAF also carried out courses for referee instructors and assessors. The objectives of these courses were to evaluate the instructors and assessors and also introduce them to new methodologies to assist in their delivery of forthcoming courses. During a five-day course, practical and theoretical sessions include video and match analysis, online reporting, editing of clips, methodology and teaching practices, as well the use of the CAF content management system (CMS).

CAF referees identified that further moves towards uniformity and standardisation are required. The changes and advancements to the development programme and the refereeing system more widely have been embraced. However, referees believe that there are still areas that can be further developed by CAF and FIFA: "the footballing authorities can move towards a more standardised approach to refereeing by studying how football is changing and create a system that will suit the modern way of football." Referees believed that those in positions of governance need to adapt to the changing nature of the game, although this is challenging given that referees identified that "the organisation of the game is different in CAF".

Although referees believed that the structure of the game in CAF is different from other confederations, the opinion was mixed concerning whether referees could be trained and developed in the same way across confederations and at FIFA level. One referee believed that "all referees must work under the same conditions and with the same means. This does not happen … we need to change and adapt to cultural values." However, another referee believed that culture can be disregarded, and that it is possible to train referees in the same way due to the shared game values that they possess: "it is very much possible for referees to be trained and developed in the same way, no matter which country they are from because they have the same goals." It was also identified that referees in Europe have some advantages to referees from CAF in particular, due to the wide variety of countries that players in European leagues come from: "referees in Europe have got vast experience – when compared to referees from CAF – because they deal with professionals from different countries." Referees believed that online resources such as a "FIFA referee platform" or "online information dissemination to help create a link which everyone has access to" are methods that can be employed further in order to move towards greater uniformity and provision for referees.

CONCACAF

According to the CONCACAF General Secretary Ted Howard (CONCACAF, 2016), refereeing has been one of the most critical functions for the CONCACAF confederation. On 7 March 2016, CONCACAF appointed FIFA technical instructor Brian Hall as Director of Refereeing. One of the main goals for the new Director of Refereeing was to "develop a professional workforce that enables CONCACAF referees to achieve and maintain a consistent, high-quality performance on the field" (CONCACAF, 2016).

In 2017 CONCACAF created the Program of Refereeing Excellence (PORE), supported by the CONCACAF federations, which was run in collaboration with the Mexican Football Federation (FMF) in the first year of operation. The programme allowed up to 14 referees from 13 CONCACAF member associations to be trained by professional Mexican referees and FIFA instructors and officiate matches in the FMF's lower divisions. Organised in two, four-week sessions, the programme provided additional tools for top international referees, and prepared officials with "crucial professional experience to deliver excellence on the field" (CONCACAF, 2018).

Special sessions on technical, physical, psychological, and nutritional aspects were conducted. The cooperation with other confederations in the first edition of this course was high, with instructors from other confederations involved, such as Shamsul Maidin, AFC Head of Refereeing Department and FIFA instructor, and referee Ahmed Darwish from the United Arab Emirates. With the support of this programme, three of the match officials who were part of this programme were promoted to the FIFA referee lists in 2018.

CONACAF referees believed that the training that they have received is appropriate and has permitted their development and promotion to the FIFA list: "the amount and kind of training was very appropriate. I would have liked more on-field training with players, but logistically, that can be difficult to organise." Referees also believed that the assessment systems in place within the confederation were not as supportive and effective as they might have been, and therefore did not assist particularly with their development:

> The assessment system is not adequate since there are not enough assessors at each level that actually officiated at the level they are assessing … it is not constructive, and the referees receive more constructive feedback from the management than from the assessors.

Despite the issues identified with the assessment system, referees did feel supported by both CONCACAF and FIFA. Referees identified that their training and development is appropriate, due principally to the increased contact between CONCACAF and FIFA and the improving relationships between them. Referees also believed that it is possible to train referees in the same way, irrespective of country or culture, although it was identified that there are some differences which should be considered:

> In terms of evaluating situations and in the interpretation of law, it can be the same. The teaching material should be consistent all around the world. With that being said, each country has their own points of emphasis and challenges which will change the development of the referees. For example, in the US, the second division is well below in skill level from the first division. Therefore, referees are tried out in the MLS with a limited amount of experience and exposure to top level football.

Referees identified that there are differences in the competition levels between leagues in different countries. Therefore, it was argued that the delivery of material and the dissemination of resources required uniformity and consistency in order to be effective, although this was perceived to be less of a concern at FIFA or confederation level:

> There needs to be a consistent message that is sent from the top and continues all the way down to the grassroots level. This requires communication and a sharing of resources. This happens for the most part from what I have seen. There also needs to be qualified instructors and assessors working with referees both on and off the pitch. More education is needed to raise the level of the instructors and assessors. Domestically, this tends to be more of an issue than at FIFA or the confederation level.

The differences between confederations and contributory factors to any perceived differences experienced by referees were personified with the explanation regarding the structure of CONCACAF:

> the majority of CONCACAF is made up of small island nations. Many of them do not have a true professional league, but they do have FIFA referees. Therefore, many of the referees that are in MLS have more experience than many of the CONCACAF FIFA referees.

Clearly some of these factors are unique to CONCACAF, but this example does highlight the challenges FIFA and confederations face when attempting to move towards uniformity.

CONMEBOL

Since 2017, CONMEBOL have been holding periodic five-day preseason courses for elite referees, with the aim of evaluating and training all the international referees, prior to the next season of the CONMEBOL tournaments. The participants were evaluated through the official FIFA physical tests. Topics included in the programme were positioning, disputes, use of hands, offside, and teamwork, among other aspects. During the training, practical exercises were undertaken with players simulating specific situations of play, with the aim of referees achieving a good interpretation of the situation and subsequent decision. Referees also received specific talks on competitions concerning ethical principles.

In addition, CONMEBOL develop training courses for young, talented referees, organised by the APF and the CONMEBOL Referees Commission, to give support to the different talent development programmes of the member federations. This selection of referees is made using the criteria that those referees will officiate in the first division of their respective local leagues, as well as demonstrating the potential to become FIFA referees in the coming years.

The training courses designed and developed by CONMEBOL have received mixed comments from the referees. One referee identified that both FIFA and CONMEBOL have been supportive throughout their officiating career, something which has also been commented upon positively by referees from other confederations. One referee identified that "the support of CONMEBOL and FIFA was always the best in terms of training and also support for the process of my training", whilst another referee believed that the support and training provided was "very constructive". Nevertheless, referees also believed that there was still work that was required in order for uniformity to be achieved. For example, despite the developments on training provided by both FIFA and CONMEBOL, referees believed that there were differences that existed in their regions: "I think that the South American refereeing philosophy is getting standardised."

The introduction of the training courses has been well received by CONMEBOL referees, although they recognise that there are still issues in the development of referees in their confederation. The level of competition in the different regions and in different competitions was identified as a concern in referee progression: "there are differences, like the level of play in the match, the competitive level and the expectations that are generated from different tournaments." In order for referees to develop further in the confederation, additional innovation is required. Referees acknowledged that there were possible areas which could be focused upon, such as information sharing: "there are methods, for example to transmit experiences from referees that have an international path to the new ones. Information sharing for example." Moreover, another referee stated that professionalisation was the change which would bring the greatest level of progression for referees in the confederation: "refereeing should be professionalised, so that the referee can dedicate himself fully … if professional refereeing were adopted … it would be much more effective and uniform."

UEFA

UEFA's comprehensive education programme for referees was divided into the education of match officials through the national associations under the UEFA Referee Convention guidelines and structure, and the education of FIFA match officials through other activities. These activities include referee courses, talents and mentors, Centre of Refereeing Excellence (CORE), observers' education, incidents from UEFA matches (FAME), publications, directives and guidelines, and fitness monitoring.

UEFA Referee Convention Guidelines and Structure

The UEFA Referees Convention was approved on 23 March 2006, and aimed to:

> improve the quality of match officials and refereeing specialists at all levels of the game by incorporating specific rules aimed at standardising and enhancing their status and education as well as the way refereeing is organised within the different UEFA member associations.

The UEFA Referee Convention specified the criteria for promotion and relegation and match appointments. It also detailed the skills, responsibilities, and knowledge of teaching methodologies to be a national referee instructor in one of the UEFA nation members. Furthermore, the convention explained the ideal or proposed structure, formal training, and assessment of the education, training, and development of elite and grassroots referees, assistant referees, referee observers, referee instructors, and mentors.

This convention also established the main duties of the elite referee managers and the refereeing department administration rules (UEFA, 2012, pp. 1–20).

UEFA Centre of Refereeing Excellence (CORE) and Other Activities

UEFA created the Centre of Refereeing Excellence (CORE) in September 2010. This training programme was part of UEFA's education delivery and was controlled by the Referees Committee and subcommittees. Held once every two years, all CORE courses included a ten-day introductory session, followed by an eight-day consolidation course, both held on separate occasions. All 55 UEFA members were invited to send a young (25–30 years old) referee and two assistants who had the potential to reach the FIFA panel, once every two years; referees were also given the opportunity to officiate at matches in Switzerland and France.

UEFA also organised other courses for specific groups of referees: elite referees (summer and winter), new FIFA referees, referees and assistants in the talent programme, women referees, futsal referees, and both male and female referees and assistants selected for UEFA tournaments, such as EURO finals, and u21, u19, and u17 finals. UEFA courses involved all or some of the following topics: plenary sessions – presentations, group discussions, analysis of match incidents, fitness training and testing, psychological preparation, lifestyle advice, input from coaches, players, etc.; integrity and media guidelines and administrative matters.

UEFA referees were supportive in their comments concerning the training and development that they receive from UEFA and FIFA; referees find the development opportunities provided by "FIFA and UEFA more complicated and professional" than that provided by their national associations. Referees elaborated that "at the local level it [training] could be more structured, following UEFA guidelines". Perhaps the biggest concern for UEFA and FIFA is the differences that referees perceive to exist between countries in this confederation and also the differences in the level of competition.

Referees identified that there were "significant differences [across different competitions] in many aspects – audience, media attention, atmosphere in the stadium, responsibility and pressure". Moreover, referees have stated that they have appreciated differences between referees from different national associations when they meet for UEFA or FIFA training events. Referees identified that there are differences in the performances of referees from different countries: "two years ago … I started with UEFA on a mini tournament, with other international referees from other countries, from the east part of Europe. And at this time, I saw a very big difference." The most notable difference identified was the variances in preparation and training: "it's different, in Portugal we train three times a week with all the referees and one time alone, but I saw in Bulgaria for example the referees always train two times a week and alone, it's different." However, contrary to the differences in training across different countries that have been identified by the referees, they also recognise that standardisation within competitions is a vitally important, if difficult, matter to resolve: "inside the same competition it's very important the referees adopt the same conduct, the same management; but it's different inside [each] country, it's impossible for the referee to change completely."

OFC

OFC is the smallest football confederation. In 2008, their referee development received a boost thanks to the FIFA-funded RAP. Under the umbrella of the FIFA Development Office, the programme had two objectives. First, it aimed to prepare referees for international competitions, particularly the FIFA World Cup. Second, it aimed to develop referees in each of the member associations (MAs) through scouting and education.

According to FIFA and AFC instructor Subkhiddin M. Salleh, one of the objectives in those courses was to help referees to build confidence: "during FIFA tournaments we noticed referees outside of Europe were really quiet, they felt that they were behind referees from UEFA, and we need to take this idea away from referees," he added (OFC, 2016).

Confederation Participation in the FIFA World Cup

The FIFA World Cup is the highest level of international competition. The participation of the officials in the FIFA World Cup can indicate the quality of the referees from a given country, although it could also be influenced by other variables. According to FIFA, the referees are selected for the World Cup based especially on "their personality and their quality in football understanding, by being able to read the game and the teams' tactical approaches towards each game" (FIFA, 2014a). In the last three FIFA tournaments the distribution of the participation of the officials was as shown in Table 16.3.

Table 16.3 Participation of referees and assistant referees in the last three FIFA World Cups

FWC	*AFC (47*)*	*CAF (54*)*	*CONCACAF (36*)*	*CONMEBOL (10*)*	*OFC (11*)*	*UEFA (55*)*
Russia 2018	16	15	14	21	5	39
Brazil 2014[1]	14	13	13	17	4	29
Africa 2010	12	12	12	18	6	30

Source: Authors' own elaboration based on data from FIFA.

*number of member associations per confederation.

1 www.fifa.com/worldcup/news/referee-trios-and-eight-support-duos-representing-different-countries-ap-2261111

In Figures 16.1, 16.2, and 16.3, the ratio of participation per confederation is demonstrated. This has been calculated by considering the number of referees by confederation and the number of member associations per confederation. For example, CONMEBOL has ten member associations, if they sent 21 referees to the 2018 FIFA World Cup, it means a 210 per cent rate of participation. On the contrary, CAF, with 54 member associations, and 15 referees in the 2018 FIFA World Cup, has a ratio of 27.78 per cent.

We can observe that the highest percentage of participation by confederation is CONMEBOL, followed by UEFA, CONCACAF, AFC, CAF

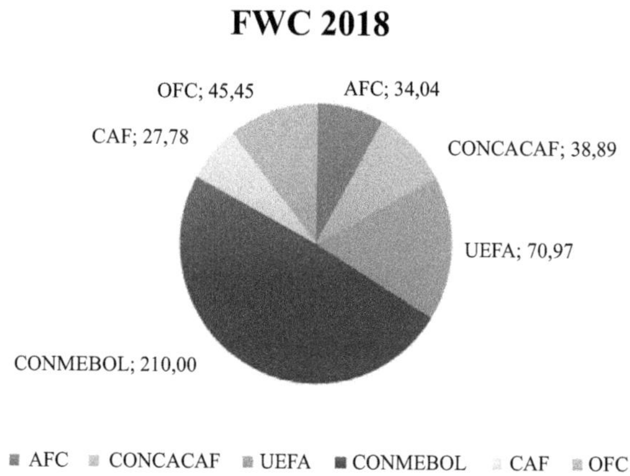

Figure 16.1 Confederation and referee participation in the Football World Cup 2018

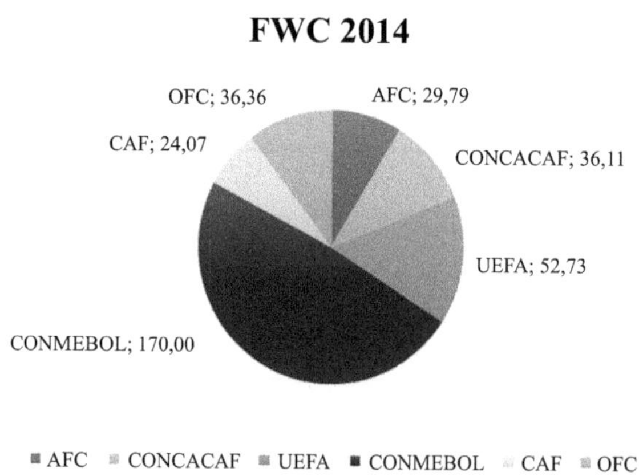

Figure 16.2 Confederation and referee participation in the Football World Cup 2014

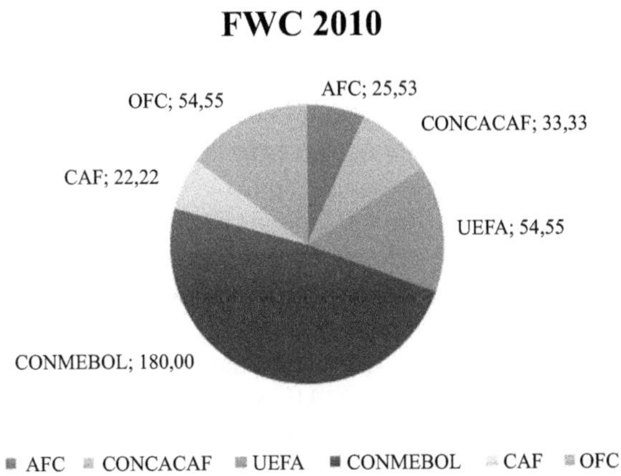

Figure 16.3 Confederation and referee participation in the Football World Cup 2010

and, finally, OFC. The arrival of VAR increased the participation of CONMEBOL and UEFA officials drastically over the rest of confederations, perhaps due to the initial training and positive attitude towards trials by member associations and the domestic competitions therein.

Conclusions

The training of elite football officials around the world has received an important expansion with the implementation of the FIFA Referee Assistance Programme, launched by FIFA President Joseph Blatter in 2007. For some confederations such as OFC, the RAP has been decisive in preparing the elite referees, but also in promoting young referees to international categories. The RAP allowed FIFA to create a homogeneous structure for training in all the confederations and a uniformity in the contents, methodology, and teaching resources for the education of football officials.

The FUTURO III, as part of the RAP, aimed at the education of football referee instructors, has been the main link in the learning chain between FIFA referee instructors and elite officials around the world. It is evident that the period between 2011 and 2017 has seen an increased importance attached to the education of elite football referees. One factor instrumental in the developments discussed in this chapter is the implementation of the VAR system, although there are still concerns around the use of the system, the protocols employed, and the different approaches being taken in different leagues/competitions.

Despite the successful period in the implementation of the RAP and FUTURO III programmes, and the general satisfaction about training that both instructors and referees have shown, it has been observed that there are still issues that referees perceive related to how uniform delivery, performance, and training of referees has become. Undoubtedly, this has improved, but referees are mixed in their responses when asked about uniformity in training and performances between competitions in particular, and these beliefs and observations appear to transcend the confederations, pointing to wider training- and development-related concerns which require attention by the confederations and FIFA.

All confederations have developed programmes for the preparation of elite referees and for the promotion of younger referees to the FIFA list, with UEFA, in particular, demonstrating variety in the organisation of teaching programmes for their referees. Recently, some confederations such as AFC, with the Academy Program, and CONCACAF, with the Program of Refereeing Excellence (PORE), have shown a structured plan for the preparation of talented and future international referees. Less detailed are the CAF and OFC programmes. Another scenario for the training of elite referees are the seminars prior to international competitions. It has been observed that these seminars have been influenced by the FIFA Head of Department, regarding the differences in teaching for seminars and the rest of FIFA courses for referee instructors.

The feeling that the FUTURO III and RAP courses and seminars in competitions had a different learning approach was also part of a lack of autonomy and freedom of the referees and referee instructors during their participation in seminars, specifically when different interpretations about a situation occurred. It is very likely that this learning process, typical of the authoritarian leadership style, plus a lack of teaching experience during the VAR experiments, was one of the consequences of the confusion which occurred in the management of some decisions during the FIFA Club World Cup in 2016 and 2017, exposing experienced elite referees in such a regrettable experience (Begley, 2017) (see also Chapter 11).

Finally, it was observed that there was no relationship between the number of member associations and the number of participants during the FIFA World Cup. The confederation that benefitted most in the FIFA World Cup was CONMEBOL. During the last three FIFA World Cups, and with just ten member associations (MAs), CONMEBOL sent an average of 19 officials, much more than AFC, with 14 officials and 47 MAs; CAF, with 54 MAs and 13 participants; OFC, with 11 MAs and 5 participants; and CONCACAF, with 36 MAs and 13 participants. Just UEFA (55 MAs), with an average of 33 participants, exceeded CONMEBOL participation.

Further research and work in this area should focus on how uniformity can be achieved, both within confederations and between confederations, and also within FIFA training and competitions. In addition, further work could identify the issues related to cultural difference, consider how training might be implemented to counter any cultural implications, and reflect how any training-based intervention could be achieved. In order to do this, further data collection should be undertaken with FIFA coaches and trainers, as well as those coaches and trainers who operate within the different confederations and at national association level. The proposed data collection would focus on moving towards uniformity across confederations and competitions, and how this might be achieved, particularly given the cultural differences which have been identified here.

This chapter has sought to examine the preparation, training, and performance of FIFA referees, also operational at confederation and national association level. We have focused on the training referees receive from their confederations and national associations, the dissemination and training provided by FIFA related to the consistency and uniformity in the interpretation of the Laws of the Game, the courses that are provided for referees to achieve this, and whether uniformity and standardisation are maintained across this training and development framework. Referees have expressed that there are still differences in the training, development, and performance across their confederations and at FIFA level, and therefore this requires further consideration in terms of the training and courses provided, the role of culture within these training programmes, and how any such programmes might be adapted further to move towards increased uniformity and standardisation.

Notes

1 The 2018 Russia World Cup was different because of the implementation of VAR, and referees did specific VAR courses. www.fifa.com/worldcup/news/referee-trios-and-eight-support-duos-representing-different-countries-ap-2261111

References

AFC (2012). Asian Referees Development Programme. [Video interview]. Retrieved from www.youtube.com/watch?v=bWE0dimNo3I

AFC (2015). AFC Asian Cup Australia 2015. Referee Technical Report. Retrieved from www.the-afc.com/about-afc/departments/referees/tech-reports/

AFC (2016a, March 30). AFC Referees Committee pledges to elevate refereeing. Retrieved from www.the-afc.com/news/afcsection/afc-referees-committee-pledges-to-elevate-refereeing-30273

AFC (2016b, August 9). AFC Referees Committee pledges to elevate refereeing. Retrieved from www.the-afc.com/news/afcsection/course-for-afc-development-group-referees-2016-held-31978

AFC (2017). Refereeing experts workshop report. 25–26 April 2017. Kuala Lumpur. Retrieved from www.the-afc.com/about-afc/departments/referees/tech-reports/

Armenteros, M., Benítez, A.J., & Curca, D.G. (2010). The trivia game as tool for learning and evaluating Football referees and assistant referees. In *Proceedings of World Conference on Educational Multimedia, Hypermedia and Telecommunications* (pp. 137–142). Chesapeake, VA: AACE.

Armenteros, M., Benítez, A.J., Fernández, M., De la Vega, R., Sillero-Quintana, M., & Sánchez Cid, M. (2019). Collaborative learning methods and multimedia tools for the education and training of instructors: The case of FIFA referee technical instructors. *International Journal of Information and Learning Technology 36*(pending). doi:10.1108/IJILT-07-2017-0061

Armenteros, M., Benítez, A.J., & Sillero, M. (2009). *Advantages of applying video to interactive auto-evaluation: A model used for the training of referees and instructors of the Fédération Internationale of Football Association (FIFA) and the Union of European Football Associations (UEFA).* Paper presented at the Internet and Multimedia Systems and Applications, Cambridge (United Kingdom).

Armenteros, M., & Curca, D. (2008). *Use of educational hypermedia for learning Laws of Game. FIFA Multimedia Teaching Materials.* Paper presented at the XII Conference on Educational Multimedia, Hypermedia and Telecommunications (ED-MEDIA 2008), Vienna, Austria.

Armenteros, M., Domínguez, A., Fernández, M., & Benítez, A.J. (2016). The video database for teaching and learning in Football refereeing. *International Journal of Social, Behavioural, Educational, Economic, Business and Industrial Engineering, 10*(9), 3041–3046.

Armenteros, M., Liaw, S.-S., Fernández, M., Díaz, R.F., & Sánchez, R.A. (2013). Surveying FIFA instructors' behavioural intention toward the multimedia teaching materials. *Computers & Education, 61*(1), 91–104. doi:http://dx.doi.org/10.1016/j.compedu.2012.09.010

Armenteros, M., Liaw, S.-S., Sánchez-Franco, M.J., Fernández, M., & Sánchez, R.A. (2017). Analysis of FIFA referees and assistant referees' motivational factors towards the multimedia teaching materials. *Education and Information Technologies, 22*, 841–872.

Begley, E. (2017, July). Confederations cup 2017: VAR confusion, Russian hospitality & Ronaldo complaints. *BBC.* Retrieved from www.bbc.com/sport/football/40447230

CAF (2018, September 18). CAF young referees' course. Retrieved from www.cafonline.com/en-US/NewsCenter/News/NewsDetails?id=PNGKI7%2fMaGmPZ%2bfiJ7zYnw%3d%3d

CONCACAF (2016, March 7). CONCACAF appoints Brian hall as director of refereeing. Retrieved from www.concacaf.com/article/concacaf-appoints-brian-hall-as-director-of-refereeing

CONCACAF (2018, April 9). CONCACAF and FEMEXFUT announce the second edition of the program of referee excellence. Retrieved from www.concacaf.com/article/concacaf-and-femexfut-announce-the-second-edition-of-the-program-of-referee-excellence

FIFA (2007, December 15). Approval for Refereeing Assistance Programme and upper altitude limit for FIFA competitions. Retrieved from www.fifa.com/about-fifa/who-we-are/news/approval-for-refereeing-assistance-programme-and-upper-altitude-limit–660747

FIFA (2012). *FIFA statutes.* Budapest: Author.

FIFA (2014a, January 15). Referee trios and support duos appointed for 2014 FIFA World Cup. Retrieved from www.fifa.com/worldcup/news/referee-trios-and-eight-support-duos-representing-different-countries-ap-2261111

FIFA (2014b, February 5). Seminars announced for 2014 FIFA World Cup referees. Retrieved from www.fifa.com/worldcup/news/seminars-announced-for-2014-fifa-world-cup-referees-2275045

FIFA (2018, October 26). First FIFA VAR course for Member Associations held. Retrieved from www.fifa.com/development/news/y=2018/m=10/news=first-fifa-var-course-for-member-associations-held.html

FIFA (n.d.). *Football development: Referees: Missions and goals*. Retrieved from www.fifa.com/development/education-and-technical/referees/index.html

Giulianotti, R., & Robertson, R. (2004). The globalization of football: a study in the glocalization of the 'serious life'. *The British Journal of Sociology, 55*(4), 545–568.

Giulianotti, R., & Robertson, R. (2009). *Globalization and football*. London: SAGE Publications Ltd.

IFAB (2018). Laws of the game. Retrieved from www.theifab.com/laws

Mascarenhas, D., O'Hare, D., & Plessner, H. (2006). The psychological and performance demands of association football refereeing. *International Journal of Sport Psychology, 37*(2), 99–120.

Nevill, A., Webb, T., & Watts, A. (2013). Improved training of football referees and the decline in home advantage post WW2. *Psychology of Sport and Exercise, 14*(2), 220–227.

OFC (2016, October 26) Referee course lifting standards. Retrieved from www.oceaniafootball.com/referee-course-lifting-standards/

Slack, L.A., Maynard, I.W., Butt, J., & Olusoga, P. (2013). Factors underpinning football officiating excellence: perceptions of English premier league referees. *Journal of Applied Sport Psychology, 25*(3), 298–315.

UEFA. (2012). *UEFA referee convention on referee education and organisation: Edition 2012*. Nyon: UEFA.

Webb, T. (2014a). The emergence of training and assessment for referees in Association Football: Moving from the side-lines. *The International Journal of the History of Sport, 31*(9), 1081–1097.

Webb, T. (2014b). Elite refereeing structures in England: a perfect model or a challenging invention? *Soccer & Society, 18*(1), 47–62.

Webb, T. (2017). *Elite soccer referees: Officiating in the premier league, La Liga and Serie A*. London: Routledge.

Webb, T., Dicks, M., Thelwell, R., & Nevill, A. (2018). The impact of referee training: Reflections on the reduction of home advantage in association football. *Soccer & Society, 19*(7), 1024–1037. doi:10.1080/14660970.2016.1267626

Webb, T., & Thelwell, R. (2015). "He's taken a dive": Cultural comparisons of elite referee responses to reduced player behaviour in association football. *Sport, Business and Management: an International Journal, 5*(3), 242–258.

Webb, T., Wagstaff, C.R.D., Rayner, M., & Thelwell, R. (2016). Leading elite association football referees: challenges in the cross-cultural organization of a geographically dispersed group. *Managing Sport and Leisure, 21*(3), 105–123.

17 Didactic Teaching Strategies for the VAR

An Approach to Gamification and 21st-Century Skills

Marta Fernández Ruiz, Miguel Ángel Betancor and Manuel Armenteros

Introduction

Football is a sport, and a game. Systems of rules, goals, and classification charts are elements that football professionals in general and referees in particular may be familiar with. The main goal of referee training is to control and ensure the correct development of the matches according to the Laws of the Game. The authority of referees, because it is imposed, must be recognized and therefore respected by all the active and passive participants of the sport; the education of referees is a clear reflection of their identity and socio-sports image (Betancor, 1998). The introduction of technology into football, especially the recent implementation of the Video Assistant Referee (VAR), aims to further support fair play and respect on the field of play.

All in all, referees serve the game, but can games serve referees? This work aims to assess games from the instructional point of view through gamification, with a special focus on the new generations of referees and the implementation of the VAR. Learning to officiate a match no longer implies only technical, management, and fitness skills, now undergoing a process of global digital transformation. The development of new digital skills in football referees will require prompt and systematic work by referee committees to provide referee training with new methodologies and technologies in different areas such as communication, objectivity, emotional control, anticipation, and integrity. This chapter also introduces the concept of "digital referees", namely, those who should possess, in addition to knowledge about the rules of the game, a set of digital skills. The "digital referee" is like the regulator of refereeing sports actions and behaves as an evaluator as he or she observes, orders, interprets, and helps in making the right decision, with the help of technology. In this respect, this work explores the feasibility of expanding the training curriculum of referees by implementing a new skill set that integrates media and digital literacy along with soft skills. Through educational gamification guidelines and examples of sports gamified applications, a didactic proposal is developed for the instructional design of VAR refereeing courses.

Gamification

Gamification is the use of game design elements in non-game contexts (Deterting et al., 2011). According to Kim et al. (2018), gamification in learning is different from serious games, as serious games are games developed to achieve learning goals in the real world. Game players learn while they are interacting with the game and reach the goals when they successfully complete the missions in the game. This means serious games, with real-world problems, are "implemented within games". By contrast, the purpose of gamification is to create real-world environments that support learning and problem-solving. This means it is "implemented within the real world".

Following Kim et al., gamification may be useful for learning and instruction because it can increase learners' engagement by means of its fun and playful features. Engagement is defined as the "simultaneous experience of concentration, interest, and enjoyment in the task at hand" (Shernoff, 2013, p. 12), and is considered to be an important factor influencing academic success (Newmann, 1992), motivation (Shernoff & Hoogstra, 2001), and social learning outcomes (Klem & Connell, 2004). Engagement is closely tied to the flow theory. According to Csikszentmihalyi (1975), there are four mental states: anxiety, apathy, boredom, and flow. Flow is the optimal state in educational contexts as it plays an important role in learning outcomes. Csikszentmihalyi (1990) defines flow as a mental state of total absorption in an activity that is both challenging and enjoyable. For Csikszentmihalyi (2004), individuals engage in an activity when the activity is challenging and performable enough. If the activity is too difficult for an individual's skill, the individual may feel anxious and quit the activity. If the activity is too easy, the individual tends to feel bored and quit the activity as well. Therefore, it is important to reach the correct balance between challenge level and the learners' skills. Csikszentmihalyi (1997) identified eight components of flow: the challenges match the individual's skills; the goal of the task is clear; there exists concentration and focus; the individual has control over the task; there exists direct feedback; the individual loses his or her self-consciousness; the activity is worthy by itself; and the individual becomes unaware of the passage of time.

Gamified systems for learning have been widely implemented in different education levels and areas of knowledge (Ibáñez et al., 2014; Lister, 2015; Marín et al., 2015; Tóth and Lógó, 2018; Urh et al., 2015). Kapp (2013, pp. 224–237) provides guidelines for the implementation of structural gamification in education:

- Clear goals: clear goals add a purpose, a focus (helping to keep the flow state), and measurable results to learning activities. Goals are useful for students as they allow them to know what they need to complete a task and what the result will be. It is important to keep the goals specific and unambiguous.

- Rewards and status: students can be rewarded for achieving these goals, as it helps them to avoid frustration and provides them with a sense of completeness and progress. Structural gamification provides visible notification of knowledge and mastery of different topics.
- Incremental goals and challenge: goals shall be incremental. This means it is possible to create a sequence of mini-goals that cover specific pieces of the curricular content. Each mini-goal should be harder to complete than the previous one to keep the learning activity engaging enough (and not to break the flow state). The goals and rewards must convey a sense of difficulty and risk. The difficulty can be linked to the number of points that need to be achieved, finding hidden badges or levelling up. Achieving the final level within a gamification environment must be equated with mastering the skills needed to complete a whole course.
- Progress: letting the students know their progress for an instructional activity can increase their motivation towards achieving the learning goals. Progress can be shown with a progress bar, or with an avatar moving forward on a board.
- Real-time feedback: immediate feedback guides actions, keeps students oriented towards the goals, and allows corrections before the course finishes.
- Transparency: transparent structural gamification designs provide data to all the students and participants in the same way. They also let students know exactly how to get rewards. Transparent systems make it possible to track progress and portray statistics about student behaviour and right and wrong decisions.
- Time: learning efforts can be distributed in multiple short sessions, and each session will be focused on a task. Distributed practice allows students to retain memorized information over long periods of time because this spacing provides deeper processing of the learning achievements.

Following Kapp, for these guidelines to be effective, it is necessary to use a series of game design elements:

- Rules: provide context and guidelines for students to progress towards learning-oriented goals and keep everything in balance. For example, the level of difficulty of a task can be controlled by introducing rules that add time limits and goals to complete. Rules make other elements of structural gamification (badges, rewards, scores) meaningful. When adding rules, it is important to keep them simple, to make learners know what they must do, and to ensure they do not cheat.

- Reward structure: to compensate the students for performing boring tasks (scores, badges, levelling up, unveiling of new contents) and giving them feedback on interesting tasks.
- Rankings: it is important to let the students choose who they want to compete with. Rankings can be created by territories or for specific groups.
- Scores: they can be used to reward progress and correct answers, to get social status, to unlock contents, or to spend as currency to get virtual or physical goods. It is important to design the score system in advance so that students can know what learning activities are most important. In this sense, score systems can be powerful tools for guiding the learning process.
- Coins/currency: a special type of score. Using coins implies there will be (virtual) objects or rewards that can be bought. In this sense, it is important to ensure that these objects are significant to the student.
- Badges: visible symbols of recognition.
- Level-ups: involves moving from one area to another while the difficulty increases. In structural gamification, it means earning enough points to go to the next content area, winning a new badge, and dominating a part of the learning curriculum. Levels enable instructors to control the progress of students, as they let them establish checkpoints. For students, knowing how many levels there are helps them to know how far they are from the learning process and how far they should go. It is important to tie each level to a specific learning objective.
- Share at the social level: students should be provided with tools to share their achievements with their colleagues so that they can congratulate them.

In the field of sports, and according to Tóth and Lógó (2018), the increase of smartphones and wearable technologies has led to the opportunity to track and analyse sport performance with gamified tracking applications that aim to keep users motivated. In this sense, running applications such as Strava and Nike+ Run are reward based. Experience points, trophies, or badges help users to understand the key points of their workout. These kinds of applications usually provide training calendars to track progress, allowing users to know how many of the goals are completed and how much more work is left. They also enable users to compare their statistics to others' and share these results among their contacts.

As per the specific context of football refereeing, there are examples of serious games and interactive digital applications developed to promote the understanding and the uniformity of criteria when learning and interpreting the rules of the game of football (Armenteros et al., 2019), but gamification strategies for learning and education in real environments have not been found yet.

21st-Century Skills and the Digital Referee

The arrival of information and communication technologies (ICT) such as the Internet, smartphones, social networks, and video games has led to working and performing day-to-day tasks with the aid of multimedia, audio-visual, multiscreen, and interactive devices. These new media and technological ecosystems demand new literacies that go beyond reading and writing. In this sense, media and digital literacy is key to participating in any activity nowadays, along with soft skills, which are those interpersonal skills that are linked to emotional intelligence (Ritter et al., 2018). This set of skills has commonly been sorted into what van Laar et al. (2017, p. 583) call 21st-century skills and is strongly tied to ICT. These skills are:

- Technical: the ability to understand the characteristics of new devices and applications, and to efficiently interact, access, and navigate through digital environments.
- Information management: the skill to use ICT to search for information and find and retrieve data from a wide range of sources. It also encompasses the ability to judge the usefulness and accuracy of information for a specific purpose.
- Communication: being able to convey information and ideas effectively to different audiences using multiple media and online formats.
- Collaboration: the skill to use devices and digital environments to create social networks to exchange information, discuss and reach agreements, and make decisions with respect to others in order to achieve a common goal.
- Creativity: being able to create new ideas or process familiar ideas in an innovative way to transform them into a product or service.
- Critical thinking: the skill to use ICT to make judgements and choices about the information gathered, and to communicate ideas with the aid of well-founded arguments and reasoning.
- Problem-solving: the ability to use ICT to understand and assess a problem situation in combination with the use of knowledge to find a solution to the problem.

VAR technology is growing within this new digital and media-based ecosystem, which means referee skills should match this set of 21st-century skills.

A Didactic Proposal for VAR Instruction

Based on the digital transformation all aspects of society in general and football refereeing in particular are experiencing, this work suggests a series of didactic teaching strategies aimed at covering both refereeing performance and technological issues in an engaging way. Goals, skills, contents, and methodology will be assessed in this respect.

Goals, Skills, and Contents

Goals are achievements that referees need to reach along their learning process. A VAR-oriented instructional design should contemplate goals such as becoming familiar with the Video Operation Room (VOR), understanding and highlighting important aspects of video output, mastering the match situations where the VAR shall intervene, having control over the procedure to follow after the referee on the pitch asks for an on-field review, and knowing and respecting the roles of the whole refereeing team.

Goals can be clear and well sequenced, but if they are not totally aligned with the desired skills, the skills might not be acquired. According to Murguía (2006), there are three main skills referees must attain: perceiving the game situation and making a judgement; comparing and interpreting the situation with the rules of the game; and sanctioning the situation by applying the correct rules. Additionally, from the psychological point of view, skills such as concentration, regulation of stress levels, self-confidence, positive thinking, and relaxation techniques play a key role in the performance of referees and need to be further assessed in the context of the VAR (see Chapter 18). Considering these factors and the digital global context in which the VAR has arisen, new skills are proposed to be taken into account for the referees' educational curriculum:

- Familiarity with the VOR and its components.
- Mastery in reading (motion) pictures.
    - Understanding of the laws of perspective, being able to take the correct references to draw vanishing points, being able to correctly read offside lines. Failing in this respect can lead to incorrect offside decisions.
    - Ability to analyse, in a multiscreen environment, relevant camera shots from video output provided by the replay operators.
    - Familiarity with the possibilities of managing the analysis of video output (frame-by-frame analysis, close-ups, speeding up or slowing down).
    - Ability to judge the accuracy of these shots to assist the referee on the field of play.
- Knowledge of the situations in which the referee shall request or be recommended an on-field review (minimum intervention, maximum benefit): goal situations, potential penalty situations, potential red card situations, and mistakes in player identification.

- Knowledge of the procedure when signalling the review with the VAR.
- Knowledge of the specific language to communicate with the referee.
- Knowledge of the different devices and channels available to establish communication with the refereeing team on the field of play.
- Mastery of the roles each member of the video review team plays: VAR, AVAR 1, AVAR 2, AVAR 3, and replay operators. Some situations that could harm video assistance in football refereeing are lack of mutual support, evasion of responsibility of one of the parties, lack of agreement in the decision-making, and invasion of one of the parties into another's ground.

These skills can be placed in relation to the 21st-century skills presented by van Laar et al. (2017). Table 17.1 shows this relationship.

Contents should match the goals and enhance the development of the desired skills. In instructional design, there are three types of contents: conceptual content, attitudinal content, and procedural content. Conceptual content consists of facts, phenomena, or concepts students can learn. In the context of refereeing, conceptual contents could be the rules of the game updated with VAR requirements, the new referees' signals (on-field review), the roles of the different members of the refereeing team, and the different video signals available to check at the VOR. Traditionally,

Table 17.1 Relationship between new skills to be added to the VAR refereeing curriculum and 21st-century skills

Skills proposed for referees' VAR educational curriculum	*21st-century skills (van Laar et al. 2017)*
Familiarity with the VOR	Technical
Mastery in reading video output	Technical, information management, critical thinking
Knowledge of the situations in which the VAR shall intervene	Information management, critical thinking, collaboration
Knowledge of the procedure when signalling the review with the VAR	Communication, collaboration
Knowledge of the specific language to communicate with the referee	Communication, collaboration
Knowledge of the devices and channels available to establish communication with the referees on the field of play	Communication
Mastery of the roles each member of the video review team plays	Communication, collaboration

Source: Authors' own elaboration.

attitudinal contents are used to work the values, but more recurrent is the work of the attitudes, especially in issues such as how to use the VAR system in unclear situations, how not to alter the veracity of the refereeing performance, and how long the delay of a restart should be so as not to affect the atmosphere of the match. Procedural contents deal with how to execute learned and internalized actions. Procedures appear in a sequential and systematic way. They require the repetition of actions that lead students to gain proficiency in the required skills. This implies the sequences of review and decision-making in the context of different game situations.

Method: Designing the Course as a Game

Up to now, referee instruction has consisted mainly of learning activities and tools that are implemented both on the field of training and outside it. On-field training is focused mainly on the development of practical sessions that cover the technical (positioning, foul recognition, identification of offside situations), management (players, restarts, free kicks), and fitness (physical exercises) areas of refereeing. These activities can be captured by video cameras and applications and shown to referees so that they can see their performance through a screen with the additional feedback of refereeing instructors (Schweizer et al., 2011). Off-the-field training consists mainly of theoretical sessions organized by refereeing organizations, but also of the use of applications (2D or 3D animations, games, quizzes, video tests) and learning management systems to enable referees to test their knowledge on the rules of the game, or to analyse videos of real match situations sorted by topics such as fouls, offsides, teamwork, and positioning to gain more insight on how to interpret and manage these situations (Armenteros et al., 2016). Through these learning activities referees can work with conceptual, attitudinal, and procedural contents.

These learning activities and tools could be enriched with the integration of the VAR educational requirements and the implementation of the proposed new goals, skills, and contents. A method to achieve this in an engaging way could be the addition of gamification strategies. Going back to Kapp (2013), and in refereeing contexts, learning activities could be framed through ruled systems that provide clear sequences of goals, levels of difficulty, rewards, and immediate feedback.

As per the implementation of goals, learning activities could be divided into small goal-oriented tasks. These small tasks could be classified by their level of complexity or demand of effort. Table 17.2 shows an example of the division of goals into smaller ones through different levels.

Levels could also be structured by adding a rule setting a time restriction to complete the judgement and management of a VAR match situation.

Table 17.2 Distribution of learning goals into mini-goals in a structure based on levels

Level	*Technical goals*	*Information management goals*	*Communication and collaboration goals*
1	Identify the media and technological structure enabling the video assistance process.	Identify the main tools to analyse video output (close-ups, frame-by-frame players, modification of the speed, addition of infography).	Identify the situations where VAR is required to intervene. Identify the roles each member of the VAR team has (role-play).
2	Understand how the media and technological structure work to support video assistance refereeing.	Analyse the accuracy of the camera shots available according to specific match situations.	Correctly use the devices provided to communicate with the on-field refereeing team.

Source: Authors' own elaboration.

This way referees may work on the acceleration of the decision-making process. The time restriction can be increased as levels are successfully completed. It is also possible to increase the difficulty of the learning tasks by increasing the complexity of the match situations as referees progress through the different levels. Video analysis tools, 2D and 3D simulations, and learning management systems can be used to support these methods. Figure 17.1 shows the implementation of the time restriction and the increasing complexity of the match situations to manage.

Progress through the levels can be both reported to the referees and rewarded. Current learning applications can track the progress made by their users and represent that advance in progress bars. The bars could show the referee's progress or the lack of it in relation to other referees or to the learning goals. These applications may also show statistics and create reward systems such as badges and rankings that could be visible and shared in (online) refereeing communities.

The possibilities that video feedback applications offer in the field of training for immediate feedback could be reinforced with VAR-oriented activities and with ICT and information management exercises. In the

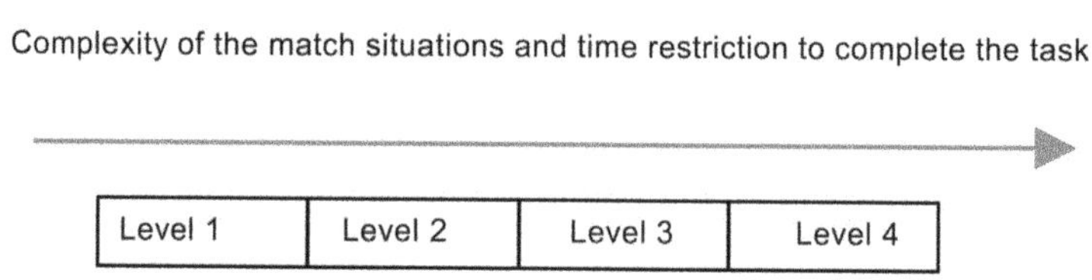

Figure 17.1 Complexity of the match situations and time restriction to complete the task

same way, the immediate feedback provided by off-the-field learning tools such as video tests, quizzes, or discussion activities can also be used to let referees know about their failures and to access the explanations to these failures so that they can improve.

Learning activities for referees can be distributed over time, and with the aid of learning management systems and other applications, instructors can control this distribution, programming specific dates and completion times for the activities (such is the case of quizzes and video tests).

An important consideration when designing a gamified course is taking into account the student's user experience. In this context, the refereeing learning experience can be assessed by designing and delivering user experience questionnaires, focus groups, surveys, or in-depth interviews to get to know what is more encouraging and motivating for a specific group of referees (for example, some users prefer to win, others prefer to solve problems, others become more engaged by working in groups, or by collecting, exploring, or role-playing, and they might be more or less proficient in the use of different technologies and devices).

Conclusions

Based on previous learning activities and tools used for the instruction of referees and also on sport gamified applications, the purpose of this chapter is to propose a set of didactic teaching strategies for the on-field and off-field training of the new generations of VAR referees. These strategies will make use of ICT, media literacy, and soft skills as contents and gamification as a methodology to structure the development of learning tasks.

Gamified instructional designs that include clear goals, increasing levels of difficulty, balancing rule systems, rewards, and social components enable students to create a habit and make their progression easily understandable. Further research on user and learning experience in the context of gamification and football refereeing will be done to reinforce and settle this proposal.

References

Armenteros, M., Benítez, A., De la Vega, R., Sillero-Quintana, M., & Sánchez Cid, M. (2019). Collaborative learning methods and multimedia tools for the education and training of instructors. The case of FIFA referee technical instructors. *The International Journal of Information and Learning Technology*, 36(5), 395–409.

Armenteros, M., Domínguez, A., Fernández, M., & Benítez, A. (2016). The video database for teaching and learning in football refereeing. *International Journal of Social, Behavioral, Educational, Economic, Business and Industrial Engineering*, 10(9), 3041–3046.

Betancor, M. (1998). Bases para la formación del árbitro de baloncesto. Doctoral dissertation. Las Palmas de Gran Canaria University. Las Palmas de Gran Canaria, Spain.

Csikszentmihalyi, M. (1975). *Beyond boredom and anxiety*. San Francisco: Josey-Bass.

Csikszentmihalyi, M. (1990). *Flow: The psychology of optimal experience*. New York: Harper Perennial.

Csikszentmihalyi, M. (1997). *Finding flow: The psychology of engagement with everyday life*. New York: Basic Books.

Csikszentmihalyi, M. (2004). *Good business: Leadership, flow, and the making of meaning*. New York: Viking.

Deterting, S., Dixon, D., Khaled, R., & Nacke, L. (2011). From game design elements to gamefulness. *15th International Academic MindTrek Conference on Envisioning Future Media Environments* (pp. 9–15). Tampere: ACM.

Ibáñez, M., Di-Serio, Á., & Delgado-Kloos, C. (2014). Gamification for engaging computer science students in learning activities: A case study. *IEEE Transactions on Learning Technologies*, 7(3), 291–301.

Kapp, K. (2013). *The gamification of learning and instruction fieldbook: Ideas into practice*. San Francisco, CA: John Wiley & Sons.

Kim, S., Song, K., Lockee, B., & Burton, J. (2018). *Gamification in learning and education. Enjoy learning like gaming*. Cham: Springer.

Klem, A., & Connell, J. (2004). Relationships matter: Linking teacher support to student engagement and achievement. *Journal of School Health*, 74(7), 262–273.

Lister, M. (2015). Gamification: The effect on student motivation and performance at the post-secondary level. *Issues and Trends in Educational Technology*, 3(2), 1–22.

Marín, V., López, M., & Maldonado, G. (2015). Can gamification be introduced within primary classes? *Digital Education Review*, 27, 55–68.

Murguía, C. (2006). Elaboración y estandarización de un instrumento de medición del autocontrol de los árbitros de fútbol mexicanos. Doctoral dissertation. Iberoamerican University. México DF.

Newmann, F. (1992). *Student engagement and achievement in American secondary schools*. New York: Teachers College Press.

Ritter, B., Small, E., Mortimer, J., & Doll, J. (2018). Designing management curriculum for workplace readiness: Developing students' soft skills. *Journal of Management Education*, 42(1), 80–103.

Schweizer, G., Plessner, H., Kahlert, D., & Brand, R. (2011). A video-based training method for improving soccer referees' intuitive decision-making skills. *Journal of Applied Sport Psychology*, 3(4), 429–442.

Shernoff, D. (2013). *Optimal learning environments to promote student engagement*. New York: Springer.

Shernoff, D., & Hoogstra, L. (2001). Continuing motivation beyond the high school classroom. *New Directions in Child and Adolescent Development*, 93, 73–87.

Tóth, Á., & Lógó, E. (2018). The effect of gamification in sport applications. 9th *IEEE International Conference on Cognitive Infocommunications* (pp. 69–74). Budapest.

Urh, M., Vukovic, G., Jereb, E., & Rok, P. (2015). The model for introduction of gamification into E-learning in higher education. *Procedia. Social and Behavioral Sciences*, 197, 388–397.

van Laar, E., Van Deursen, A., Van Dijk, J., & De Haan, J. (2017). The relation between 21st-century skills and digital skills: A systematic literature review. *Computers in Human Behavior*, 72, 577–588.

18 Psychological Response of the Referee to the Introduction of VAR

Ricardo de la Vega and Juan Pedro Fuentes

Introduction

Over the course of a football match, the referee must have the necessary skills to make complex decisions in a very short space of time and with high psychological and physiological demands (Fernández-Elías, Gómez-López, De la Vega, & Clemente-Suárez, 2017; Schweizer, Plessner, Kahlert, & Brand, 2011). The set of variables that can influence referees' decision-making has been the object of different studies and is of great validity (MacMahon, Helsen, Starkes, & Weston, 2007; Morillo, Reigal, Hernandez-Mendo, Montana, & Morales-Sánchez, 2017). We can distinguish, in general terms, two basic sources of information that can influence and affect decisional judgement: endogenous and exogenous (Kurtz, Shapcott, Kaiser, Schmiedt, & Schmid, 2017). Regarding the former, sports psychology, as a behavioural science, has focused its interest on knowing how cognitive, emotional, and psychophysiological processes (Cokely & Kelley, 2009; Larkin, Mesagno, Berry, & Spittle, 2018) can affect decisional judgement. Regarding the exogenous variables, the studies have focused on analysing their influence on athletes – including referees (Raab, 2002) – as well as on establishing the bases to avoid the possible biases that may arise, increasing the probability of attaining greater accuracy in the decisions that are made (Hutzler & Bar-Eli, 2013).

In this sense, this chapter aims to analyse how the implementation of the VAR can influence the behaviour of referees, studying the possible endogenous and exogenous changes that may occur and on which, undoubtedly, different research will be carried out in the coming years.

The Complexity of Decision-making of Referees in the Real World: The Need for a VAR

The continuous decision-making demanded of referees is of great interest for sports psychology, since it is known that in the interaction with the environment, the judgements that are taken can be seriously altered, with the accuracy and the speed with which they are taken being essential (De la Vega et al., 2017a). To analyse the possible effects of VAR implementation at this level, it is pertinent to start from a model that allows us to understand how decision-making is carried out, emphasizing those variables that, in the case of refereeing, are of greater importance (Raab & Helsen, 2015).

In the field of psychology, the study of decision-making has focused, mainly, on research into the processes that are at its base, such as perception (Put, Baldo, Cravo, Wagemans, & Helsen, 2013) and memory (MacMahon & Ste-Marie, 2002). From the first approach, the study of the decision-making processes has taken a traditional perspective of "closed systems analysis" (Starkes, Helsen, & Jack, 2001). In the attempt to know how a player, a coach, or referee makes decisions during the development of a football match, the ecological model, as an alternative to the traditional approach, offers important advantages (Araújo, Davids, & Hristovski, 2006). From this perspective, the different possibilities of action are known as affordances (Gibson, 1979), and are defined through the relationships established between the person and their environment, emphasizing the importance of the perceptual component (Turvey & Shaw, 1999).

The incorporation of the VAR can imply, from the ecological model, new adaptive forms of interaction of the referee with game situations. According to the ideas presented by Araújo et al. (2006), there are two central aspects that should focus our interest: first, the importance of identifying the informational variables that are used to guide behaviour (anticipatory signals that indicate that VAR assistance should be requested, generating *emergent behaviour patterns*); second, it's important to analyse the temporal evolution of behaviours. The decision-making behaviour must be based on perception of key information sources from their actions to the variable external environment (Araújo, Davids, & Serpa, 2005).

A clear example of the importance of understanding this referee–media interaction process in the analysis of decision-making is the study of some interesting and probably implicit biases that may arise during a match. For example, Schwarz (2011), in a study carried out with the analysis of 12,902 league games played between 1963 and 2006, found a compensatory bias in refereeing behaviour, consisting of, when a penalty has been awarded to one of the two teams, there is a much higher probability than normal that this will be compensated with another penalty in favour of the opposing team. This bias, which is increased in the case of awarding the first penalty to the visiting team, is an example of the implicit processes that arouse great interest in sports psychology and that could be corrected with the implementation of the VAR system. On the other hand, Plessner and Betsch (2001), carried out an experimental study with referees and footballers, and found that the previous information they had about actions watched through recorded plays had a significant influence with respect to the decisions that were made about the awarding of penalties. Consequently, different authors, such as Bar-Eli, Plessner, and Raab (2011), defend the importance of studies on refereeing and decision-making in sport being carried out following the principles and biases of studies conducted from social psychology (Plessner, Brand, & Schweizer, 2007).

Due to the importance of improving decisional judgements, there is a growing interest in proposing forms of referee training that, in general, seeks to complement an approach based on training in digital contexts such as tablets, computers, virtual reality, etc. (Schweizer et al., 2011); and even in the laboratory (Thomas, Gallagher, & Lowry, 2003), regarding the advantages and difficulties of training in contexts and situations of real simulation (Brand, Schmidt, & Schneeloch, 2006). MacMahon et al. (2007) examined sport expertise as a function of role (seven elite referees versus 34 youth academy soccer players), and they found that referees were better than players in a video-based decision-making task.

The implementation of the VAR, in this sense, involves a new refereeing challenge that involves the search for ways of improvement in which, precisely, digital media is already incorporated in the analyses that must be carried out in real time, which we believe should involve a support to the development of studies of the forms of training and the later utility of the work with these formats (Schweizer et al., 2011).

Once the basic characteristics in which the VAR is inserted are presented, as a tool to help in making correct refereeing decisions in football, as well as the main challenges of future research, the analysis of the main endogenous psychological variables will be considered, which may be affected in the context of refereeing performance.

Referee's Psychological Response to the VAR System: Anxiety, Self-confidence, and Motivation

In the study of the skills and strategies that the referee has to have in order to face the difficulties that may arise as a game unfolds, interesting research has been carried out that extends the models used from the field of basic and applied psychology (Kaissidis, Anshel, & Porter, 1997; Young-Kil & Jae-Youn, 2016). Louvet, Campo, and André (2015) conducted a study with a sample of 222 football referees, of whom 66 were professionals, analysing the direct and indirect effects of variable situations, such as the perception of the importance of the game, cognitive and somatic anxiety, and self-confidence; and dispositions, such as the anxiety trait, self-esteem, and some personality variables such as perfectionism and pessimism, as determinants of the coping strategies used. In their results, they found that these could be predicted by the perception of the importance of the match, by the intensity of the cognitive and somatic anxiety they had before the game, as well as by the perfectionism of the referee, which emphasizes the importance of analysing how the referee may respond to these levels with the implementation of the VAR system.

In this sense, the possibility of being evaluated could mean that the perception one has regarding the usefulness of the VAR may be different. Wolfson and Neave (2007), in a study conducted through the application of surveys to a total of 42 English referees, found that, faced with errors, the strategies most indicated for coping were: trying to learn from those situations in which they had made the wrong call, mentally reviewing the situations, as well as self-instructions aimed at pointing out that they had done the best they could.

In a recent study, Yun and Jeon (2016) analysed the psychological responses of 35 referees in the Korean football league just after committing an error of judgement. In their results they found eight main categories of psychological response in these situations: increased anxiety; reflexion on wrong calls; concern about reprimand; feeling apologetic; psychological fragmentation; feeling helpless; concern about reputation; and acknowledgement of wrong judgement. In the case of the VAR system being introduced, it is urgent to carry out new research that will allow us to go deeper into these levels of psychological response, knowing if the VAR acts to modulate this response and optimize the subsequent psychological and behavioural performance, or otherwise causes an increase in the doubts and negative thoughts, in negative emotions and in greater errors in the later decisional judgements.

Related to the coping strategies of situations, the level of anxiety prior to the development of a match can negatively affect the referee's performance. Johansen and Haugen (2013) assessed by questionnaire the anxiety of male and female referees in the first and second division of the Norwegian league, finding that those in the top flight showed greater levels of anxiety than their second division counterparts. In addition, those who had a perception of competence similar to the average or lower also manifested higher levels of anxiety, which is in line with the work carried out in the study of the relationships between anxiety and sports performance (De la Vega, Ruiz-Barquín, Aguayo, & Márquez, 2017b).

On the other hand, Pedrosa and García-Cueto (2016) carry out a relevant study with all the professional referees serving in the Spanish League (first and second divisions), in which they analyse the burnout syndrome, their anxiety and stress response, as well as their perception of social support. The results indicate that the presence of burnout syndrome is similar to that of other athletes (Gustafsson, Kenttä, & Hassmén, 2011), standing at around 2.4 per cent of the sample, the most experienced being those with the highest risk of developing the syndrome, which would go against results obtained in other studies in the context of work in which, with greater experience, there would be a lower risk associated with developing the burnout syndrome (Hamama, 2012). To these data we should add the positive correlation of the syndrome with the scores obtained in anxiety and stress, and negative with social support (Pedrosa & García-Cueto, 2016). The introduction of the VAR system means, in this sense, the challenge of carrying out future research in which it is analysed if the burnout syndrome decreases in prevalence when being perceived as an aid, analysing its differential effects at the level of anxiety and stress.

Regarding the importance of the referees' self-confidence, although there are no studies at the moment in relation to the VAR, previous studies on self-confidence and refereeing in football point in the direction of emphasizing the importance of this variable to adapt to situations and pressures that professional football presents (Folkesson, Nyberg, Archer, & Norlander, 2002). In a recent study (Furley & Schweizer, 2016), self-confidence was analysed in an experiment based on the non-verbal behaviour (NVB) of referees when making committed decisions. Viewers find the NVB less safe in ambiguous decisions, with there also being a greater chance of players and managers disputing their decisions. In this sense, it is important to study whether the referees perceive the VAR system as an aid to their work, reinforcing their self-confidence and conveying to the players, in this sense, greater security and control of the game.

If we consider how the introduction of the VAR system can affect motivational level, following the theory of cognitive evaluation (Deci & Ryan, 1985), there would be four fundamental aspects that should be taken into account to know how it may affect each referee: (1) When the referees perceive that they have participated voluntarily in the decisions to introduce the VAR system, as well as its putting into practice, their intrinsic motivation increases. On the contrary, if what they have is a perception of control, that is, a feeling of being controlled, it will decrease. (2) In relation to the perception of competence, when the referees perceive greater control and capacity of choice, the intrinsic motivation is likely to increase (Deci & Olson, 1989). In this case, it is important to carry out new research that can ratify these theoretical approaches. (3) Extrinsic factors, such as VAR, which are perceived as informative with respect to perceived competence, increase intrinsic motivation. In the case of being perceived as tools of behavioural control, intrinsic motivation will decrease (Moreno & Martínez, 2006). (4) In the referees oriented to the task, the effect of the VAR would be negative on the intrinsic motivation, while in those oriented towards the ego, it will also be able to affect at the level of their perceived pressure and on their self-esteem (Moreno & Martínez, 2006).

Leadership and VAR

Among the factors that are perceived to have greater influence within the skills and competencies that a football referee should have, those which are immersed in the correct running of the match stand out. Conflictive situations, attempts on the part of players and managers to pressure them, the relevance of the points at stake, as well as a whole host of other elements, can affect the referee's performance. Slack, Maynard, Butt, and Olusoga (2013) collected data from interviews with 15 English Premier League referees, finding that they perceived that success in the correct running of the game was based on a combination of physical and psychological abilities and their management of the environment. In particular, the aspects that were given greater importance were mental strength, preparation of the matches, improvement of the level of performance, the opportunities that were offered to them to thrive in their refereeing career, the personal characteristics of each one, and, finally, their physical qualities.

Similar results appear in the analysis of research conducted on these same aspects in different sports. Mascarenhas, Collins, and Mortimer (2005) emphasize the importance of the strategies of verbal and non-verbal communication used during rugby matches, later analysing the factors that are identified as keys for adequate decision-making in conflict situations that require control and management of game situations (Mascarenhas, O'Hare, & Plessner, 2006). These results, together with those found in other sports such as hockey or handball (Pizzera & Raab, 2012), reinforce the approach of the existence of a series of common factors related to excellence in the refereeing of sports competitions.

With regard to a more detailed analysis of the psychological skills found in their study, those that are of greater importance and that should be studied in greater depth with the implementation of the VAR system would be "mental strength" and "personal characteristics". "Within mental strength, the ability to face the pressures of competition, striving for, resilience, a solid self-confidence, a tenacious attitude and sports intelligence" (Slack et al., 2013) would be included. With regard to "personal characteristics", those that stand out as the most relevant by professional referees themselves are, in order of importance, being accessible, honest, humble, reserved, and a sense of humour. These competences and abilities, from the perspective of sports psychology, can and should be strengthened and improved, so that referees are offered the possibility of having an optimal capacity to adapt to the demands of today's football.

With regard to the mastery of verbal and non-verbal communication skills in the management of the game, these have been shown to be greatly important because of their influence on the perception of control that the players have and the effect on the self-confidence of the referee (Furley & Schweizer, 2016), with the impact of non-verbal rather than verbal language being more relevant due to its influence (Lex, Pizzera, Kurtes, & Schack, 2014). These communication mechanisms have a component of deliberate control and unconscious control regulated by the autonomic nervous system, hence the importance of each referee training them and gaining awareness of the handling of these processes (Matsumoto, Frank, & Hwang, 2013). At the same time, it is relevant to emphasize that people, in our case, players, have a high capacity to interpret and adapt to the messages they receive at the implicit level of the referees – and vice versa – with training in the control of the conflicting situations in which, for the most part, the use of the VAR is requested, being necessary (Furley & Schweizer, 2016). In this regard, although the referees are aware of the importance of controlling the verbal and non-verbal communication processes, they also show their difficulties in knowing how to train and improve, perceiving that they are difficult skills to learn and to teach (Cunningham, Simmons, Mascarenhas, & Redhead, 2014). In line with the ideas defended by Furley and Schweizer (2016), applying the postulates of Ekman (2003) to refereeing, the referee tries to maintain a coherent line in his way of acting throughout a match. Given their perception of a possible error in a decision made, non-verbal language reflects that inadequacy, being captured by the players, especially when the non-verbal language shows a lack of confidence. The confidence in the VAR as an instrument that helps in decision-making can be very useful to increase the level of coherence that is to be maintained in the game.

Discussion: Is Psychological Training for Referees Necessary?

Over recent decades, the incorporation of psychology at the level of basic and applied research has been increasing (Fortin-Guichard, Boudreault, Gagnon, & Trottier, 2017). It is considered essential to control and adapt mentally to the highly demanding conditions which an athlete must handle, being considered a determinant for the optimization of performance (Weinberg & Gould, 2011). For its part, the implementation of psychological training is well established in the preparation of the most important events and competitions internationally (Gould, Guinan, Greenleaf, Medbery, & Peterson, 1999); it is essential to know what its benefits can be with a suitable training programme that has been designed on the basis of scientific knowledge.

In the analysis of the intervention programmes that have been applied to date with football referees, interest has been growing in recent decades (Catteeuw, Helsen, Gilis, & Wagemans, 2009). Authors such as Weinberg and Gould (2011) argue that a mental training programme should incorporate exercises for the systematic and planned improvement of concentration, the regulation of stress levels, self-confidence, and motivation of the athlete – in our case the referee. Recent research has shown the importance of mental training in football referees (Mathers & Brodie, 2011), introducing strategies for the training of visualization, positive thinking, relaxation techniques, and pre-match workout routines.

Giske, Haugen, and Johansen (2016) point out, in research carried out with elite referees compared with others at sub-elite level in Norwegian football, how those of the first group report a greater use and importance given to psychological training with respect to the sub-elite group (in the case of both men and women). The most used techniques in psychological training were visualization, followed by concentration and self-talk techniques, also emphasizing that 69 per cent of the sample performed psychological training, with the percentage in the group of elite referees being 87 per cent. In this sense, the refereeing level and competitive demands, such as the highest level of perceived stress in the highest refereeing categories (Johansen & Haugen, 2013), would also arouse greater interest in, and motivation for, the practice of psychological training, which would explain the differences in quantity and quality of mental practice among the different groups, supporting results previously found in other research (Behncke, 2004).

Regarding psychological training programmes carried out with football referees, the scientific literature is scarce, although the work of Slack, Maynard, Butt, and Olusoga (2015) does stand out. In it, they implement and evaluate, over a period of eight months, the effectiveness of a Mental Toughness Education and Training Program (MTETP) developed with three national football referees. This programme follows the recommendations proposed by Gucciardi and Gordon (2009) on the importance of adapting mental training to the specific conditions of sport and athletes, and is developed through the implementation of four individual and two group workshops for the development of mental strength and refereeing performance. In these workshops, relevant content such as coping strategies for pressure, resilience, self-confidence, setting objectives, training in relaxation control techniques or management of attention focus and self-talk, among others, are presented, with the referees showing a high level of satisfaction with the implementation of the programme, as well as improvements in the scores obtained in the refereeing psychological performance indicators used. Regarding the implementation of the VAR, it should be noted that one of the benefits found with the implementation of the MTETP is that the referees say they improve their response to stoppages in the game and the moments when the decisions taken are conflictive and of a high level of difficulty (Slack et al., 2015).

Conclusion

By way of conclusion, it seems important to develop new research in the context of the football referee to analyse the extent to which benefits are obtained from psychological training in coping with situations such as those exposed by the VAR. This line of work follows a growing interest, and under this perspective of strengthening referees' performances it seems that important training and research challenges may arise in the coming years (García-Aranda, 2003). It will be particularly important to adapt refereeing performance to such a demanding, and even changing, context as the current one; it is vital to conceptualize the referee as a high-level athlete subject to similar processes of influence in which sports psychology can make valuable contributions (De la Vega et al., 2017a).

References

Araújo, D., Davids, K., & Hristovski, R. (2006). The ecological dynamics of decision making in sport. *Psychology of Sport and Exercise*, 7, 653–676. doi:10.1016/j.psychsport.2006.07.002.

Araújo, D., Davids, K., & Serpa, S. (2005). An ecological approach to expertise effects in decision-making in a simulated sailing regatta. *Psychology of Sport and Exercise*, *6*(6), 671–692.

Bar-Eli, M., Plessner, H., & Raab, M. (2011). *Judgement, decision making and success in sport*. Chichester: Wiley-Blackwell.

Behncke, L. (2004). Mental skills training for sports: A brief review. *Athletic Insight: The Online Journal of Sport Psychology*, *6*(1), 1–19.

Brand, R., Schmidt, G., & Schneeloch, Y. (2006). Sequential effects in elite basketball referees' foul decisions: An experimental study on the concept of game management. *Journal of Sport and Exercise Psychology*, *28*, 93–99. doi:10.1123/jsep.28.1.93.

Catteeuw, P., Helsen, W., Gilis, B., & Wagemans, J. (2009). Decision-making skills, role specificity, and deliberate practice in association football refereeing. *Journal of Sports Sciences*, *27*, 1125–1136. doi:10.1080/02640410903079179.

Cokely, E.T., & Kelley, C.M. (2009). Cognitive abilities and superior decision making under risk: A protocol analysis and process model evaluation. *Judgement and Decision Making*, *4*(1), 20–33.

Cunningham, I., Simmons, P., Mascarenhas, D., & Redhead, S. (2014). Skilled interaction: Concepts of communication and player management in the development of sport of cials. *International Journal of Sport Communication*, 7, 166–187. doi:10.1123/IJSC.2013-0098.

De la Vega, R., Carrasquilla, D., Ortiz, E., Ruiz, R., & Armenteros, M. (2017a). Influence of reaction time and movement in the performance of football referees. *Acción Psicológica*, *14*(1), 17–26. doi:10.5944/ap.14.1.19255.

De la Vega, R., Ruiz-Barquín, R., Aguayo, E., & Márquez, S. (2017b). Restoration of confidence and perception of coaches following sports injury. *Cogent Psychology*, *4*, 1. doi:10.1080/23311908.2017.1312047.

Deci, E.L., & Ryan, R.M. (1985). The general causality orientations scale: Self-determination in personality. *Journal of Research in Personality*, *19*, 109–134. doi:10.1016/0092-6566(85)90023-6.

Deci, E.L., & Olson, B.C. (1989). Motivation and competition: Their role in sports. In J.H. Goldstein (Ed.), *Sports, games, and play* (2nd ed., pp. 83–110). Hillsdale, NJ: Erlbaum.

Ekman, P. (2003). *Emotions revealed*. New York: Henry Holt & Co.

Fernández-Elías, V., Gómez-López, M., De la Vega, R., & Clemente-Suárez, V. (2017). Physical demands, heart rate response and performance of talent soccer referees. *Medicina Dello Sport*, *70*(4), 447–456. doi:10.23736/50025-7826.17.03076-9.

Folkesson, P., Nyberg, C., Archer, T., & Norlander, T. (2002). Soccer referees' experience of threat and aggression: Effects of age, experience and life orientation on outcome of coping strategy. *Journal of Aggressive Behavior, 28*, 317–327. doi:10.1002/ab.90028.

Fortin-Guichard, D., Boudreault, V., Gagnon, S., & Trottier, C. (2017). Experience, effectiveness, and perceptions toward sport psychology consultants: A critical review of peer-reviewed articles. *Journal of Applied Sport Psychology, 30*(1), 3–22. doi:10.1080/10413200.2017.1318416.

Furley, P., & Schweizer, G. (2016). Nonverbal communication of confidence in soccer referees: An experimental test of Darwin's Leakage Hypothesis. *Journal of Sport & Exercise Psychology, 38*(6), 590–597. doi:10.1123/jsep.2016-0192.

García-Aranda, J. (2003). Another step forward in preparing referees. *FIFA Magazine, 79*, 36.

Gibson, J.J. (1979). *The ecological approach to visual perception*. Hillsdale, NJ: Lawrence Erlbaum Associates.

Giske, R., Haugen, T., & Johansen, B.T. (2016). Training, mental preparation and unmediated practice among soccer referees: An analysis of elite and sub-elite referees´reported practice. *International Journal of Applied Sports Science, 28*(1), 31–41. doi:10.24985/ijass.2016.28.1.31.

Gould, D., Guinan, D., Greenleaf, C., Medbery, R., & Peterson, K. (1999). Factors affecting Olympic performance: Perceptions of athletes and coaches from more and less successful teams. *Sport Psychologist, 13*(4), 371–394. doi:10.1123/tsp.13.4.371.

Gucciardi, D.F. & Gordon, S. (2009). Revisiting the performance profile technique: Thheoretical underpinnings and application. *The Sport Psychologist, 23*, 93–117. doi:10.1123/tsp.23.1.93.

Gustafsson, H., Kenttä, G., & Hassmén, P. (2011). Athlete burnout: An integrated model and future research directions. *International Review of Sport and Exercise Psychology, 4*(1), 3–24. doi:10.1080/1750984X.2010.541927.

Hamama, L. (2012). Burnout in social workers treating children as related to demographic characteristics, work environment, and social support. *Social Work Research, 36*(2), 113–125. doi:10.193/swr/svs003.

Hutzler, Y., & Bar-Eli, M. (2013). How to cope with bias while adapting for inclusion in physical education and sports: A judgement and decision-making perspective. *Quest, 65*(1), 57–71. doi:10.1080/00336297.2012.727372.

Johansen, B.T., & Haugen, T. (2013). Anxiety level and decision-making among Norweigan top-class soccer referees. *International Journal of Sport and Exercise Psychology, 11*(2), 1–12. doi:10.1080/1612197X.2013.773665.

Kaissidis, R.A., Anshel, M.H., & Porter, A. (1997). Personal and situational factors that predict coping strategies for acute stress among basketball referees. *Journal of Sports Sciences, 15*(4), 427–436. doi:10.1080/026404197367218.

Kurtz, P., Shapcott, K.A., Kaiser, J., Schmiedt, J.T., & Schmid, M.C. (2017). The influence of endogenous and exogenous spatial attention on decision confidence. *Scientific Reports*, 7, 6431. doi:10.1038/s41598.017.06715w.

Larkin, P., Mesagno, C., Berry, J., & Spittle, M. (2018). Exploration of the perceptual-cognitive processes that contribute to in-game decision-making of Australian football umpires. *International Journal of Sport and Exercise Psychology, 16*(2), 112–124. doi:10.1080/1612197X.2016.1167760.

Lex, H., Pizzera, A., Kurtes, M., & Schack, T. (2014). Influence of players' vocalisations on soccer referees' decisions. *European Journal of Sport Science, 15*(5), 1–5. doi:10.1080/17461391.2014.962620.

Louvet, B., Campo, M., & André, A. (2015). Déterminants psychologiques des strategies de coping des arbitres de football (Psychological determinants of coping strategies of soccer referees). *Movement & Sport Sciences*, *87*, 63–77. doi:10.1051/sm/2014015.

MacMahon, C., Helsen, W.F., Starkes, J.L., & Weston, M. (2007). Decision-making skills and deliberate practice in elite association football referees. *Journal of Sport Sciences*, *25*(1), 65–78. doi:10.1080/02640410600718640.

MacMahon, C., & Ste-Marie, D.M. (2002). Decision-making by experienced rugby referees: Use of perceptual information and episodic memory. *Perceptual and Motor Skills*, *95*(2), 570–572. doi:10.2466/PMS.95.6.570-572.

Mascarenhas, D., Collins, D., & Mortimer, P. (2005). Elite refereeing performance: Developing a model for sport science support. *The Sport Psychologist*, *19*, 364–379. doi:10.1123/tsp.19.4.364.

Mascarenhas, D., O'Hare, D., & Plessner, H. (2006). The psychological and performance demands of association football refereeing. *International Journal of Sport Psychology*, *37*, 99–120.

Mathers, J.F. & Brodie, K. (2011). Elite refereeing in professional soccer: A case study of mental skills support. *Journal of Sport Psychology in Action*, *2*(3), 171–182. doi:10.1080/21520704.2011.609018.

Matsumoto, D., Frank, M., & Hwang, H.S. (2013). *Nonverbal communication: Science and applications*. Los Angeles, CA: Sage.

Moreno, J.A., & Martínez, A. (2006). Importancia de la Teoría de la Autodeterminación en la práctica físico-deportiva: Fundamentos e implicaciones prácticas (Importance of the Self-determination Theory on physical-sporting practice: Fundamentals and practical implications). *Cuadernos De Psicología Del Deporte*, *6*(2), 39–54.

Morillo, J.P., Reigal, R.E., Hernández-Mendo, A., Montana, A., & Morales-Sánchez, V. (2017). Decision-making by handball referees: Design of an ad hoc observation instrument and polar coordinate analysis. *Frontiers in Psychology*, *8*, 1842. doi:10.3389/fpsyg.2017.01842.

Pedrosa, I., & García-Cueto, E. (2016). Burnout síndrome in elite referees: The Spanish Professional Soccer League (LFP) under analysis. *Revista Iberoamericana De Diagnóstico Y Evaluación*, *42*(2), 59–68. doi:10.21865/RIDEP42-59.

Pizzera, A., & Raab, M. (2012). Perceptual judgements of sports officials are influenced by their motor and visual experience. *Journal of Applied Sport Psychology*, *24*, 59–72.

Plessner, H., & Betsch, T. (2001). Sequential effects in important referee decisions: The case of penalties in soccer. *Journal of Sport and Exercise Psychology*, *23*, 200–205. doi:10.1123/jsep.23.3.254.

Plessner, H., Brand, R. & Schweizer, G. (2007). A social cognition approach as the basis for the development of a decision training for referees in soccer. *Journal of Sport & Exercise Psychology*, *29*, S10–S10.

Put, K., Baldo, M.V.C., Cravo, A.M., Wagemans, J., & Helsen, W.F. (2013). Experts in offside decision making learn to compensate for their illusory perceptions. *Journal of Sport & Exercise Psychology*, *35*(6), 576–584. doi:10.1123/jsep.35.6.576.

Raab, M. (2002). T-ECHO: Model of decision making to explain behaviour in experiments and simulations under time pressure. *Psychology of Sport and Exercise*, *3*(2), 151–171. doi:10.1016/S1469-0292(01)00014-0.

Raab, M., & Helsen, W. (2015). How experts make decisions in dynamic, time-constrained sporting environments. In J. Baker, & D. Farrow (Eds.), *Routledge Handbook of Sport Expertise* (pp. 64–73). Oxon: Routledge.

Schwarz, W. (2011). Compensating tendencies in penalty kick decisions of referees in professional football: Evidence from the German Bundesliga 1963–2006. *Journal of Sports Science, 29*(5), 441–447. doi:10.1080/02640414.2010.538711.

Schweizer, G., Plessner, H., Kahlert, D., & Brand, R. (2011). A video-based training method for improving soccer referees' intuitive decision-making skills. *Journal of Applied Sport Psychology, 23*(4), 429–442. doi:10.1080/10413200.2011.555346.

Slack, L.A., Maynard, I.W., Butt, J., & Olusoga, P. (2013). Factors underpinning football officiating excellence: Perceptions of English Premier League referees. *Journal of Applied Sport Psychology, 25*, 298–315. doi:10.1080/10413200.2012.726935.

Slack, L.A., Maynard, I.W., Butt, J., & Olusoga, P. (2015). An evaluation of a Mental Toughness Education and Training Program for early-career English football league referees. *The Sport Psychologist, 29*, 237–257. doi:10.1123/tsp.2014-0015.

Starkes, J., Helsen, W., & Jack, R. (2001). Expert performance in sport and dance. In R. Singer, H. Hausenblas, M. T. Turvey, & Shaw, R. E. (1999). Ecological foundations of cognition: I. Symmetry and specificity of animal–environment systems. *Journal of Counsciousness Studies, 6*(11–12), 95–110.

Thomas, J.R., Gallagher, J., & Lowry, K. (2003, June). Developing motor and sport expertise: Meta-analytic findings. Communication to the Conference of the North American Society of Psychology of Sport and Physical Activity (NASP-SPA), Savannah, GA.

Turvey, M.T. & Shaw, R. (1999). Ecological foundations of cognition I: Symmetry and specificity of animal-environment systems. *Journal of Consciousness Studies, 6*, 95–110.

Weinberg, R.S., & Gould, D. (2011). *Foundations of sport and exercise psychology*. Champaign, IL: Human Kinetics.

Wolfson, S., & Neave, N. (2007). Coping under pressure: Cognitive strategies for maintaining confidence among soccer referees. *Journal of Sport Behavior, 30*(2), 232–247.

Young-Kil, Y., & Jae-Youn, J. (2016). Psychological experience and coping strategies for K-League referees just after the moment of wrong judgement. *Korean Journal of Sport Science*, 27(4), 957–969.

Yun, Y.-K., & Jeon, J.-Y. (2016). Psychological experience and coping strategies for K-League referees just after the moment of wrong judgment. *Korean Journal of Sport Science*, 27(4), 957–969.

Glossary

Art McNally GameDay Central (AMGC) A room named in honour of Art McNally, the long-time official and officiating executive, 42 feet long and 36 feet wide, full of displays to review the games, and placed in New York where NFL officials monitor the games.

Assistant Referee (AR) In football, the Assistant Referee is a member of the referees team who helps the referee to control the match. This commonly includes such matters as: inspecting the field, the balls used, and player equipment, determining if problems with equipment or bleeding have been resolved, monitoring the substitution process, maintaining back-up records of time, goals, or misconduct.

Assistant Video Assistant Referee (AVAR) A current or former referee/assistant referee appointed to assist the Video Assistant Referee (VAR).

Associazione Italiana Arbitri (AIA) The AIA is in charge of selection, training, and representation of Italian football referees. The AIA is part of the Federazione Italiana Giuoco Calcio (FIGC) and has as one of its most important tasks the appointment of the referees who manage all the matches of the Italian football championships developed under the control or with the authorization of the FIGC. At the same time, the AIA is present in the Italian territory through the regional committees, the autonomous provincial committees, and the refereeing sections.

Attacking possession phase (APP) It's the "starting point" of the reviewable period of play for a goal, penalty incident, or DOGSO offence. It is the start of the attacking move which led to the goal, penalty incident, or DOGSO offence and, if relevant, how possession of the ball was gained at the start of that phase of play (but not a restart which began the attack).

Automatic Goal Detection (AGD) The AGD System is a measurement system that is capable of automatically detecting the scoring of a goal during a football match. It also called Goal-Line Technology (GLT).

Back judge Focuses on tight end and action surrounding him. Responsible for players on the end of the lines, counts defensive players on field, keeps track of 40- or 25-second game clock and all television breaks.

BNC Stands for Bayonet Neill-Concelman, a type of radio-frequency and other signals connector for coaxial cables. It is widely used to carry analogue and digital video signals.

CCU Stands for "Camera Control Unit", a remote device to control and configure camera parameters related to exposition (iris, shutter speed, signal gain) or colour design (white balance), to free television production camera operators from these tasks.

Codification Codification "is the process by which data is systematically transformed and aggregates into units that allow an accurate description of the relevant characteristics of the content" (Holsti, cited by Rivoire, 2013). Rivoire, C. (2013). Retrieved from: https://repositori.upf.edu/bitstream/handle/10230/22222/TFM_CorentineRivoire.pdf?sequence=1&isAllowed=y.

Confederation Body responsible for football in a continent. The six confederations are AFC (Asia), CAF (Africa), CONCACAF (North, Central America and Caribbean), CONMEBOL (South America), OFC (Oceania), and UEFA (Europe).

Conselho de Arbitragem (CA) It is the supervisory body and is responsible for defining the guidelines and coordination, planning, and administration of refereeing activity throughout the Portuguese territory.

Digital literacy Ability to retrieve, evaluate, and create contents and pieces of information through the use of information and communication technologies.

Digital referee Referee who combines a set of digital skills with knowledge about the Laws of the Game. He or she behaves as an evaluator as he or she observes, orders, interprets, and helps in making the right decision with the help of technology.

DOGSO Denying an obvious goal-scoring opportunity.

Engagement Experience of concentration, interest, and enjoyment in the process of doing a task.

European Club Association (ECA) It is an independent and individual body that represents the European football teams, defending and protecting their interests. Founded in 2008, it is the only body of its kind to be recognized by UEFA.

Favourites It was the service's primary way for users to show their acknowledgement, laughter, etc. created in 2006. On 2015, Twitter replaced favourites with likes.

Federação Portuguesa de Futebol (FPF) It is the governing body of football in Portugal, based in Lisbon. It organizes the First Division of Portugal and the Selection of Soccer of Portugal. The FPF is composed of 22 territorial federations.

Fédération Internationale de Football Association (FIFA) It is the highest football institution and includes 211 associations (2018) or football federations from different countries. It was founded in 1904 and is headquartered in Zurich, Switzerland. FIFA is responsible for the organization of the most important international football competitions, such as the FIFA World Cup, which was created in 1930, and the Women's World Cup, which began in 1991.

Federazione Italiana Giuoco Calcio (FIGC) It is the highest organization in Italian football. The FIGC coordinates the professional football championships (Lega Serie A, Lega B, Lega Pro), non-professional competitions (from the Lega Nazionale Dilettanti "Serie D" to provincial championships) as well as the different divisions of futsal and the Serie A and Serie B women's soccer championships. It also directs and coordinates the activities of the Italian national team and the different youth sections. The following organizations are part of the FIGC: the Lega Nazionale Professionisti Serie A, the Lega Nazionale Professionisti B, the Lega Italiana Calcio Professionistico, the Lega Nazionale Dilettanti, the Associazione Italiana Calciatori, the Associazione Italiana Allenatori Calcio, and the Associazione Italiana Arbitri.

Field judge In NFL matches, the field judge is one of the referees who controls the game. The field judge will line up 25 yards deep in the defensive backfield, and his responsibilities include watching the game clock and watching for pass interference penalties, illegal use of hands by defensive players, proper count of players on the field, and the spot where a player goes out of bounds on his side of the field.

FIFA World Cup The FIFA World Cup is the biggest single-event sporting competition in the world and is contested by the senior men's national teams from the 211 member associations of FIFA (2018). The current format of the tournament involves 32 (48 approved in March 2019) teams competing for the title at venues within the host nation(s) over a period of one month – this phase is often called the final competition. A qualification phase, the preliminary competition – which currently takes place over the preceding three years – is used to determine which teams qualify for the tournament together with the host nation(s).

Flow theory Psychological theory developed by Mihály Csíkszentmihályi consisting of the identification of a mental state of total immersion in a feeling of focus, involvement, and enjoyment in the process of an activity.

Football Association Cup (FA Cup) It is the oldest English football competition. The top four levels (Premier League and the three divisions of the Football League) are automatically eligible. Non-league football teams (football leagues which play outside the top leagues) are also eligible provided they have played in either the FA Cup, FA Trophy, or FA Vase.

Fußball-Bundesliga It is a professional association and the most important football league in Germany. 18 teams participate from August to May using the promotion and relegation system.

Gamification The use of game design elements such as rules, goals, rewards, and rankings in non-game contexts.

GIFs Stands for "Graphic Interface Format". It has an efficient compression algorithm, so it is widely used on social networks to interchange brief animations.

Goal-line technology (GLT) Electronic system which immediately informs the referee when a goal has been scored, i.e., the ball has wholly passed over the goal line in the goal.

Hashtag Hashtags are labels formed by one or more key words that are written without spaces and are preceded by the hash symbol: #. These tags serve to group the contents on the same subject, making it easier to search.

Hawk-Eye Technology applied in some sports such as tennis and cricket to know the trajectory of the ball through a computer system. In this way, the judges can use it to decide the validity of some dubious plays.

Head linesman In the NFL, primarily responsible for ruling on offside, encroachment, and actions pertaining to the scrimmage line prior to or at snap. Generally, keys on closest setback on his side of the field. On pass plays, the linesman is responsible to clear his receiver approximately seven yards downfield as he moves to a point five yards beyond the line. The linesman's secondary responsibility is to rule on any illegal action taken by defenders to delay the receiver moving downfield. Has full responsibility for ruling on sideline plays on his side, e.g., pass receiver or runner in or out of bounds. Together with the referee, the linesman is responsible for keeping track of the number of downs and is in charge of mechanics of his chain crew in connection with its duties.

Instant Replay Technician (IRT) Similar to RO in sports like basketball, but he/she works only for referee decision aid.

Instructional design Systematic development of specifications aimed at delivering instructional products and experiences enabling the acquisition of knowledge in an efficient and engaging way.

Laws of the Game (LoG) The Laws of the Game (LoG) are a set of rules governing the play of football in order to keep the game "fair" and "safe" and apply in every match in every confederation, country, town, and village throughout the world. The first "universal" football laws were drawn up in 1863, and for a law to be changed, the IFAB must be convinced that the change will benefit the game, focusing on: fairness, integrity, respect, safety, the enjoyment of the participants, and encouraging participation from everyone, regardless of background or ability.

Lega Serie A or Lega Nazionale Professionisti Serie A It is the body that is responsible for leading the most important Italian football club tournaments at the national level, such as the Serie A, the Coppa Italia, the Italian Supercoppa, the Primavera 1 Championship, the Italia Primavera Cup, and the Primavera Supercopa.

Lega Serie B or Lega Nazionale Professionisti B It is the body that has the task of managing the Italian Serie B championship, the Spring 2 Championship and the Spring 2 Supercoppa.

Line judge In NFL matches, the line judge assists the head linesman at the other end of the line of scrimmage, observing for possible offsides, encroachment, and other fouls before the snap. As the play progresses, he is responsible for the action near his sideline, including whether a player is out of bounds, and for counting offensive players.

Live trials The "live" trials of VAR were developed during the period between December 2016 and March 2018. The referee had the option to review and change the original decision using replay footage directly ("on-field" review) or indirectly ("VAR only" review).

Malaysia Super League It is the men's top professional football division of the Malaysia football league system. Twelve teams participate in these leagues, where the top two teams in the Malaysia Premier League division replace the two lowest-placed in the Malaysia Super League.

Media literacy Ability to read, create, and critically analyse media contents (audio-visual, multimedia, transmedia, crossmedia).

Mentions The mention is the way for Twitter users to refer to someone or something. To do this, before the name is placed the symbol of @. Thanks to mentions it is easy to interact with other users. In addition, people or companies that have a Twitter account can measure their presence.

National Football League (NFL) The NFL is an American sport organization founded in 1920. The NFL regular season begins on the weekend following the first Monday of September and ends in December or early January. Each one of the 32 teams plays 16 games during a 17-week period.

NFL side judge The NFL side judge takes care about clock malfunctions, counts defensive players on the field, and signals the referee when time expires at the end of each quarter.

OBV (Outside Broadcast Van) Vehicle with a TV room inside with live production facilities.

Offline trials The "offline" trials of VAR were developed between March 2016 and December 2016. In this phase, there was no contact between the VAR and the referee, so there was no impact on the game. "Offline" matches were used for VARs, AVARs, and ROs to practise using the VAR setup, assessing whether an incident was a "clear and obvious error" or "serious missed incident", and gaining experience of using different angles and replay speeds.

On-field review (OFR) When the referee decides to view the replay footage in the referee review area (RRA).

Picture In Picture (Effect) Video effect that shows one image superimposed over another, so two cameras can be seen simultaneously.

Playlist A related and ordered collection or list of video clips waiting to be emitted.

Portable electronic performance and tracking system (EPTS) Tracking systems are used to assess players' performance, and to monitor the position of the players and the ball. They can be used in conjunction with accelerometers, gyroscopes, heart rate monitors, and other devices that measure physiological parameters or medical information.

Red card (RC) Disciplinary action when a player is required to leave the field for the remainder of the match having committed a sending-off offence (indicated by a red card).

Referee NFL The NFL official leader of a seven-official crew who is identified with a white cup. This official has the authority on disputed rulings and general responsibilities, such as overseeing everything related to the officials, maintaining the pace of the game, signalling all fouls, and counting the offensive players on the field.

Referee Review Area (RRA) In matches using VARs there must be at least one RRA where the referee undertakes an "on-field review" (OFR). The RRA must be in a visible location outside the field of play and clearly marked.

Replay Operator (RO) The RO is a person with technical knowledge in video replay who assists the VAR in the video operation room (VOR). He or she should also know about refereeing.

Retweet A retweet is a republication of a tweet. That is, it allows a user to share in their own profile a publication of another.

Review Assistant (RA) The RA is a person who assists the referee in the referee review area.

Review Operator Person with technical knowledge that assists the VAR in the area of the Video Operation Room (VOR).

Russian FIFA Confederation Cup A quadrennial international men's football tournament with 16 matches played in a group stage and a later knockout stage. Organized by FIFA and held in Russia from 17 June to 2 July 2017.

Serious games Games developed with the purpose of educating in a specific area.

Silent check The silent check is the process of checking a decision/incident by the VAR, but no communication with the referee takes place (no clear error identified).

Sky Italia Sky Italia is an Italian digital television platform owned by Sky Private limited company, which is a pan-European British communication company, owned by Comcast and based in London and founded by Rupert Murdoch. This platform has a wide range of sports broadcasts, highlighting the football competitions in Italy and in other European countries, as well as the UEFA Champions League and the UEFA Europa League.

SMART TV Also named "connected TV", a conventional TV set provided with a connection to the Internet.

Soft skills Interpersonal skills that are linked to emotional intelligence that help people to develop good social relationships and working performances. These skills include communication, teamwork, and problem-solving.

Standard Definition TV or SD TV PAL or NTSC television system, as opposed to HD (high definition). For many years, space resolution of TV signals were only 576 lines in a PAL system and 480 in NTSC per frame. Images registering and representation was poor. In HD the resolution per frame is 1080. It is expected that upcoming 4K's resolution per frame will be 3840 (UHDTV1 Standard).

Tally, or Tally Light A coordinating TV production system that uses red lights to keeping the attention on which camera or video signal is selected to be aired in each moment. If a camera operator sees this light on in their display, they know their camera is hot, and every movement they make can be seen instantly by the audience.

The International Football Association Board (IFAB) The International Football Association Board (IFAB) was founded in 1886 by the four British football associations (The FA, Scottish FA, FA of Wales, and Irish FA) as the worldwide body with sole responsibility for developing and preserving the Laws of the Game. FIFA joined the IFAB in 1913.

TIM Cup or Coppa Italia Football competition held in Italy in a single match, except for the semi-finals, which are played over two matches, in which the Italian soccer clubs of Serie A, Serie B, Serie C, and Serie D participate. The champion club of this competition has the right to participate in the UEFA European League and dispute the Supercoppa against the winner of Serie A.

Time Code A figure made by four groups of two numbers that is assigned to each frame of each video feed registered. It stands for hours, minutes, seconds, and frames. In the PAL system, after "01:59:59:24", the next frame is always "02:00:00:00", because it is designed for 25 frames per second. In sport broadcasting, all cameras are usually recorded in different recording devices with synchronized TC.

TMO (Television Match Official) In rugby, system that is used by the referee to clarify some situations. An official can be provided by the match organizer and usually HB provides video feeds.

Triax A type of coaxial cable to carry television and video signals. It is a short form of triaxial. These cables allow camera signals to be carried to the CCU, and TV program signals from the CCU to the camera, and intercom bi-directional signals simultaneously. Nowadays single-mode optical fibre is replacing this technology.

UEFA Champions League The UEFA Champions League (abbreviated as UCL) is an annual club football competition organized by the Union of European Football Associations (UEFA) and one of the most prestigious tournaments in the world disputed by the top-division European clubs.

UEFA Europe League Previously "UEFA Cup" and first "Inter-Cities Fairs Cup", it is considered the second most important football clubs competition in Europe. It is played annually.

Umpire referee In NFL matches, the umpire referee maintains control at the LOS by watching for holding and blocking infractions, reviews players' equipment, counts the number of offensive players on the field, and marks off penalty yardage.

Union of European Football Associations (UEFA) UEFA is the highest representative of football in the European continent. It is one of the six confederations of the Fédération Internationale de Football Association (FIFA). It is responsible, among many other aspects, for organizing competitions of European teams, such as the UEFA European Championship or the League of Nations, as well as competitions of European clubs such as the UEFA Champions League and the UEFA Europa League.

VAR system VAR system refers to all VARs and broadcasting technologies involved in the use of video to help referees. It involves an audio communication system, a protocol, and a team of referees that help the referee on the pitch.

Video Assistant Referee (VAR) A current or former referee appointed to assist the referee by communicating information from replay footage only in relation to a "clear and obvious error" or "serious missed incident" in one of the reviewable categories.

Video Operation Room (VOR) The room/area where the Video Assistant Referee (VAR), Assistant VAR (AVAR) and Replay Operator (RO) work. They have independent access to, and control of, the broadcaster's video replay footage, and it may be in/near to the stadium or at a more distant location.

Viral Viral content is distributed in a massive way through social networks, so that it multiplies and spreads like a computer virus in a short space of time.

World Rugby (WR) Rugby Union governing body. New name for the International Rugby Football Board and International Rugby Board.

Yellow card (YC) Official sanction which results in a report to a disciplinary authority; indicated by showing a yellow card; two cautions in a match result in a player being dismissed (sent off).

Index

Locators in *italics* refer to figures and those in **bold** to tables, though where concurrent with related text these are not distinguished from principal locators.

www.ingramcontent.com/pod-product-compliance
Ingram Content Group UK Ltd.
Pitfield, Milton Keynes, MK11 3LW, UK
UKHW022004180926
475701UK00003B/192
* 9 7 8 1 0 3 2 0 8 3 8 5 8 *